Chronotherapeutics

UNIVERSITY OF HERTFORDSHIRE
The School of Pharmacy

Chronotherapeutics

Edited by
Peter Redfern
BPharm, PhD, DSc, FRPharmS

Professor of Pharmacology
Department of Pharmacy and Pharmacology,
University of Bath, UK

London • Chicago   **Pharmaceutical Press**

Published by the Pharmaceutical Press
Publications division of the Royal Pharmaceutical Society of Great Britain

1 Lambeth High Street, London SE1 7JN, UK
100 South Atkinson Road, Suite 206, Grayslake, IL 60030-7820, USA

© Pharmaceutical Press 2003

 is a trade mark of Pharmaceutical Press

First published 2003

Text design by Barker/Hilsdon, Lyme Regis, Dorset
Typeset by Type Study, Scarborough, North Yorkshire
Printed in Great Britain by TJ International, Padstow, Cornwall

ISBN 0 85369 488 5

All rights reserved. No part of this publication may be reproduced, stored in a retrieval system, or transmitted in any form or by any means, without the prior written permission of the copyright holder.

The publisher makes no representation, express or implied, with regard to the accuracy of the information contained in this book and cannot accept any legal responsibility or liability for any errors or omissions that may be made.

A catalogue record for this book is available from the British Library

Contents

Preface xii
About the editor xiv
Contributors xv

1   **An introduction to circadian rhythms and their measurement in humans**   1
    1.1   Constancy and rhythmicity   1
    1.2   Endogenous and exogenous components of a rhythm   3
        1.2.1   Constant routines   3
        1.2.2   Fractional desynchronisation   5
        1.2.3   Differences between variables in the nature and strength of their exogenous component   6
        1.2.4   Variants of 'constant routines'   7
    1.3   Some properties of the body clock in humans   8
        1.3.1   Free-running rhythms   8
        1.3.2   Zeitgebers in humans   9
    1.4   Widespread effects of the body clock   11
        1.4.1   Metabolism   12
        1.4.2   Mental and physical performance   12
        1.4.3   Sleep   15
    1.5   Field studies   16
        1.5.1   Problems with 'constant routines' and the forced desynchronisation protocols   16
        1.5.2   The choice of a marker for field studies   16
        1.5.3   'Purifying' field data   19
    1.6   Summary   27
        Acknowledgements   27
        References   28

2   **The biological clock: location, neuroanatomy and neurochemistry**   31
    2.1   Introduction: the circadian timing system   31
    2.2   The suprachiasmatic nuclei (SCN)   32
        2.2.1   SCN neurochemistry   32
    2.3   Clock inputs   32

- 2.3.1 The retinohypothalamic tract (RHT) 33
- 2.3.2 The geniculohypothalamic tract (GHT) 36
- 2.3.3 The raphe-hypothalamic tract (RaHT) 39
- 2.3.4 Converging photic and nonphotic signals 41
- 2.4 Clock outputs 41
- 2.5 Comment 45
- References 45

3 The molecular biology of the mammalian circadian clock 53
- 3.1 Introduction 53
- 3.2 Molecular aspects of the circadian clock 54
- 3.3 Which photopigment is mediating circadian photoreception? 63
- 3.4 Role of clock genes in peripheral tissues 65
- 3.5 Relevance of the molecular clock mechanism to human biology 68
- 3.6 Conclusion 69
- Acknowledgements 69
- References 69

4 Rhythms and pharmacokinetics 75
- 4.1 Introduction and objectives 75
- 4.2 Rhythms in pharmacokinetics 76
  - 4.2.1 Chronopharmacokinetic studies in humans 76
  - 4.2.2 Rhythms, physicochemical properties and formulation of drugs 85
  - 4.2.3 Rhythms and routes of drug administration 86
- 4.3 Mechanisms of chronopharmacokinetics 88
  - 4.3.1 Rhythms in drug absorption 88
  - 4.3.2 Rhythms in drug distribution 90
  - 4.3.3 Rhythms in hepatic drug metabolism 94
  - 4.3.4 Rhythms in drug excretion 100
- 4.4 Conclusions 103
- References 103

5 Rhythms and pharmacodynamics 111
- 5.1 Principles of drug action 111
- 5.2 Rhythms in receptors and their signalling pathways 111
  - 5.2.1 β-Adrenoreceptors in the heart 111
  - 5.2.2 β-Adrenoreceptors in the CNS 116
  - 5.2.3 5-HT-receptor subtypes in the CNS 116
- 5.3 Rhythms in enzyme systems 117
  - 5.3.1 The renin–angiotensin system 117
  - 5.3.2 The nitric oxide–cyclic GMP system 118
  - 5.3.3 The synthesis of cholesterol 120

	5.4	Rhythms in ion channel activity 121
	5.5	Summary and conclusions 122
		Acknowledgements 123
		References 123

6 Rhythms and therapeutics of diseases of the gastrointestinal tract   127

- 6.1 Introduction   127
- 6.2 Gastrointestinal motility rhythms   127
  - 6.2.1 Pharmacological and therapeutic implications   130
  - 6.2.2 Other rhythms influencing drug bioavailability   131
- 6.3 Rhythms in gastric acid secretion   134
  - 6.3.1 Pharmacological implications   138
  - 6.3.2 Therapeutic implications   143
- 6.4 Seasonal rhythms in peptic ulcer disease   144
- 6.5 Rhythms in gastric mucosal defence   145
  - Acknowledgements   147
  - References   148

7 Rhythms and therapeutics of asthma   153

- 7.1 Introduction   153
- 7.2 Chronobiology of Asthma   154
  - 7.2.1 Circadian variation in lung function   154
  - 7.2.2 Mechanisms of nocturnal worsening of asthma   159
- 7.3 Chronotherapeutic approach to asthma treatment   167
  - 7.3.1 Theophylline   169
  - 7.3.2 Oral β_2-adrenoreceptor agonists   171
  - 7.3.3 Inhaled β_2-adrenoreceptor agonists   172
  - 7.3.4 Oral and inhaled corticosteroids   176
  - 7.3.5 Leukotriene modifiers   182
  - 7.3.6 Sodium cromoglicate and nedocromil   184
  - 7.3.7 Acetylcholine muscarinic receptor (ACh mR) antagonists   184
- 7.4 Summary   185
  - References   185

8 Rhythms in therapeutics of cardiovascular disease   193

- 8.1 Introduction   193
  - 8.1.1 Chronobiology of the cardiovascular system   193
  - 8.1.2 Rhythms in cardiovascular events   194
- 8.2 Chronopharmacology of hypertension   195
  - 8.2.1 β-Adrenoreceptor antagonists   195
  - 8.2.2 Calcium channel blockers   196
  - 8.2.3 Angiotensin-converting enzyme (ACE) inhibitors   198

		8.2.4	Other antihypertensives 200

        8.2.4   Other antihypertensives   200
        8.2.5   Duration of the antihypertensive effect   202
  8.3   Chronopharmacology of coronary heart disease   202
        8.3.1   Oral nitrates   202
        8.3.2   β-Adrenoreceptor antagonists   203
        8.3.3   Calcium channel blockers   205
  8.4   Summary   205
        References   206

9 Rhythms, pain and pain management   211
  9.1   Introduction   211
  9.2   Chronobiology of pain   212
        9.2.1   Rhythms in endogenous opioid peptide levels   212
        9.2.2   Rhythms in experimental pain   213
        9.2.3   Pain rhythms in clinical situations   213
  9.3   Chronopharmacology of analgesics   219
        9.3.1   Local anaesthetics   219
        9.3.2   Anti-inflammatory agents   221
        9.3.3   Opioid analgesics   224
        9.3.4   Placebo   228
  9.4   Guidelines for chronotherapeutic use   228
  9.5   Conclusions   229
        References   230

10 Rhythms of cancer chemotherapy   235
  10.1   Introduction   235
  10.2   Genetic and molecular biology   235
  10.3   Circadian coordination, carcinogenesis and tumour outcome   238
  10.4   Biological rhythms in host and/or tumours   240
         10.4.1   Animal data   240
         10.4.2   Human data   243
  10.5   Chronotoxicity and chronopharmacology in animals   245
         10.5.1   Chronopharmacology of single agents   245
         10.5.2   Chronopharmacology of combination treatment   250
  10.6   Chronotoxicity and chronopharmacology in humans   251
         10.6.1   Fluoropyrimidine derivatives   251
         10.6.2   Other drugs   261
  10.7   Disruption of spontaneous rhythmicity of the host with tumour development or tumour treatment   262
         10.7.1   Influence of tumour development on host circadian rhythmicity   262
         10.7.2   Influence of anticancer treatment on host circadian rhythmicity   264

	10.8	Chronoefficacy 265
		10.8.1 Animal data 265
		10.8.2 Human data 268
		10.8.3 Organisation of chronotherapy centres within the EORTC Chronotherapy Group 271
	10.9	Conclusion 273
		References 274

11 Chronopharmaceutical drug delivery 283

- 11.1 Introduction 283
- 11.2 Time-delayed tablet formulations 285
  - 11.2.1 Osmotic control 285
  - 11.2.2 Gellable barrier layers using hydrophilic polymers 287
  - 11.2.3 Erodible barrier layers 289
  - 11.2.4 Tablets that rupture 290
  - 11.2.5 Three-dimensional printing techniques 291
- 11.3 Time-delayed pellet formulations 291
  - 11.3.1 Pellet formulation with a sigmoidal release profile 291
  - 11.3.2 Pellets that rupture 292
- 11.4 Time-delayed capsule formulations 293
  - 11.4.1 Plug separation from permeable capsules 293
  - 11.4.2 Impermeable capsules with separable components 295
  - 11.4.3 Impermeable capsules fitted with erodible plugs 297
  - 11.4.4 Pressure-controlled systems 300
  - 11.4.5 Hydrophilic sandwich capsule 300
- 11.5 The potential use of pH-based systems using gastroresistant coatings 302
- 11.6 Conclusions 303
- References 303

12 Circadian rhythm abnormalities 309

- 12.1 Normal variations in the phase of circadian rhythms 309
- 12.2 Abnormal variations in the phase of circadian rhythms 312
- 12.3 Occupational health and long-haul travellers; jet lag 314
  - 12.3.1 Differences between individuals 315
- 12.4 Occupational health and night workers; shift-workers' malaise 317
  - 12.4.1 Accidents and short-term effects 318
  - 12.4.2 Long-term effects 319
  - 12.4.3 Differences between individuals 320
  - 12.4.4 Adjustment of the body clock 321

x Contents

12.5 Treatment of circadian rhythm disorders 323
 12.5.1 Treatment of jet lag 323
 12.5.2 Treatment of shift-workers' malaise 329
 12.5.3 Treatment of clinical problems associated with circadian rhythm disorders 331
12.6 Summary 336
Acknowledgements 337
References 337

13 Drug effects on suprachiasmatic nuclei function 343

13.1 Introduction 343
 13.1.1 What are the challenges? 344
 13.1.2 Innervation of the SCN 344
 13.1.3 Generation of the endogenous period 345
 13.1.4 Rhythm amplitude 346
 13.1.5 Phase response 346
13.2 Methodology 347
 13.2.1 *In vitro* methodology 347
 13.2.2 *In vivo* methodology 348
 13.2.3 SCN c-*fos* induction 350
13.3 Neurotransmitter and drug effects on rhythms 350
 13.3.1 Excitatory amino acids 350
 13.3.2 Serotonin 352
 13.3.3 GABA 357
 13.3.4 Melatonin 358
 13.3.5 Acetylcholine 359
 13.3.6 Purinergic 359
 13.3.7 Histamine 360
 13.3.8 Dopamine 361
13.4 Conclusion 361
Acknowledgements 362
References 362

14 The effect of ageing on melatonin secretion: clinical aspects 369

14.1 Melatonin secretion and control 369
 14.1.1 Synthesis 369
 14.1.2 Noradrenergic control 369
 14.1.3 Light control 371
 14.1.4 Circadian rhythm 371
 14.1.5 Metabolism 372
14.2 Pharmacokinetics and toxic effects 372
14.3 Melatonin and ageing: physiological effects 374
 14.3.1 Fetal period and infancy 374
 14.3.2 Melatonin in the elderly 374

	14.4	Effects of melatonin on sleep 381
	14.5	Melatonin as a resynchronising agent 383
	14.6	Melatonin as a free-radical scavenger 384
	14.7	Conclusion 385
		References 386

15 Chronobiological mechanisms in seasonal affective disorder   393

- 15.1 Introduction   393
- 15.2 Diagnosis of SAD and light therapy   393
- 15.3 Biological rhythms in SAD   396
  - 15.3.1 Seasonality and brain serotonin function   396
  - 15.3.2 Photoperiod and melatonin secretion   399
  - 15.3.3 Circadian phase shift   400
- 15.4 Conclusion   403
- Acknowledgement   403
- References   403

Index   409

Preface

Chronobiology – the study of how plants and animals measure time and of how time of day and the seasons affect biological processes – is not a new science. For as long as the human race has observed and recorded biological processes there has been evidence of systematic diurnal and seasonal variation. The concept of a 'biological clock' is newer and has only gained scientific respectability in the last half century. The demonstration of the exact location of a self-sustaining oscillator in the suprachiasmatic nucleus of the mammalian hypothalamus was seminal. Subsequent research has elucidated the neural and hormonal connections through which the oscillator is controlled and through which, in turn, the oscillator signals information about time to the rest of the organism.

The full armamentarium of the molecular biologist has also been brought to bear; as a result we now have a much clearer and more detailed understanding of how cells can generate and sustain a reliable time signal. What has also emerged strongly in recent decades is the extent to which disease processes and their pharmacological treatment can be influenced by time of day.

Chronotherapeutics – the application of chronobiological principles to the treatment of disease – is still in its infancy, but the infant is robust and growing rapidly. Remarkable strides have been made, particularly in oncology, in cardiovascular, respiratory and gastrointestinal medicine, in pharmacokinetics and in drug delivery.

The intention behind this book is to promote and encourage these developments by reaching out to scientists and practitioners who are expert in their own field but to whom the intricacies and delights of chronobiology may be relatively new. There have been many excellent books written on chronobiological topics by and for experts in the field. This book has certainly been written by experts and I am grateful to friends and colleagues in the relatively small world of chronobiology who have rallied to yet another call on their valuable time.

I hope therefore that this book will be of interest to scientists and practitioners alike who are not primarily chronobiologists but who are interested in becoming better informed about a branch of science that is at once fascinating and of growing importance.

Peter Redfern
April 2003

About the editor

Peter Redfern studied Pharmacy at the University of Nottingham, receiving his BPharm degree in 1962. He stayed in Nottingham to carry out research in pharmacology under the supervision of the late James Crossland and was awarded his PhD in 1965. After a brief spell in the Portsmouth School of Pharmacy, he joined the fledgling University of Bath as lecturer in pharmacology in 1968. He has remained in Bath ever since and is now Professor of Pharmacology. Throughout his career he has been active in teaching and research. As a founder and now co-director of the Pharmacy Consortium for Computer Aided Learning (PCCAL) he has been at the forefront of the use of new technologies in learning and teaching in the pharmaceutical sciences. His research in psychopharmacology and circadian rhythms in the nervous system has led to over 200 refereed publications, as well as books, reviews and contributed chapters. He was awarded his DSc by the University of Bath in 1991. He is a member of the Institute for Learning and Teaching and a Fellow of the Royal Pharmaceutical Society of Great Britain.

Contributors

Denis Beauchamp PhD
Professor of Microbiology, Université Laval, School of Medicine and Centre de recherche en infectiologie, Centre Hospitalier Universitaire de Québec, Québec, Canada

Rodolfo Costa PhD
Professor of Genetics, Department of Biology, University of Padova, Padua, Italy

Christian Focan MD, PhD
Head of Department of Internal Medicine & Oncology, Cliniques Saint-Joseph, Liège, Belgium and EORTC Chronotherapy Group, Brussels, Belgium

Chiaki Fukuhara PhD
Assistant Research Professor, Neuroscience Institute, Morehouse School of Medicine, Atlanta, Georgia, USA

Martine L Garabette PhD
Medical writer and editor, Department of Pharmacy and Pharmacology, University of Bath, Bath, UK

David J Kennaway PhD
Associate Professor, Department of Obstetrics and Gynaecology, Adelaide University Medical School, South Australia

Gaston Labrecque PhD
Chronopharma Inc., Cap-Rouge, Québec, Invited Professor, Facultie de Pharmacie, Université de Montréal, Montréal, and Centre de recherche en infectiologie, Centre Hospitalier Universitaire de Québec, Québec, Canada

Esther L Langmack MD
Assistant Professor of Medicine, National Jewish Medical and Research Center, Denver, Colorado, USA

Björn Lemmer MD, PhD
Professor of Pharmacology and Toxicology, Director, Institute of Pharmacology and Toxicology, Ruprecht-Karls-University of Heidelberg, Mannheim, Germany

Richard J Martin MD
Professor of Medicine, Head, Pulmonary Division, National Jewish Medical and Research Center, Denver, Colorado, USA

David S Minors PhD
School of Biological Sciences, University of Manchester, Manchester, UK

John G Moore MD
Professor of Medicine, University of Utah; and Chief, Gastrointestinal Section, Department of Veterans Affairs (VA) Salt Lake City, Health Care System (VASLCHCS), Salt Lake City, Utah, USA

Alexander Neumeister MD
Senior Investigator, NIMH, Mood and Anxiety Disorders Program, Bethesda, Maryland, USA

Peter Redfern BPharm, PhD, DSc, FRPharmS
Department of Pharmacy and Pharmacology, University of Bath, Bath, UK

Howard N E Stevens BPharm, PhD, CChem, FRSC, FRPharmS
Pfizer Professor of Exploratory Drug Delivery, Department of Pharmaceutical Sciences, University of Strathclyde, Glasgow, Scotland, UK

Gianluca Tosini PhD
Associate Professor of Anatomy and Neurobiology, Neuroscience Institute, Morehouse School of Medicine, Atlanta, Georgia, USA

Yvan Touitou MD
Professor, Department of Biochemistry and Molecular Biology, Faculty of Medicine Pitie-Salpetriere, Paris, France

Marie-Claude Vanier BPharm, MSc
Associate Professor of Clinical Pharmacy, Faculté de pharmacie, Université de Montréal, Montréal, Québec, Canada

James M Waterhouse DPhil
Research Institute for Sport and Exercise Sciences, Liverpool John Moores University, Liverpool, UK

Klaus Witte MD
Institute of Pharmacology and Toxicology, Ruprecht-Karls-University of Heidelberg, Mannheim, Germany

Mauro Zordan PhD
Assistant Professor of Genetics, Department of Biology, University of Padova, Padua, Italy

1

An introduction to circadian rhythms and their measurement in humans

James M Waterhouse, Peter Redfern and David S Minors

1.1 Constancy and rhythmicity

One of the most important concepts in physiology is that of 'homeostasis', the principle by which the internal environment of the body is maintained within narrow limits in spite of the tendency for an individual's environment and activities to produce change. For example, core body temperature is normally kept within the range 35–40°C, even when the subject is active or asleep, and when the environmental temperature can change over a far wider range. Thermoregulatory reflexes ensure that such comparative constancy is maintained in health. Other reflex mechanisms are important in controlling other physiological and biochemical variables.

For the past fifty years or so, there has been the development of means of measuring these variables repeatedly and conveniently in subjects who are living normally: that is, active in the daytime and sleeping at night. These measurements have revealed that physiological and biochemical variables show daily rhythms that are in synchrony with the sleep–activity cycle of the individual. A selection of these variables is shown in Figure 1.1, and the list could be extended almost indefinitely. The rhythms observed tend to fall into one of two groups. In the first are those that peak during the daytime and are associated with the activity phase of the individual; core temperature, mental, physical and gastrointestinal activities, blood pressure and heart rate, and the secretion of adrenaline (epinephrine) all fit into this group. The second group, where rhythms show a peak during nocturnal sleep, is smaller and includes secretion of several hormones, among which are growth hormone, cortisol and melatonin.

2 Circadian rhythms and their measurement in humans

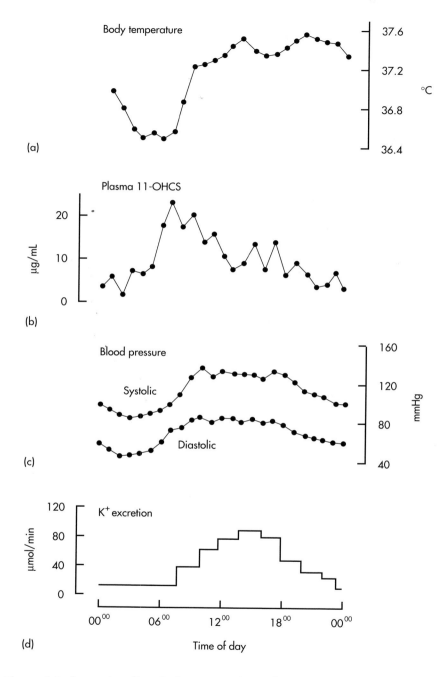

Figure 1.1 Some circadian rhythms in a subject sleeping from 00.00 to 08.00. (11-OHC, 11-hydroxycorticosteroids.) Reproduced from Redfern et al. (1991).

1.2 Endogenous and exogenous components of a rhythm

It is reasonable to suppose that such rhythms result from behavioural changes associated with the sleep–activity cycle. During the daytime, we are physically and mentally active, and we eat and drink in an environment that is busy and stimulating. At night, we rest and recuperate during sleep; growth hormone and cortisol both promote growth of tissues and, in the absence of food intake and insulin secretion, maintain plasma glucose at the same time as fat is metabolised.

Two main lines of indirect evidence indicate that this exogenous cause cannot be a full explanation of the position. First, after long-haul flights (to the east or west and crossing several time zones, rather than in a north–south direction), individuals suffer temporarily from 'jet lag' (Waterhouse *et al.*, 1997), a group of symptoms that includes fatigue during the new daytime, and yet an inability to sleep well in the new night-time; falls in the abilities to concentrate and to perform complex mental and physical tasks; and loss of appetite, coupled with feeling bloated after a meal and an irregularity of bowel habits. The second line of evidence is that similar symptoms – 'shift-workers' malaise' (Waterhouse *et al.*, 1992) – occur in those working at night. (These two areas are considered in more detail in Chapter 12.)

Such results indicate that it is not possible for the body to adjust its sleep–activity cycle immediately. This leads to the idea, developed independently in studies in other animals and plants, that some internal mechanism might be involved: the 'body clock'.

1.2.1 Constant routines

There are two methods by which a more direct assessment can be made of the role of such a body clock in generating the observed rhythm of core temperature. The first of these requires the direct effects of the environment and sleep–wake cycle to be reduced as much as possible. A protocol to do this is called a 'constant routine' (Mills *et al.*, 1978; Czeisler *et al.*, 1985). In this, the subject is required to stay awake and sedentary (or, preferably, lying down and relaxed) for at least 24 hours in an environment of constant temperature, humidity and lighting; to engage in similar activities throughout, generally reading or listening to music; and to take identical meals at regularly spaced intervals. When this protocol is undertaken, it is observed (Figure 1.2) that the rhythm of core temperature does not disappear, even though its amplitude becomes decreased. Three deductions can be made from this result.

4 Circadian rhythms and their measurement in humans

1. The component of the temperature rhythm that remains must arise from within the body. It is described as an 'endogenous' rhythm and it is this component that is attributed to the 'body clock'.
2. An effect of the environment and lifestyle does exist, however, since the two curves differ; this component of the rhythm is termed 'exogenous'. The exogenous effects for core temperature are mainly physical activity (which raises core temperature during the daytime) and sleep (which lowers it at night). Since the effect of exogenous influences is to obscure the role of the endogenous, clock-driven component in a set of measured data, these influences are known as 'masking effects'.
3. In subjects living normally, these two components are in phase, both raising core temperature in the daytime and both lowering it at night. (The implications of this finding will be considered in Chapter 12, which considers abnormal circadian rhythms.)

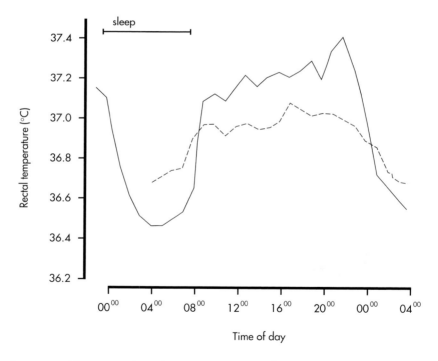

Figure 1.2 Mean circadian changes in core (rectal) temperature measured hourly in eight subjects: living normally and sleeping from 00.00 to 08.00 (solid line); and then woken at 04.00 and spending the next 24 h on a 'constant routine' (dashed line). Reproduced from Minors and Waterhouse (1981).

1.2.2 Fractional desynchronisation

The second method for assessing the role of a body clock in generating the observed rhythm of core temperature is that of 'fractional desynchronisation' (Kleitman and Kleitman, 1953; Boivin *et al.*, 1997). This approach is based on the observation that the body clock is not able to adjust to an imposed lifestyle whose period differs substantially from 24 h (see later). For example, if a subject lives on 27 h 'days' (with 9 h of sleep and 18 h of activity each 'day'), the endogenous component of the circadian rhythm retains a period close to 24 h; as a result, 8×27 h 'days' equal 9×24 h endogenous cycles. This length of time is called a 'beat cycle'. One result of this protocol is that the endogenous component of the rhythm moves continually out of phase with the sleep–wake cycle and then back into phase. If temperatures are measured throughout a beat cycle, then they can be averaged in one of two ways. First, if the results from nine 24 h cycles are averaged into a single 24 h rhythm, any phase of this average rhythm is mixed with all phases of the imposed (27 h) sleep–wake cycle. That is, provided that the sleep–wake cycle is similar day by day, any effects directly or indirectly due to it will be cancelled out, and the average rhythm observed will represent the endogenous component of the measured rhythm (see Figure 1.3). Second,

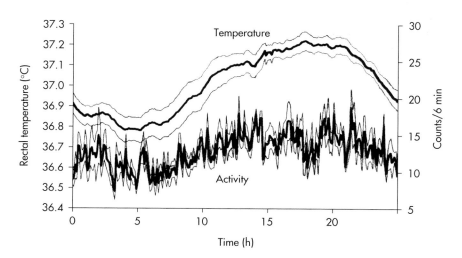

Figure 1.3 Mean (±SE) rectal temperature and activity (measured by means of a 'wrist actimeter') in a group of subjects living on 27 h 'days' but retaining a circadian period of 24 h due to the presence of adequate external time cues (zeitgebers) of this period. Results educed at 24 h. Reproduced from Waterhouse *et al.* (1999c).

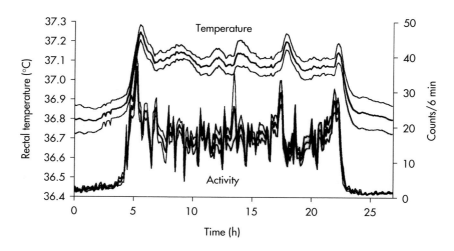

Figure 1.4 Mean (±SE) rectal temperature and wrist activity in a group of subjects living on 27 h 'days' but retaining a circadian period of 24 h due to the presence of adequate zeitgebers of this period. Results educed at 27 h. Reproduced from Waterhouse *et al.* (1999c).

if the temperatures from eight 27 h cycles are averaged into a single 27 h rhythm, any phase of this average rhythm is mixed with all phases of the endogenous (24 h) cycle. That is, any effects directly or indirectly due to the body clock will be cancelled out, and the average rhythm observed will represent the exogenous component of the measured rhythm (see Figure 1.4).

1.2.3 Differences between variables in the nature and strength of their exogenous component

The above principle, that a measured rhythm consists of an endogenous, clock-driven component and an exogenous component driven by the individual's lifestyle and environment, is a general one and applies to all variables. In practice, however, differences exist between variables. The first differences are those between the aspects of the subject's sleep–wake cycle and environment that contribute to the exogenous component of a measured rhythm. To take some examples:

- Blood pressure, heart rate and airway calibre are affected similarly to core temperature, being raised by physical and mental activity and lowered by sleep.
- Moderate levels of activity and sleep appear to have less effect upon

the secretion of melatonin, but its secretion is suppressed by bright light (Lewy *et al.*, 1980).
- The levels of several hormones are modified by food ingestion; and antidiuretic hormone (ADH) release is affected by fluid intake.

The second difference between variables is in the relative size of the endogenous and exogenous components of a rhythm. Some variables appear to have much larger exogenous than endogenous components: for example, heart rate, blood pressure, the excretion of water by the kidneys, and the secretion of growth hormone by the pituitary gland (which is promoted by sleep). By contrast, the excretion of potassium in the urine and, particularly in sedentary subjects studied indoors, the secretion of melatonin and cortisol into the plasma have smaller exogenous than endogenous components. Core temperature has endogenous and exogenous components of similar magnitude (compare the two curves in Figure 1.2).

1.2.4 Variants of 'constant routines'

The requirement of a 'constant routine' is to minimise the exogenous component of a rhythm experimentally; exactly how this is done depends upon the variable being measured. For melatonin, whose secretion is suppressed by bright light, it becomes necessary to take blood samples in dim light. Lewy's group (Lewy *et al.*, 1999) has tended to use the time at which melatonin secretion starts its nocturnal rise as a phase marker, and this is therefore called the DLMO (dim light melatonin onset). The time of cessation of melatonin secretion has also been used; when the excretion of melatonin in the saliva or urine is considered, the method of fitting a cosine curve to values obtained over the course of 24 h and estimating the time of peak of the fitted curve has been employed.

For studies of the circadian rhythms of the hormones associated with the control of carbohydrate and fat metabolism, Van Cauter's group (Van Cauter *et al.*, 1989) has used a constant intravenous infusion of glucose in place of eating identical, evenly spaced meals. The rationale for this procedure is that eating identical meals does not guarantee a constant uptake of nutrients into the bloodstream since it depends upon digestion and absorption from the gut, both of which themselves show circadian variation. Intravenous infusion of a glucose solution overcomes this problem.

1.3 Some properties of the body clock in humans

1.3.1 Free-running rhythms

When individuals are studied in environments where there are no time cues – in an underground cave or a specially constructed 'isolation unit', for example – all rhythms continue and, to a very considerable extent, show the same phase relationships with each other as when subjects are living normally. The example of the sleep times of a subject studied in such conditions is shown in Figure 1.5. This observation confirms the endogenous origin of the rhythms, and they are said to be 'free-running'. However, whereas rhythms show a period of exactly 24 h (the same as that of the solar day) when the subject is studied in a normal environment, when they are measured under such temporal isolation their period is closer to 25 than 24 h. It is for this reason that they are called *circadian* (from the Latin: 'about a day'). More recent measurements (Carskadon *et al.*, 1999), using a forced desynchronisation protocol, indicate that the free-running period of the body clock in humans is about 24.3 h. This is shorter than the values found previously; the earlier values were influenced by the light cycle to which the individual was exposed as a result of the self-selected sleep–wake cycle.

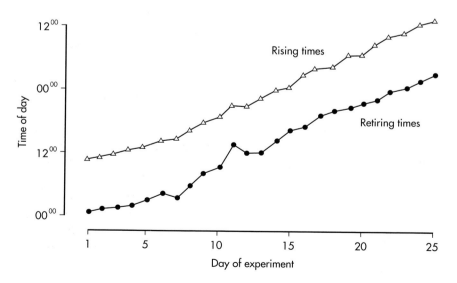

Figure 1.5 Successive times of retiring and rising in a subject living alone in a time-free environment. Reproduced from Waterhouse *et al.* (1990).

Clearly, a clock that loses several minutes per day is useless, even though this is an error of only about 1%, much smaller than for many other physiological and biochemical variables. For the body clock to be of value, and for individuals to show the same timing of rhythms that is necessary for the data of Figure 1.1 to be obtained, the circadian rhythms and the body clock need to be adjusted continually for them to remain synchronized to a solar (24 h) day. Synchrony is achieved by *zeitgebers* (German for 'time-giver'), which are rhythms resulting, directly or indirectly, from the environment. A zeitgeber is said to 'entrain' the body clock to a period equal to that of the zeitgeber. There is a limit to the range of entrainment that can be produced by a zeitgeber, which depends upon the zeitgeber itself and its strength. Normally, the body clock in humans cannot be entrained to a period as short as 21 h or as long as 27 h; this becomes the basis for the method of fractional desynchronisation (see Section 1.2.2).

1.3.2 Zeitgebers in humans

In different mammals, rhythms of the light–dark cycle, of food availability–unavailability, of activity–inactivity, and of social influences act singly or in combination as zeitbegers. In humans living normally, some combination of these zeitgebers is normally present. However, the evidence strongly suggests that the most important zeitgeber is the light–dark cycle, combined with the rhythm of melatonin secretion. Other putative zeitgebers, including physical activity and meal times, have been found to be comparatively unimportant (see, for example, Beersma and Hiddinga, 1998; Mrosovsky, 1999).

1.3.2.1 Light

The effect of light depends on the time at which it is presented (Honma *et al.*, 1987; Czeisler *et al.*, 1989; Minors *et al.*, 1991; Dawson *et al.*, 1993). It has been shown that

- Pulses of bright light (that is, light of an intensity found outdoors rather than domestically), centred in the 6 h 'window' immediately after the trough of the body temperature rhythm (the trough normally being about 05.00, see Figure 1.1), produce a phase advance.
- Pulses centred in a window 6 h before the temperature minimum produce a phase delay.
- Pulses centred away from the trough by more than a few hours have little effect (Figure 1.6).

10 Circadian rhythms and their measurement in humans

This nonparametric relationship between the time of light exposure and the phase shift of the body clock that it produces is called a phase–response curve (PRC). More recent work has shown that weaker light pulses (as found with domestic lighting) also affect the clock and can produce smaller phase shifts (Boivin *et al.*, 1996; Waterhouse *et al.*, 1998). This is important since many humans living in industrial and postindustrial societies work indoors and have very little exposure to natural daylight.

When light acts as a zeitgeber, it is believed that the information passes directly from the retina to the suprachiasmatic nuclei (SCN) of the

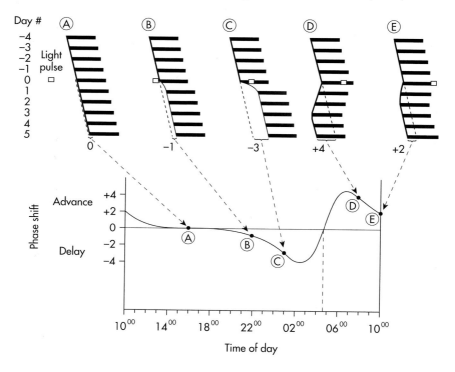

Figure 1.6 Illustrating the derivation of a phase–response curve. Upper panel: The effect of pulses of light given at different phases of the free-running sleep–wake cycle upon the subsequent phase of this cycle. A–E represent separate experiments performed using the same subject under identical experimental conditions. It can be seen that the light pulse (represented by the open rectangle) induces either a delay (B, C), an advance (D, E) or no change (A), depending on the phase in the cycle at which the light is administered. Lower panel: The size of the advance or delay is plotted against the (circadian) time of light administration, to produce the phase–response curve. The temperature minimum in this illustration is at about 04.30. Based on Moore-Ede *et al.* (1982).

hypothalamus via the retinohypothalamic tract. The origin of this input from the retina is still being researched actively. There is evidence to suggest that the retina contains a subgroup of receptors whose visual pigment is based on vitamin B_2 rather than the more conventional vitamin A_1-based opsins, but the nature of the receptor that contains this pigment and its distribution in the retina are not yet clear (Miyamoto and Sancar, 1998; Lasko *et al.*, 1999; Visser *et al.*, 1999). It has even been suggested, but not confirmed in later studies, that information about the light–dark cycle can originate from the skin (Lindblom *et al.*, 2000).

1.3.2.2 Melatonin

Ingestion of melatonin has also been shown to adjust the phase of the body clock. Again, according to the time of ingestion, it can phase-advance, phase-delay or have no effect upon the body clock; that is, there is a PRC for melatonin ingestion (Lewy *et al.*, 1998). The shifts tend to be the inverse of those produced by light: melatonin ingestion in the afternoon and early evening advances the body clock and in the night and early morning delays it. Receptors for melatonin have been found in the SCN (Reppert *et al.*, 1988; Liu *et al.*, 1997). Light inhibits melatonin secretion, the amount of inhibition being dependent on light intensity (Lewy *et al.*, 1980), and this makes the phase-shifting effects on the body clock of these two zeitgebers reinforce each other. Thus, bright light in the early morning, just after the temperature minimum, advances the phase of the body clock, not only directly via the PRC to light but also indirectly, as it suppresses melatonin secretion and so prevents the phase-delaying effect that melatonin would have exerted at this time (Lewy and Sack, 1996).

1.4 Widespread effects of the body clock

The effects of the hypothalamic SCN permeate the body because they act upon core temperature, plasma hormone concentrations, the outflow of the sympathetic nervous system, and the sleep–wake cycle, all of which exert widespread effects and originate in, or have connections with, the hypothalamus.

These effects of the body clock act upon the body as a whole in two ways. First, they promote the 'ergotropic' actions of the body (those involving physical and mental activity and their associated biochemical and cardiovascular changes) in the daytime, and separate them from the 'trophotropic' actions (those involving recovery and restitution during a period of inactivity), which are promoted at night. The second role is to enable preparations to be made for the switches from the active to the sleeping phase, and vice versa. Such profound neurophysiological changes cannot be brought about rapidly, but require instead ordered reductions or increases in activities of many biochemical and physiological functions, and this takes time. Individuals have to be prepared for going to sleep and for waking up.

Of particular importance are the rhythms of metabolism (Waterhouse *et al.*, 1999a), mental performance (Folkard, 1990), physical performance (Reilly *et al.*, 1997; Youngstedt and O'Connor, 1999) and sleep (Akerstedt and Gillberg, 1981).

1.4.1 Metabolism

During the daytime, particularly in the morning, the sensitivity of the tissues to the hormone insulin is high. This, coupled with the fact that we eat then, causes most daytime energy to come from the metabolism of glucose, and the extra energy to be stored in the muscles and liver as glycogen. Insulin is also lipogenic, that is, it has the property of causing the fats that we eat to be laid down in adipose tissue. At night, by contrast, the sensitivity of the body tissues to insulin is reduced and growth hormone, whose secretion is associated with the early part of sleep, further reduces the effectiveness of insulin. As a result, fatty acids are released from the adipose tissue stores during the night and they, rather than glucose, are metabolised then. Also, there is an increased release of cortisol towards the end of the night (see Figure 1.1), which conserves glucose by promoting gluconeogenesis (Waterhouse *et al.*, 1999a).

1.4.2 Mental and physical performance

Most types of mental performance, mood and physical performance are phased similarly to the rhythms of core temperature and plasma adrenaline, and peak in the daytime (Figure 1.7a,b). For physical activity (Reilly *et al.*, 1997), this reflects a higher tissue metabolism and cardiovascular performance at this time. For mental performance and mood, the parallelisms that are often noted (Folkard, 1990) are those with the circadian rhythms of core temperature and plasma adrenaline.

The general parallelism between mental performance and core temperature has been investigated in more detail by studying subjects during the 'fractional desynchronisation' protocol (see Section 1.2.2).

These studies (Boivin *et al.*, 1997) have confirmed a parallelism between body temperature and mental performance; a parallelism that also exists during the trough of the circadian cycle, at a time when the subject is normally asleep. This parallelism applies also to self-chosen work rate when physical performance is considered (see Figure 1.7b). As a result of this last circadian rhythm, there will also be rhythms in the pace and intensity of exercise when it is self-selected (in a training context, for example), and in the amount of physical work performed in an occupational setting.

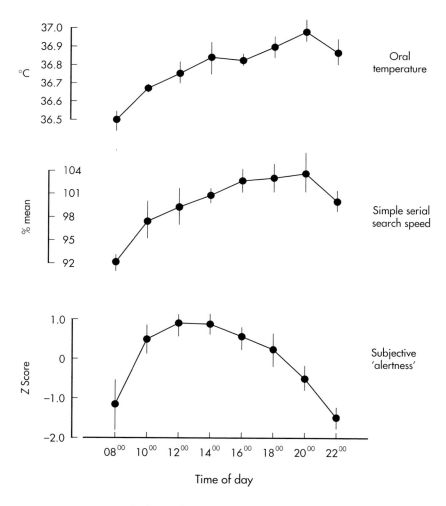

Figure 1.7a Daytime rhythms of oral temperature, a simple mental task performance and subjective alertness. (Reproduced from Folkard, 1990.)

14 Circadian rhythms and their measurement in humans

A close inspection of Figure 1.7a shows that the rhythms differ in detail when their time-courses are considered. These differences can be seen as variations in the rate at which a decrement occurs as a result of time elapsed while awake. As a general rule, mood and complex mental performance tasks show the greater speed of deterioration; simpler mental performance tasks show less, as is the case with physical performance

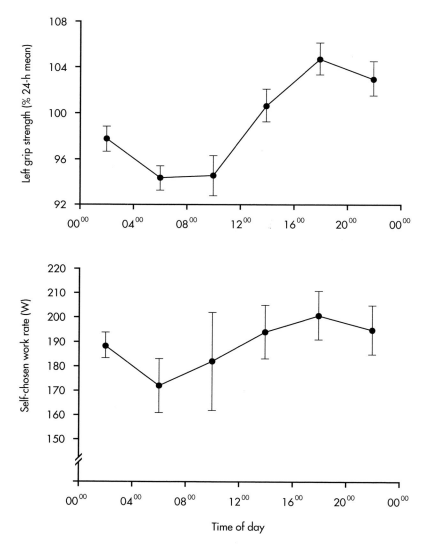

Figure 1.7 b 24 h rhythms of left grip strength and self-chosen work rate. (Based on Reilly et al., 1997.)

tasks (Folkard, 1990). This deterioration is often referred to as 'fatigue'. In the present context, it can be considered to be an exogenous component of the measured circadian rhythm. The effects of fatigue upon physical performance are less direct. Whereas there seems to be comparatively little effect upon muscle strength, performance at sports and games includes components that are affected, such as decisions regarding strategy, the processing of sensory inputs, and the desire to train maximally (Reilly et al., 1997).

Other exogenous factors for mental performance include the conditions (lighting, noise, etc.) under which the task is being performed. The fractional desynchronisation protocol (Boivin et al., 1997) has confirmed the presence of this deterioration due to time spent awake, and that it is more marked for complex tasks than for simple tasks.

1.4.3 Sleep

The ease of getting to sleep and staying asleep shows a circadian variation that is linked to the rhythm of core temperature (Akerstedt and Gillberg, 1981). Sleep onset is easiest to initiate at the temperature trough or when temperature is falling rapidly; it is most difficult to initiate at the peak or when the temperature is rising rapidly. Spontaneous waking is most likely to occur when core temperature is high or rising, particularly some 3–6 h after the temperature minimum. By contrast, it is least likely if core temperature is low or falling quickly, about 3–6 h before the temperature minimum. Bearing in mind that the minimum of core temperature is normally around 05.00 (see Figure 1.1), then the conventional times of sleeping – between 23.00 and 08.00 – can be seen to be synchronised most appropriately with the circadian rhythm of core temperature, a fact that has been stressed by several authors (see, for example, Dijk and Czeisler, 1995).

The time of greatest rate of fall of the temperature rhythm follows shortly after the time of onset of melatonin secretion. Melatonin promotes a fall of core temperature, possibly by causing cutaneous dilatation. Whatever the reason, melatonin is often considered to be a 'natural hypnotic' (see Haimov and Lavie, 1997). Moreover, the links between increasing sleep propensity, declining core temperature and rising melatonin secretion are believed to be functionally important (Lack and Lushington, 1996; Reid et al., 1996; Murphy and Campbell, 1997; Shochat et al., 1997).

1.5 Field studies

1.5.1 Problems with 'constant routines' and the forced desynchronisation protocols

Although the constant-routine protocol (and its variants) is often regarded as providing the 'gold standard' when describing a circadian rhythm (and inferring the output of the body clock), it suffers possible limitations. By its very nature, the protocol is too demanding for field conditions, for experiments with animals, and for individuals who cannot give informed consent. It is also inappropriate for use over extended periods of time, such as on consecutive days after a time-zone transition or during night work. There is also the problem that, since sleep is prohibited, variables that are affected by the amount of time spent awake (many aspects of mood and mental performance, for example) cannot be assessed satisfactorily. (A related protocol – 'control days' – was developed by Moog and Hildebrandt (1989), in which subjects underwent a constant routine except that they were allowed to sleep at night. Although this protocol overcame one of the major limitations to the use of constant routines on successive days, the direct effects of sleep are not eliminated, and this might reduce its value.)

Loss of complete sleep periods cannot be levelled against the forced desynchronisation protocol (even though their length changes from 8 h, see Section 1.2.2). However, the protocol is such that it can only give an average value over the course of a beat cycle, and the other problems associated with constant routines apply to this method also.

In other words, the above methods are laboratory-bound, and a different strategy must be adopted in field conditions and in other circumstances where constant routines are impracticable or otherwise unacceptable. For field studies, two additional problems, not present in laboratory studies, must be overcome. These are, first, the choice of a marker of the circadian system that is acceptable to the subject as well as the experimenter and, second, the ability to correct the measured data for the exogenous effects of the environment and sleep–activity cycle of the subject. This last technique is generally called 'purification'.

1.5.2 The choice of a marker for field studies

Methods that require the subject to be woken at night – taking urine or saliva samples, for example – are unsuitable, as are those that require continual venepuncture or an indwelling venous catheter. In practice, therefore, the methods that can be considered are recordings of core temperature, movement, heart rate and, possibly, blood pressure. None of these is ideal.

- The measurement of heart rate and blood pressure can be automated, but the inflation of the pneumatic cuff to measure blood pressure can disturb sleep, and the record of heart rate is dominated by exogenous effects. As a result of this, a suitable method for purifying the data (see Section 1.5.3) becomes of prime importance.
- For core temperature, rectal temperature has been shown to be successful, but the wearing of a rectal probe is unacceptable to many, particularly if they are active rather than sedentary. Tympanic, oral and mid-stream urine temperatures have also been used, but they all require the subject to be woken from sleep; a third of the potential data is missed if recordings are not made during the period of sleep. Two other methods of recording core temperature have been tried: swallowing a 'pill' – a temperature-measuring device that passes through the gut – and insulated skin temperature (generally of the axilla).

In preliminary studies, we have compared the records of rectal, gut and insulated temperatures in subjects living normally. Results (Figure 1.8a) show that there is a very strong similarity between the records of rectal and gut temperature, but that insulated skin temperature was less satisfactory. Our current interpretation of the skin data is that the insulation of the axillary probe is satisfactory (provided that the arms are not raised above the head), but that there might be an effect due to blood reaching the axilla from adjacent, noninsulated areas of skin – for example, if axillary temperature is influenced by the temperature of the blood returning from the arm. Another possible problem arises in that changes in some parts of the cutaneous vasculature are the means by which heat loss is regulated to produce the circadian rhythm of core temperature (Aschoff and Heise, 1972). That is, even at rest in a thermally neutral environment, the delivery of blood to the cutaneous vasculature of the limbs will show a circadian rhythm, peaking when core temperature is falling most rapidly and showing a trough in the early morning when core temperature is rising most quickly. The axilla temperature data (Figure 1.8a) accord with this interpretation, indicating, overall, an appropriate phase delay compared with core temperature of about 6 h.

While the use of gut temperature appears to be a satisfactory alternative to rectal temperature, the pills are expensive and recycling them is unacceptable.

- For movement, generally measured by motion sensors strapped to the wrist of the nondominant hand, the exogenous effects are dominant (Figure 1.8b); many have used data from this device to do little more than record times and amounts of high (awake) and low (asleep) activity.

However, more use can be made of this convenient source of data. The 'dichotomy' between out-of-bed activity and in-bed activity has been measured, and it has been shown that small changes from normal exist in subjects whose sleep–activity cycle is

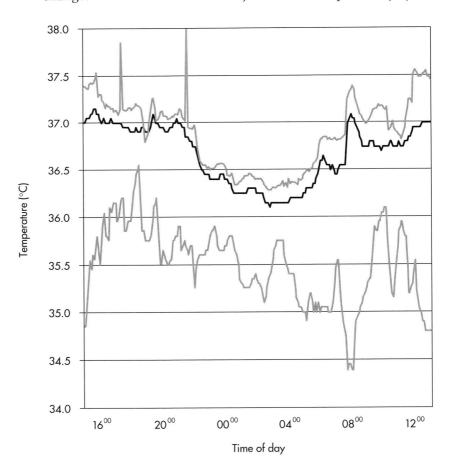

Figure 1.8 a Gut (upper grey trace), rectal (black trace) and insulated axilla (lower grey trace) temperatures in a subject living normally.

changed, as during fractional desynchronisation (Minors *et al.*, 1996a) or during night work (Minors *et al.*, 1996b), and in patients with advanced cancer (Mormont *et al.*, 2000). Figure 1.9 shows the activity profile of a subject who, while undergoing a fractional desynchronisation protocol involving 27 h 'days', was required to attempt sleep from 16.00 to 01.00. The poor sleep during this time is indicated by the activity count, which, though low, was higher than would be associated with a period of consolidated sleep. Also, in the period 02.00–08.00 – which was during the 'daytime' as far as the protocol was concerned but coincided with night by the body clock – activity was very low, indicative of a subject experiencing fatigue at this time.

1.5.3 'Purifying' field data

Whatever recording is made, if the endogenous component of a rhythm is required, it must be 'purified', that is, corrected for the direct, exogenous

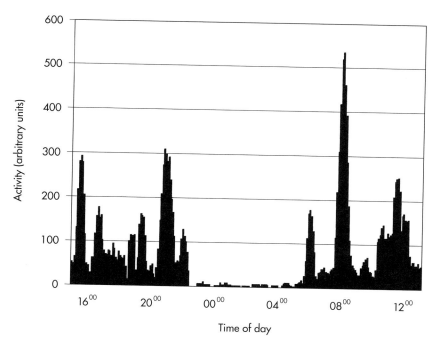

Figure 1.8 b Wrist activity. The subject slept from about 23.00 to 07.00, and went jogging outdoors at about 10.00. Unpublished data from Ben Edwards (Liverpool John Moores University).

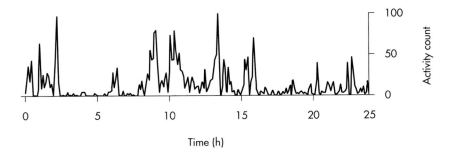

Figure 1.9 Wrist activity record of a subject living on a 27 h 'day' but also retaining a 24 h circadian rhythm due to the presence of adequate zeitgebers of this period. Sleep was taken from 16.00 to 01.00 (shown as 16–25 h on the figure), and the rest of the time was spent awake. Reproduced from Waterhouse *et al.* (1999c).

effects of the individual's sleep–activity cycle and environment. Purification methods use the raw (i.e. 'masked') data and then attempt to correct (or 'purify') these data mathematically to take into account the effects of exogenous influences. Several methods have been developed, and these have been reviewed recently (Waterhouse *et al.*, 2000).

1.5.3.1 Early models of purification

One of the earliest models used was with temperature data, and consisted of adding 0.3°C to values associated with sleep (Wever, 1985). A later version (Folkard, 1989) was applied to data collected after a time-zone transition. It consisted of estimating the exogenous effect of sleep and activity upon control data by comparing raw data with 'normative endogenous' temperature data. (Normative endogenous data were estimates of the mean temperature rhythm of a group of subjects who had remained sedentary and awake throughout a 24 h period.) The difference between raw and normative endogenous data – the exogenous component – could then be adjusted so that it was timed appropriately for the individual's lifestyle after a time-zone transition. This adjusted profile could then be subtracted from the raw data obtained after the time-zone transition in order to estimate the new phase of the endogenous component.

These methods were limited in that they assumed that the exogenous effect was due to sleep only (first method), or that it did not vary between successive days (second method). While this second assumption might be reasonable after a time-zone transition, it is suspect during night work, since sleep length usually changes and the pattern of activity when awake is generally leisure followed by work, whereas work normally comes before leisure during daytime shifts.

1.5.3.2 Recent purification methods

More recently, the methods have been improved in that they calculate the exogenous effect for each day on the basis of results from a diary (recording types of activity) or from measurements of activity (from an actimeter containing motion sensors worn on the nondominant wrist) or of heart rate (Waterhouse *et al.*, 1999b). This correction enables the method to be used whatever the subject's habits. It can also be applied to data collected on consecutive days, since no change to the subject's lifestyle is required.

Purification by categories This method (Minors and Waterhouse, 1992) describes the measured rhythm of core temperature as the sum of an endogenous component (one that is described by a sinusoid) and an exogenous component (one that can be estimated from knowledge of the activity profile). Owing to the thermal capacity of the body, the effect of activity will not be immediate, but it can be estimated from the total activity in the previous 30 min.

Different amounts of activity are allocated to separate 'categories', each then being assessed for its masking effect. The lowest category of activity (associated with sleep) is assumed to have resulted in a negative masking effect upon temperature, and the higher categories (associated with being awake and active) with positive masking effects. The number of categories of activity used can be varied but, when activity is assessed from a diary, six categories are used (labelled 'asleep', 'awake, but lying down', 'sitting, relaxed', 'sitting, doing something', 'standing' and 'walking, moving around'). Using more categories than this can cause difficulty in labelling the categories. When activity is measured by an actimeter or a heart rate monitor, the number of categories can be increased, and the allocation of a given amount of activity to a particular category can simply be decided mathematically (Waterhouse *et al.*, 1999b). It would be predicted that increasing activity when awake would produce successively greater masking effects, but this is not a requirement of the method.

There are other masking effects upon core temperature, including the effect of eating and of mental activity. These have not often been applied to purification models, though the direct effect of the light–dark cycle has been studied in healthy neonates living in a hospital ward (Weinert et al., 1997). Results indicated that the rises in temperature produced by behavioural activity (awake and alert) were greater in the light than in the dark, and that the falls during sleep were greater in the dark then the light.

Purification by intercepts   The second 'purification' model was developed from work on mice (Weinert and Waterhouse, 1998). Its basic assumption is that there is a linear effect of activity upon core temperature. Therefore, a linear regression of measured temperatures upon summed activity over the previous 30 min would enable an intercept temperature, T_{int} (one that would have existed in the absence of any activity) to be estimated. This method assumes that the masking effect increases linearly with the amount of activity, but the shape of the circadian rhythm produced from the intercept temperatures (the T_{int} data set) need not be specified.

The method was initially less successful in humans, probably because the low temperatures during sleep in humans were due not only to inactivity (which was taken into account by the intercept method) but also to a further fall caused by a period of consolidated sleep that was not taken into account by the method. (This problem does not arise with rodent data, since there are not extended periods of sleep in this species.) This outcome necessitated a correction of the temperature data collected from humans during the times allocated to sleep. After this correction (Waterhouse et al., 2000), the method was as successful in dealing with data from humans as with those from mice.

Some studies (Van Dongen and Duindam, 1997) indicate that it might be the change in posture associated with sleep, rather than sleep per se, that is responsible for the extra fall in temperature associated with sleep in humans. If sleep itself does not produce a large fall in temperature, then this argues for the value of the 'control days' protocol of Moog and Hildebrandt (1989) described above, when posture and activity were controlled but sleep was permitted.

Comparison of purification by categories and by intercepts, and other purification methods   The two purification models (purification by categories and purification by intercepts) are compared in Table 1.1 with regard to their assumptions and incidental outcomes, that is, outcomes apart from estimates of the endogenous component of the circadian rhythm and the size of the masking effect. It will be noted that the two models are complementary insofar as the incidental outcomes of one

Table 1.1 A comparison of the assumptions and some incidental outcomes of two 'purification' models

	Purification method	
	Category	Intercept
1. Assumptions of the method		
A. Shape of endogenous component	Sinusoidal	—
B. Effect of activity upon temperature	Phase-independent	Linear
2. Incidental outcomes of the method	Is effect of activity linear?	Is endogenous component sinusoidal? Is masking independent of phase?

Based on Waterhouse *et al.* (2000).

model enable the assumptions of the other to be tested, and vice versa. Results from the two methods have been compared (Waterhouse *et al.*, 2000) using data collected from a forced desynchronisation protocol. Figure 1.10 shows an example of raw temperature data (a), the activity record (d), and the data after purification by these methods (b, c). In summary, the methods produced very similar results with regard to the phase and amplitude of the purified data, and both were far superior to the use of the original raw data, which was considerably affected by the imposed sleep–wake schedule. For details of these comparisons, together with a discussion of the assumptions on which the methods were based and a more detailed comparison with other variants of purification, the reader is referred to the original review (Waterhouse *et al.*, 2000).

There are links between the method of purification by categories and methods that have been used by others. Some have adopted the practice of correcting only sleep temperatures, which, in effect, is using two categories of activity, 'awake' and 'asleep'. A more elaborate method for correcting sleep data has been described that does not treat the masking effect of sleep as being constant throughout the sleep period (Carrier and Monk, 1997).

How accurate are results from purified data? The answer to this question is not known, if only because it has not been possible so far to measure directly the output of the body clock in humans. If it is assumed

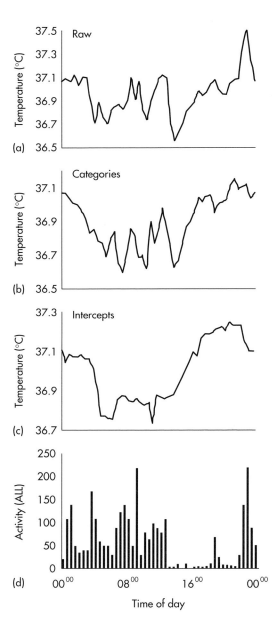

Figure 1.10 Temperature and activity data from a single subject in bed from 13.00 to 22.00. From top downwards: raw temperatures; temperatures purified by the categories method (16 categories used, based on the activity record); temperatures purified by the intercepts method (corrected for sleep, see text); activity profile. Reproduced from Waterhouse *et al.* (2000).

that this output can be inferred from the 'gold standard' of the measured circadian rhythm during constant routines, then the issue becomes one of comparing constant-routine and 'purified' data.

In earlier studies (Minors and Waterhouse, 1992; Minors et al., 1994), it was shown that, after a simulated time-zone transition, the shifts measured using raw data differed significantly from those measured by constant routines in the same subjects. By contrast, shifts measured after purification were not significantly different from the constant-routine results (however, there exists in this result the possibility of a Type II error). It has also been shown (Minors and Waterhouse, 1992) that, during a fractional desynchronisation protocol, the daily phase of the circadian rhythm of purified core temperature showed similar shifts to those of melatonin excretion in urine (a rhythm that, under the conditions of low light intensity, would be predicted to show minimal masking effects). However, this similarity was absent when raw, masked temperature data were considered instead.

Purification by covariance analysis The purification methods described so far were devised either pragmatically or with biological principles in mind, and were described mathematically only afterwards. As a result, certain mathematical considerations, including effects of serial correlation and problems arising from the use of iterative processes, had not been considered when deriving the methods.

A further method that is essentially statistically based, purification by covariance analysis, is being developed (Waterhouse et al., 2001). This model also investigates the combined effects of time of day and spontaneous activity on changes in core temperature, using as the covariate activity recorded by the wrist monitor and summed over the previous 0–18 min (A18), 18–30 min (A30) and 30–42 min (A42). Moreover, a more complex, nonlinear model for describing the effect of spontaneous activity upon core temperature, was used. The model can be described as follows.

$$\text{Temperature} = a_0 + a_1(A18) + a_2(A18)^2 + a_3(A18)^3 + a_4(A30) + a_5(A30)^2 + a_6(A30)^3 + a_7(A42) + a_8(A42)^2 + a_9(A42)^3$$

where the unknown parameters $a_0, a_1, a_2, \ldots a_9$ were estimated using analysis of covariance (ANCOVA).

Such a model can describe an immediate effect upon core temperature of activity (via a significant linear term: a_1, a_4, a_7), or a slow initial effect that rises increasingly (via a significant positive quadratic term: a_2,

a_5, a_8) and ultimately peaks with a further increase in activity (via a significant negative cubic term: a_3, a_6, a_9). The method also enables the temperature profile to be calculated in the absence of effects due to activity (the purified, endogenous circadian component).

The model has, so far, been tested on data obtained from subjects undergoing a forced desynchronisation protocol (Waterhouse *et al.*, 2001), and enabled the 'purified' temperature rhythm shown in Figure 1.11 to be produced. It also produced results that had been predicted from current chronobiological theory and had previously been found in other studies using the other two purification methods (see Weinert and Waterhouse, 1998). Investigation of the relationship between activity and the temperature rise it produced indicated that a linear description was normally a reasonable approximation, lending support to the model of purification by intercepts (see Table 1.1). However, with higher levels of activity, there was evidence that the rise in temperature produced showed a maximum (the a_3, a_6 and a_9 coefficients, those

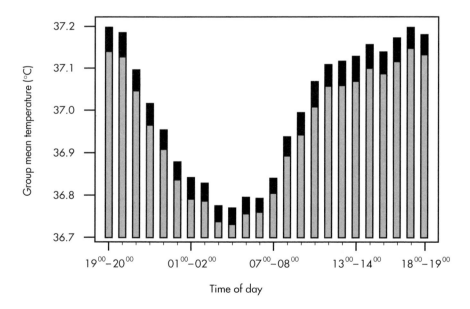

Figure 1.11 Mean hourly temperatures from a group of subjects living on 27 h 'days' but retaining a 24 h circadian rhythm due to the presence of adequate zeitgebers of this period. The data were analysed by ANCOVA and the hourly means have been divided into the portion that could be accounted for by the activity (black bars) and that which was independent of it ((grey) bars). Reproduced from Waterhouse *et al.* (2001).

describing the cubic terms in the above equation, were negative), presumably reflecting the recruitment of thermoregulatory reflex mechanisms at this stage.

1.6 Summary

Our current understanding of circadian rhythms indicates that measured rhythms tend to be impure reflections of the output of the endogenous circadian oscillator, as a result of exogenous or masking effects. The size and nature of these depend upon the variable under consideration. The body clock appears to be a poor time keeper unless it is continually adjusted by zeitgebers, rhythmic influences from the environment and lifestyle. The main zeitgeber in humans is the light–dark cycle, in concert with the circadian rhythm of melatonin secretion. The body clock exerts its effects throughout the body, not only by virtue of its position in the hypothalamus but also because it affects core temperature, hormones and the sleep–wake cycle, which exert widespread influences.

The phase and amplitude of the output of the endogenous circadian oscillator cannot be measured directly in humans but only inferred from studies of marker rhythms, particularly core temperature and melatonin secretion. However, inferences drawn from measured data can often be misleading owing to the exogenous component of a circadian rhythm, the 'masking effect'. Protocols exist that can minimise the masking effect; these are the 'constant routine' and its variants. Whereas different types of 'constant routine' are still the best way to assess the phase and amplitude of the endogenous component of a circadian rhythm, the associated protocol is very demanding, is laboratory-bound, and is not suited to many types of investigation and subject. Purification methods are being developed, and these attempt to correct for the masking effects in data collected in field conditions from subjects who are not constrained by the process of data gathering. Without doubt, these methods make assumptions that must be challenged (see, for example, Klerman *et al.*, 1999), but we believe that the way ahead is to improve current methods and to develop new ones. It is likely that one or more of the methods currently developed will form the basis of an acceptable method for purification.

Acknowledgements

We thank T. Reilly and B. Edwards for their constructive comments on an earlier draft of this article.

References

Akerstedt T, Gillberg T (1981). The circadian variation of experimentally displaced sleep. *Sleep* 4: 159–169.

Aschoff J, Heise A (1972). Thermal conductance in man: its dependence on time of day and on ambient temperature. In: Itoh S, Ogata K, Yoshimura H, eds. *Advances in Climatic Physiology*. Tokyo: Igaku Shoin, 334–348.

Beersma D, Hiddinga A (1998). No impact of physical activity on the period of the circadian pacemaker in humans. *Chronobiol Int* 15: 49–57.

Boivin D, Duffy J, Kronauer R, Czeisler C (1996). Dose–response relationships for resetting of human circadian clock by light *Nature* 379: 540–542.

Boivin D, Czeisler C, Dijk D-J, et al. (1997). Complex interaction of the sleep–wake cycle and circadian phase modulates mood in healthy subjects. *Arch Gen Psychiatry* 54: 145–152.

Carrier J, Monk T (1997). Estimating the endogenous circadian temperature rhythm without keeping people awake. *J Biol Rhythms* 12: 266–277.

Carskadon M, Labyak S, Acebo C, Seifer R (1999). Intrinsic circadian period of adolescent humans measured in conditions of forced desynchrony. *Neurosci Lett* 260: 129–132.

Czeisler C, Brown E, Ronda J, et al. (1985). A clinical method to assess the endogenous circadian phase (ECP) of the deep circadian oscillator in man. *Sleep Res* 14: 295.

Czeisler C, Kronauer R, Duffy J, et al. (1989). Bright light induction of strong (Type 0) resetting of the human circadian pacemaker. *Science* 244: 1328–1333.

Dawson D, Lack L, Morris M (1993). Phase resetting of the human circadian pacemaker with use of a single pulse of bright light. *Chronobiol Int* 10: 94–102.

Dijk D-J, Czeisler C (1995). Contribution of the circadian pacemaker and the sleep homeostat to sleep propensity, sleep structure, electroencephalographic slow waves, and spindle activity in humans. *J Neurosci* 15: 3526–3538.

Folkard S (1989). The pragmatic approach to masking. *Chronobiol Int* 6: 55–64.

Folkard S (1990). Circadian performance rhythms: some practical and theoretical implications. *Philos Trans R Soc Lond B Biol Sci* 327: 543–553.

Haimov I, Lavie P (1997). Melatonin – a chronobiotic and soporific hormone. *Arch Gerontol Geriatr* 24: 167–173.

Honma K, Honma S, Wada T (1987). Phase-dependent shift of free-running human circadian rhythms in response to a single bright light pulse. *Experientia* 43: 1205–1207.

Kleitman N, Kleitman E (1953). Effect of non-24 hour routines of living on oral temperature and heart rate. *J Appl Physiol* 6: 283–291.

Klerman E, Lee Y, Czeisler C, Kronauer R (1999). Linear demasking techniques are unreliable for estimating the circadian phase of ambulatory temperature data. *J Biol Rhythms* 14: 260–274.

Lack L, Lushington K (1996). The rhythms of sleep propensity and core body-temperature. *J Sleep Res* 5: 1–11.

Lasko T, Kripke D, Elliot J (1999). Melatonin suppression by illumination of upper and lower visual fields. *J Biol Rhythms* 14: 122–125.

Lewy A, Sack R (1996). The role of melatonin and light in the human circadian system. *Prog Brain Res* 111: 205–216.

Lewy A, Wehr T, Goodwin F, *et al.* (1980). Light suppresses melatonin secretion in humans. *Science* 210: 1267–1269.

Lewy A, Bauer V, Ahmed S, *et al.* (1998). The human phase response curve (PRC) to melatonin is about 12 hours out of phase with the PRC to light. *Chronobiol Int* 15: 71–83.

Lewy A, Cutler N, Sack R (1999). The endogenous melatonin profile as a marker for circadian phase position. *J Biol Rhythms* 14: 227–236.

Lindblom N, Heiskala H, Hatonen T, *et al.* (2000). No evidence for extraocular light induced phase shifting of human melatonin, cortisol and thyrotropin rhythms. *Neuroreport* 11: 713–717.

Liu C, Weaver D, Jin X, *et al.* (1997). Molecular dissection of two distinct actions of melatonin on the suprachiasmatic circadian clock. *Neuron* 19: 91–102.

Mills J, Minors D, Waterhouse J (1978). Adaptation to abrupt time shifts of the oscillator(s) controlling human circadian rhythms. *J Physiol* 285: 455–470.

Minors D, Waterhouse J (1981). *Circadian Rhythms and the Human*. Bristol: Wright PSG.

Minors D, Waterhouse J (1992). Investigating the endogenous component of human circadian rhythms: a review of some simple alternatives to constant routines. *Chronobiol Int* 9: 55–78.

Minors D, Waterhouse J, Wirz-Justice A (1991). A human phase-response curve to light. *Neurosci Lett* 133: 36–40.

Minors D, Akerstedt T, Waterhouse J (1994). The adjustment of the circadian rhythm of body temperature to simulated time-zone transitions: a comparison of the effect of using raw versus unmasked data. *Chronobiol Int* 11: 356–366.

Minors D, Folkard S, Macdonald I, *et al.* (1996a) The difference between activity when in bed and out of bed. II. Subjects on 27 hour 'days'. *Chronobiol Int* 13: 179–190.

Minors D, Waterhouse J, Folkard S, Atkinson G (1996b) The difference between activity when in bed and out of bed. III. Nurses on night work. *Chronobiol Int* 13: 273–282.

Miyamoto Y, Sancar A (1998). Vitamin B2-based blue-light photoreceptors in the retinohypothalamic tract as the photoactive pigments for setting the circadian clock in mammals. *Proc Natl Acad Sci USA* 95: 6097–6102.

Moog R, Hildbrandt G (1989). Adaptation to shiftwork – experimental approaches with reduced masking effects. *Chronobiol Int* 6: 65–75.

Moore-Ede M, Sulzman F, Fuller C (1982). *The Clocks That Time Us*. Cambridge, MA: Harvard University Press.

Mormont M-C, Waterhouse J, Bleuzen P, *et al.* (2000). Marked 24 h rest/activity rhythms are associated with better quality of life, better response, and longer survival in patients with metastatic colorectal cancer and good performance status. *Clin Cancer Res* 6: 3038–3045.

Mrosovsky N (1999). Critical assessment of methods and concepts in nonphotic phase shifting. *Biol Rhythm Res* 30: 135–148.

Murphy P, Campbell S (1997). Night-time drop in body temperature: a physiological trigger for sleep onset? *Sleep* 20: 505–511.

Redfern P, Waterhouse J, Minors D (1991). Circadian rhythms: principles and measurement. *Pharmacol Ther* 49: 311–327.

Reid K, Vandenheuvel C, Dawson D (1996). Daytime melatonin administration – effects on core temperature. *J Sleep Res* 5: 150–154.

Reilly T, Atkinson G, Waterhouse J (1997). *Biological Rhythms and Exercise*. Oxford: Oxford University Press.

Reppert S, Weaver D, Rivkees S, Stopa E (1988). Putative melatonin receptors in a human biological clock. *Science* 242: 78–81.

Shochat T, Luboshitzky R, Lavie P (1997). Nocturnal melatonin onset is phase locked to the primary sleep gate. *Comp Physiol* 42: R364–R370.

Van Cauter E, Desir D, Decoster C, et al. (1989). Nocturnal decrease in glucose tolerance during constant glucose infusion. *J Clin Endocrinol Metab* 69: 604–611.

Van Dongen H, Duindam H (1997). The effect of lying down on body temperature. *Sleep–Wake Res Netherlands* 8: 29–31.

Visser E, Beersma D, Daan S (1999). Melatonin suppression by light in humans is maximal when the nasal part of the retina is illuminated. *J Biol Rhythms* 14: 116–121.

Waterhouse J, Minors D, Waterhouse M (1990). *Your Body Clock*. Oxford; Oxford University Press.

Waterhouse J, Folkard S, Minors D (1992). *Shiftwork, Health and Safety. An Overview of the Scientific Literature 1978–1990*. HSE contract research report. London: HMSO.

Waterhouse J, Atkinson G, Reilly T (1997). Jet lag. *Lancet* 350: 1611–1616.

Waterhouse J, Minors D, Folkard S, et al. (1998). Light of domestic intensity produces phase shifts of the circadian oscillator in humans. *Neurosci Lett* 245: 97–100.

Waterhouse J, Akerstedt T, Lennernas M, Arendt J (1999a) Chronobiology and nutrition: internal and external factors, *Can J Diabet Care* 23(Suppl 2): 82–88.

Waterhouse J, Edwards B, Mugarza J, et al. (1999b) Purification of masked temperature data from humans: some preliminary observations on a comparison of the use of an activity diary, wrist actimetry, and heart rate monitoring. *Chronobiol Int* 16: 461–475.

Waterhouse J, Weinert D, Minors D, et al. (1999c) The effect of activity on the waking temperature rhythm in humans. *Chronobiol Int* 16: 343–357.

Waterhouse J, Weinert D, Minors D, et al. (2000). A comparison of some different methods for purifying core temperature data from humans. *Chronobiol Int* 17: 539–566.

Waterhouse J, Nevill A, Weinert D, et al. (2001). Modelling the effect of spontaneous activity on core temperature in healthy human subjects. *Biol Rhythm Res* 32: 511–528.

Weinert D, Waterhouse J (1998). Diurnally changing effects of locomotor activity on body temperature in laboratory mice. *Physiol Behav* 63: 836–843.

Weinert D, Sitka U, Minors D, et al. (1997). Twenty-four hour and ultradian rhythmicities in healthy full-term neonates: exogenous and endogenous influences. *Biol Rhythm Res* 28: 441–452.

Wever R (1985). Internal interactions within the human circadian system: the masking effect. *Experientia* 41: 332–342.

Youngstedt S, O'Connor P (1999). The influence of air travel on athletic performance. *Sports Med* 28: 197–207.

2

The biological clock: location, neuroanatomy and neurochemistry

Martine L Garabette

2.1 Introduction: the circadian timing system

The mammalian circadian timing system (CTS) is the set of neural structures that have evolved to synchronise an organism with its environment. The components of the mammalian CTS are thus

- Photoreceptors and projections of retinal ganglion cells that form entrainment pathways
- A circadian pacemaker or pacemakers
- Efferent pathways that couple the pacemaker to effector systems displaying circadian function

An important aspect of pacemaker function is, therefore, the coordinated relay of messages to the pacemaker and the efficient output of signals to the effector regions.

Thirty years have passed since Stephan and Zucker discovered that free-running circadian rhythms of drinking and locomotor activity were severely disrupted by relatively discrete lesions to an area of the anterior hypothalamus (Stephan and Zucker, 1972). Since that time, a plethora of experiments have confirmed that the suprachiasmatic nuclei (SCN) house the mammalian pacemaker (the 'biological clock'), capable of independently regulating an animal's biological rhythms in the absence of environmental time cues.

Definite evidence that the SCN are responsible for generating and maintaining circadian rhythmicity came from the discovery of a mutant hamster that presented with a shortened free-running circadian period compared with its littermates (Ralph and Menaker, 1988). Colonies of this mutant (the *tau* mutant) were bred; heterozygous and homozygous mutants presented with free-running periods of 22 h and 20 h, respectively, compared with 24 h period of wild-type hamsters. Tissue generated from the SCN of homozygous mutants was subsequently implanted into SCN-lesioned, arrhythmic, wild-type hamsters in a series of delicate experiments. Rhythmicity was restored in the host animals and the resultant free-running period was attributable to the implanted donor SCN tissue (Ralph et al., 1990).

2.2 The suprachiasmatic nuclei (SCN)

The SCN are small, paired nuclei situated in the midline on the ventral surface of the brain above ('supra') the optic chiasm ('chiasmatic'). In all mammalian species studied to date, the SCN are readily compartmentalised into a ventral-lateral (VL) and dorsal-medial (DM) segment (or 'core' and 'shell', respectively) with regard to immunostaining of perykarya, cell size, neuropil extent, neurotransmitter content and afferent projections to the nucleus. Many fibres cross the midline between the left and right SCN, suggesting that the two nuclei do not function as separate entities, but rather as a single body (van den Pol and Tsujimoto, 1985; Leak et al., 1999). An interesting aspect of the hamster SCN, unique to this rodent species, is that the DM-SCN have been shown to join in the midline and present as a single nucleus (Card and Moore, 1984).

2.2.1 SCN neurochemistry

There are various techniques for the detection and measurement of neurotransmitters in the mammalian central nervous system, and immunocytochemical methods have been used extensively to elucidate the neurochemical make-up of the SCN (Table 2.1). Like the majority of brain neurons, SCN neurons contain high levels of γ-aminobutyric acid (GABA) and, in the VL-SCN, cells containing vasointestinal polypeptide (VIP) and gastrin-releasing peptide (GRP) co-localise with GABA. VIP also co-localises with GRP and peptide histidine isoleucine (PHI), and the number of cells expressing these co-localised proteins exhibits diurnal rhythmicity: significantly more cells co-express VIP and GRP in the mid-dark phase compared with the mid-light phase (Romijn et al., 1998). The VIP-containing neurons are considered the output cells of the clock, coordinating and interpreting signals from clock inputs.

2.3 Clock inputs

SCN afferent fibres collectively entrain the circadian pacemaker to environmental cues. The principal afferent fibres originate from the retina, transmitting light signals directly (via the retinohypothalamic

Table 2.1 The neurochemistry of the rodent[a]

VL-SCN (core)		DM-SCN (shell)	
Neurotransmitters	Inputs (ir fibres)	Neurotransmitters	Inputs (ir fibres)
Bombesin	Aspartate	Angiotensin II	Galanin
Calretinin	Glutamate	AVP	VIP
Calbindin	5-HT	Calbindin	
GABA	NPY	GABA	
GRP	PACAP	Met-enkephalin	
Neurotensin		PHI	
Nitric oxide		Somatostatin	
PHI			
Substance-P[b]			
VIP			

From Card and Moore (1984); van den Pol and Tsujimoto (1985); van den Pol (1991); Takatsuji *et al.* (1991); Abrahamson and Moore (2001); Romijn *et al.* (1998); Piggins *et al.* (2001b).
[a]Mouse and rat.
[b]Not mouse.
AVP, arginine vasopressin; DM-SCN, dorsal-medial SCN; GABA, γ-aminobutyric acid; GRP, gastrin-releasing peptide; 5-HT, 5-hydroxytryptamine (serotonin); ir, immunoreactive; NPY, neuropeptide-Y; PACAP, pituitary adenylate cyclase-activating polypeptide; PHI, peptide histidine isoleucine; VIP, vasointestinal polypeptide; VL-SCN, ventral-lateral SCN.

tract, RHT) and indirectly (via the geniculohypothalamic tract, GHT) to the SCN. In addition, the SCN receive a major projection from the midbrain raphe nuclei (raphe-hypothalamic tract, RaHT). Afferent fibres also originate from the infralimbic cortex, the septal nuclei, the substantia innominata and the ventral subiculum (Pickard, 1982; Moore, 1996).

2.3.1 The retinohypothalamic tract (RHT)

Environmental light is necessary for stable entrainment of animals to their environment, synchronising an animal to the 24 h light–dark cycle. Light exposure during the early subjective night will phase delay the clock, whereas light presented during the late subjective night will result in phase advances. A direct retinal projection to the SCN was demonstrated 30 years ago by autoradiography of anterogradely transported, tritiated amino acids injected into the posterior chamber of the rodent eye (Moore and Lenn, 1972). The pathway was thought to terminate solely in the VL-SCN (the 'retinorecipient' SCN); however, subsequent studies have demonstrated RHT terminals to be more widespread: terminals have been located in the anterior hypothalamic area, the retrochiasmatic area and the lateral hypothalamic area (Johnson et al., 1988a; Moore et al., 1995). The RHT is necessary for entrainment to light, as dissection of this pathway results in free-running rhythms, even under light–dark conditions (Johnson et al., 1988b), and electrical stimulation of hamster optic nerves in vivo mimics the response of the pacemaker to light (de Vries et al., 1994).

Since the discovery of a functional pathway, the transmitter of the RHT has been the focus of much speculation. Converging evidence has established glutamate as the principal transmitter, mediating light entrainment via a combination of NMDA (N-methyl-D-aspartate), non-NMDA (kainate and α-amino-3-hydroxy-5-methyl-4-isoxazolepropionic acid (AMPA)) and metabotropic glutamate receptors. Glutamate-immunoreactivity is present in presynaptic terminals of the RHT (van den Pol, 1991; de Vries et al., 1993) and glutamate is released by RHT neurons following optic nerve stimulation (Liou et al., 1986). Indeed, similarly to the effects of light, glutamate applied topically to in vitro brain slice preparations can delay or advance neuronal firing only when applied during the early or late subjective night, respectively, while having no effect on clock functioning during the subjective day (Shirakawa and Moore, 1994a). NMDA, AMPA, kainate and metabotropic glutamate receptor mRNA have been detected in the retinorecipient VL-SCN region (van den Pol, 1994; Mick et al., 1995; O'Hara et al., 1995; van den Pol et al., 1995; Ghosh et al., 1997), and many experiments in vitro (Cahill and Menaker, 1989; Kim and Dudek, 1991; Shibata et al., 1994; Scott and Rusak, 1996; Ding et al., 1994; Watanabe et al., 1994) and in vivo (Amir, 1992; Ohi et al., 1991; de Vries et al., 1994) corroborate the function of glutamate and its receptors during light regulation of circadian rhythms.

NMDA receptor stimulation releases nitric oxide (NO) via stimulation of the enzyme nitric oxide synthase (NOS) (Knowles et al., 1989), and NO has been implicated in the generation of circadian rhythms (Amir, 1992; Ding et al., 1994; Watanabe et al., 1994, 1995; Weber et al., 1995; Starkey et al., 2001). Light-induced phase shifts of circadian wheel-running activity are attenuated by systemic (Watanabe et al., 1995) or intracerebroventricular injection of an NOS inhibitor (Weber et al., 1995). In addition, topical application of an NOS inhibitor to in vitro slice preparations of the SCN can block phase delays produced by NMDA receptor stimulation during the early subjective night (Watanabe et al., 1994). However, mutant mice lacking either the neuronal (Kriegsfeld et al., 1999) or endothelial (Kriegsfeld et al., 2001) isoforms of NOS are still able to entrain to a standard light–dark cycle, phase-shift locomotor activity or free run in constant conditions, and light still induces the immediate early gene c-*fos* in the SCN in a similar pattern to that in wild-type animals. These data appear to contradict earlier studies and suggest that NO from SCN neurons may not be necessary for photic entrainment. Further studies are on-going to elucidate the role of NO in circadian rhythmicity. More recently, carbon monoxide has been identified as a potential transmitter in SCN neurons (Artinian *et al.*, 2001).

Evidence also suggests that substance-P may be involved in the light entrainment pathway, potentiating glutamate-evoked excitatory responses *in vitro* (Shirakawa and Moore, 1994b) and exhibiting phase-responses *in vitro* similar to the effects of light *in vivo* (Shibata *et al.*, 1992b). Indeed, fibres containing substance-P are found in the VL-SCN of the rat (Piggins *et al.*, 2001b), and intracerebroventricular injection of a substance-P receptor antagonist (neurokinin-1 receptor, NK_1) in the hamster can block light-induced phase advances of nocturnal locomotor activity (Challet *et al.*, 1998). Substance-P and glutamate can also interact at the level of SCN neurons, acting in series with substance-P operating upstream of glutamate (Kim *et al.*, 2001). Both substance-P and glutamate can mimic the effects of optic nerve stimulation *in vitro* on SCN neuronal cell firing during the late subjective night. However, while an NMDA and non-NMDA receptor antagonist cocktail can inhibit the effects of glutamate stimulation, it can also inhibit the action of substance-P. This is in contrast to an NK_1 receptor antagonist, which cannot block glutamate-induced phase shifts. Blocking substance-P can also signal 'darkness' to the SCN, as NK_1 receptor antagonists can induce significant phase advances of locomotor activity when injected during the subjective day (Challet *et al.*, 2001) at a time when glutamate and light are without effect.

Pituitary adenylate cyclase-activating polypeptide (PACAP) may also play a role in mediating the response of the CTS to light. Intense PACAP-immunoreactivity has been found in the VL-SCN, which is lost following bilateral enucleation (Hannibal *et al.*, 1998), and PACAP co-localises with glutamate in some retinal ganglion cells projecting to the SCN (Hannibal *et al.*, 2000). These two putative transmitters may, therefore, be co-released following retinal stimulation. Indeed, PACAP has been shown to be a significant component of the glutamatergic light-entrainment pathway, modulating the response of SCN neurons to glutamate in a phase-dependent manner (Chen *et al.*, 1999). Application of PACAP alone at circadian time (CT) 14 and 19 (CT12 designates the onset of activity), during the subjective night has no effect on *in vitro* firing rates of SCN neurons. At CT19, however, PACAP attenuates the glutamate-induced phase advances, whereas a PACAP-receptor antagonist (PAC1) potentiates the effect of glutamate. By contrast, PACAP potentiates glutamate-induced phase delays at CT14, and co-application of glutamate with the PAC1 receptor antagonist can block the glutamate-induced phase delay. The effect of the antagonist suggests that there is a tonic PACAPergic component to the response stimulated by exogenous glutamate. Moreover, an intracerebroventricular infusion of an anti-PACAP antibody (blocking PACAP receptors) in free-running hamsters significantly potentiates light-induced phase advances (Chen *et al.*, 1999), revealing PACAP as an endogenous component of the light/glutamatergic pathway. PACAP may also act as a daytime regulator of the circadian clock (Hannibal *et al.*, 1998; Piggins *et al.*, 2001a) as PACAP can induce large phase advances in rat SCN neuronal activity (Hannibal *et al.*, 1998) and hamster wheel-running (Piggins *et al.*, 2001a) during the mid-subjective day.

During periods when the circadian pacemaker is sensitive to the effects of light, therefore, release of glutamate and/or substance-P and PACAP at retinal afferent terminals produces a cascade of events, leading to modulation of circadian physiological and behavioural rhythms (Figure 2.1), although evidence also suggests that substance-P and PACAP may have the capability to signal 'darkness' to the SCN. Nitric oxide has also been implicated as a messenger necessary for photic entrainment downstream of NMDA receptor stimulation (Amir, 1992; Ding *et al.*, 1994; Watanabe *et al.*, 1994, 1995; Weber *et al.*, 1995; Amir and Edelstein, 1997; Starkey *et al.*, 2001), and light-induced phase shifts have also been shown to be dependent on the induction of a number of immediate early genes (c-*fos*, *egr-1*, *NGF1-A*, *NGF1-B*) (Ebling *et al.*, 1991; Sutin and Kilduff, 1992). c-*fos* and its protein product Fos are rapidly induced in the VL-SCN following light-inducing phase shifts (Rea, 1989), a response blocked by an NMDA receptor antagonist (Ebling *et al.*, 1991; Park *et al.*, 1993) and mimicked by intracerebroventricular injection of NMDA (Ebling *et al.*, 1991).

The molecular mechanisms responsible for the effect that light has on the CTS have begun to be uncovered, and this will be discussed in Chapter 3.

2.3.2 The geniculohypothalamic tract (GHT)

The SCN also receive a prominent projection from the retinorecipient intergeniculate leaflet (IGL) of the thalamus (Moore and Card, 1990). The

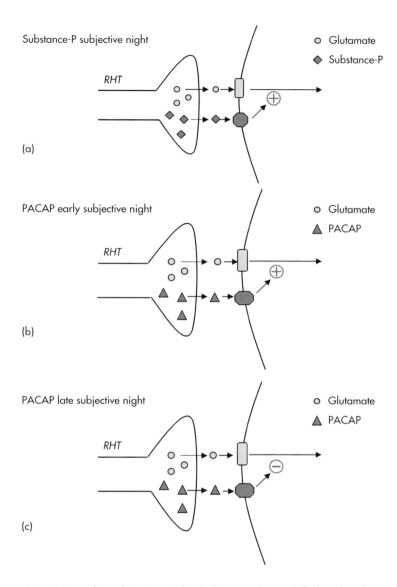

Figure 2.1 During the subjective night, light can phase shift the circadian pacemaker. (a) During the subjective night, following a light pulse, glutamate is released, which acts upon NMDA, non-NMDA and metabotropic glutamate receptors to produce a measured response. Substance-P is co-released with glutamate and can potentiate (+) the effects of glutamate. (b, c) Pituitary adenylate cyclase-activating polypeptide (PACAP), co-released with glutamate, has the ability to (b) potentiate (+) or (c) attenuate (−) the effect of glutamate, following retinal stimulation during the early and late subjective night, respectively. RHT, retinohypothalamic tract.

IGLs contain a population of neuropeptide-Y (NPY) and met-enkephalin neurons (Harrington *et al.*, 1987) that project bilaterally to the VL-SCN (Morin *et al.*, 1992). Each IGL also relays information to its contralateral nucleus via met-enkephalin-positive fibres. IGL neurons also contain GABA and neurotensin, which project to the SCN in a similar pattern to NPY and met-enkephalin-containing fibres; the majority of neurotensin molecules in GHT neurons co-localise with NPY (Morin and Blanchard, 2001).

The GHT serves a dual function: (1) to modulate photic entrainment of the CTS and (2) to relay nonphotic information to the clock. Nonphotic stimuli can induce phase shifts in activity patterns that are characterised by large phase advances or delays during the late or early subjective day, respectively. Neurotoxic lesions of the IGL can prevent both triazolam-induced (Johnson *et al.*, 1988c; Janik and Mrosovsky, 1994) and novel wheel-running-induced (Janik and Mrosovsky, 1994; Wickland and Turek, 1994; Maywood *et al.*, 1997) nonphotic phase shifts. The triazolam-induced phase shift is thought to result from an increased arousal state of the animal; indeed, the phase shifting effects of trialozam or dark pulses can be completely suppressed by immobilisation of the animals during treatment (Van Reeth and Turek, 1989). Electrical stimulation of the IGL or infusion of NPY can also mimic the effects of a nonphotic stimulus (Maywood, personal communication; Maywood *et al.*, 2002) confirming the role of the GHT and NPY as mediators of the response to a nonphotic stimulus.

An intact IGL is not necessary for photic entrainment to standard light–dark cycles as lesions of the IGL are without effect (Harrington, 1997). However, IGL lesions can alter the ability of animals to entrain to a 'skeleton' photoperiod (Edelstein and Amir, 1999), that is an environmentally relevant lighting schedule where animals are exposed to light only during periods of dawn or dusk, thus suggesting that the IGL may be a vital component of the CTS in nocturnal animals, which 'sample' environmental light in their natural habitat. Local administration of NPY can also inhibit light-induced phase advances in wheel-running activity in hamsters when administered in the late subjective night (CT19) (Weber and Rea, 1997), although NPY is without effect during the early subjective night (CT14). Interestingly, light can reciprocally inhibit NPY-induced phase shifts when presented during the subjective day (Maywood *et al.*, 2002). These experiments by Maywood and colleagues are intriguing, since light was presented to mice immediately after NPY infusion during the subjective day at a time when light has been commonly perceived not to affect circadian function.

2.3.3 The raphe-hypothalamic tract (RaHT)

The VL-SCN receive a dense serotonergic innervation from the mesencephalic raphe nuclei (Azmitia and Segal, 1978; Moore et al., 1978). This ascending serotonergic pathway also innervates the IGL (Moore and Card, 1994). The consensus of opinion has been that serotonin (5-hydroxytryptamine, 5-HT) relays nonphotic information to the pacemaker and serves to modulate photic inputs.

Controversy surrounds whether the SCN receive serotonergic input from the median (MRN) or dorsal raphe nuclei (DRN); the data seem to highlight species variations and inconsistencies relating to different methodological approaches. Early anterograde tracing studies of the raphe nuclei in rats demonstrated SCN afferent fibres originating from both the MRN and DRN (Azmitia and Segal, 1978). However, direct measurements of SCN 5-HT content of DRN- or MRN-lesioned rats suggested that the principal projection arises from the MRN; MRN lesion resulted in a significant (70%) reduction in SCN 5-HT content, whereas DRN lesion failed to affect 5-HT levels (van de Kar and Lorens, 1979). More recent retrograde tracing experiments in the rat demonstrate a projection from the DRN (Kawano et al., 1996), whereas complementary anterograde with retrograde tracing experiments suggest that the majority of fibres originate in the MRN (Moga and Moore, 1997). Likewise, in the hamster, both MRN and DRN were implicated using retrograde tracing methods (Pickard, 1982). However, subsequent experiments suggest only a MRN projection to the hamster SCN, either by functional studies examining the effects of specific lesions in either the MRN or DRN on locomotor activity (Meyer-Bernstein and Morin, 1996) or by anterograde tracing techniques (Morin and Meyer-Bernstein, 1999). More recently, it has been hypothesised that DRN input to the hamster SCN could be mediated by a DRN–MRN–SCN pathway involving a 5-HT-sensitive multisynaptic interaction between DRN and MRN neurons (Dudley et al., 1999).

Destruction of the serotonergic system by specific neurotoxic lesion results in a rapid onset of activity, delayed offset and longer duration of the nocturnal activity phase of hamsters maintained under a normal light–dark cycle (Smale et al., 1990; Morin and Blanchard, 1991; Penev et al., 1993). Similar effects have been observed in mice with electrolytic lesions of the DRN (Barbacka-Surowiak and Gut, 2001). However, under constant light conditions, normal circadian rhythmicity deteriorates rapidly in hamsters with generalised raphe nuclei lesions (Smale et al., 1990; Morin and Blanchard, 1991; Penev et al., 1993), whereas the activity phase is maintained and extended in DRN-lesioned mice (Barbacka-Surowiak and Gut, 2001). Lesioned hamsters maintained under constant darkness do not experience demonstrable changes to their circadian period, although there is an overall change in the temporal properties of their phase–response curve to light, with lesioned animals demonstrating a larger phase delay portion of the phase–response curve during the early- to mid-dark phase (Morin and Blanchard, 1991).

5-HT can phase-advance the clock when administered *in vivo* or *in vitro* during the mid-subjective day (Prosser *et al.*, 1990, 1994a; Medanic and Gillette, 1992; Starkey, 1996; Shibata *et al.*, 1992b; Cutera *et al.*, 1994, 1997; Tominaga *et al.*, 1992). This action is thought to be via stimulation of 5-HT$_{1A/7}$ receptors (Starkey, 1996; Prosser *et al.*, 1994a; Shibata *et al.*, 1992b; Cutera *et al.*, 1994, 1997; Tominaga *et al.*, 1992), activating NO production (Starkey, 1996), protein kinase A and potassium channels (Prosser *et al.*, 1994b), and may involve an increase in transcription and translation of specific proteins (Jovanovska and Prosser, 2002).

The terminal fields of the RHT and the RaHT are very closely apposed and 5-HT and its agonists can modify the response of the pacemaker to light. 5-HT, acting via 5-HT$_{1A/7}$ receptors, possibly located postsynaptically on GABA-, VIP- and AVP-immunoreactive cells in the SCN (Belenky and Pickard, 2001), can attenuate optic nerve stimulated-evoked field potentials (Liou *et al.*, 1986; Moriya *et al.*, 1996) and can reduce light-induced Fos expression in hamster SCN (Selim *et al.*, 1993; Moriya *et al.*, 1996). During the early subjective night (CT14), blockade of 5-HT$_{1A}$ receptors can also significantly increase the magnitude of light-induced phase delays *in vivo* (Smart and Biello, 2001). However, the site of action of 5-HT modulation of photic responses is not fully elucidated, and it is possible that 5-HT acts directly at the level of SCN neurons or through a circuit involving the midbrain raphe nuclei. DRN or MRN stimulation *in vivo* can also attenuate light-induced Fos-immunoreactivity in the SCN (Meyer-Bernstein and Morin, 1999).

5-HT-mediated phase shifts are also thought to involve 5-HT$_{1B}$ receptors, and 5-HT can also inhibit the effects of light on the circadian system through 5-HT$_{1B}$ heteroreceptors located on retinal afferents (Pickard *et al.*, 1996; Pickard and Rea, 1997). A reciprocal modulation of 5-HT by glutamate is also evident, since glutamate and optic chiasm stimulation *in vitro* can inhibit serotonergic-induced phase advances through both AMPA and NMDA receptors (Prosser, 2001). Moreover, 5-HT is tonically regulated by glutamate during the light phase – an effect which is absent during the dark phase, when RHT neurons are, presumably, quiescent (Garabette, 1998; Figure 2.2).

In what appears to be confounding evidence, the 5-HT system can also mediate the effects of light on the pacemaker. The 5-HT agonist quipazine can induce c-*fos* in the VL-SCN in a manner similar to that of a light pulse in the rat (Moyer *et al.*, 1997), and microinjections of the agonist into the region of the SCN can induce photic-like effects similar to those induced by systemic injections of quipazine, implicating a direct action on SCN neurons (Kalkowski and Wollnik, 1999). In addition, depletion of 5-HT significantly attenuates the induction of light-induced Fos-immunoreactivity in SCN cells, supporting the involvement of a serotonergic mediator of photic information to the SCN (Moyer and Kennaway, 2000).

There is speculation that the 5-HT-mediated effects of light may be a consequence of a direct retinal–raphe–hypothalamic tract, and that light may directly affect the activity of neurons and serotonin levels in the midbrain raphe. A bilateral retinal projection to the DRN has been identified in the rat (Villar *et al.*, 1987; Shen and Semba, 1994; Kawano *et al.*, 1996), the cat (Foote *et al.*, 1978) and the Chilean degus (Fite and Janusonis, 2001), while a projection to both the MRN and DRN has been identified in the tree shrew (Reuss and Fuchs, 2000). Although these fibres are scarce in comparison to well-established retinal projections such as the SCN and IGL, they do provide a possible link between light and the serotonergic system.

2.3.4 Converging photic and nonphotic signals

As discussed, interactions can occur between RHT–GHT and RHT–RaHT projections. The SCN afferent pathways (Figure 2.3) often relay conflicting information to the clock, but the SCN has the ability to interpret these signals and respond effectively to produce a coordinated response.

2.4 Clock outputs

Connections from the SCN to effector regions are organised topographically (Leak *et al.*, 1999; Leak and Moore, 2001). The VL- and DM-SCN each project to distinct brain regions (Table 2.2) providing areas of discrete control of circadian function. The efferent projections from the circadian pacemaker are responsible for coordinating overt rhythmicity, without which the careful assembly of pacemaker and input

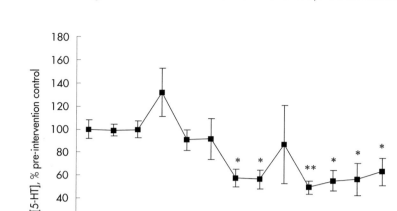
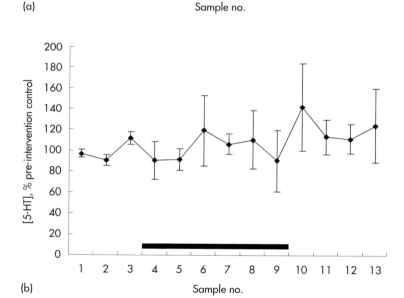

Figure 2.2 5-HT release is tonically regulated by glutamate during the light phase. The noncompetitive NMDA receptor antagonist MK801 (10 μmol/L) was infused via an *in vivo* microdialysis probe for 90 min into the SCN region of conscious rats (solid bar) maintained under a 12:12 light–dark schedule. Samples were taken every 15 min; three pre-intervention samples and 10 post-infusion samples. 5-HT concentration was measured by high-performance liquid chromatography with electrochemical detection. (a) Mid-light phase (zeitgeber time 6, ZT6; n = 5); (b) mid-dark phase (ZT18; n = 7). *$p < 0.05$, **$p < 0.01$ (From Garabette, 1998).

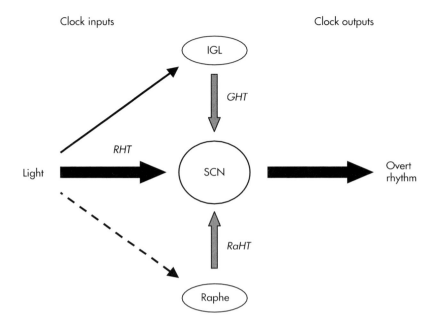

Figure 2.3 Schematic representation of the organisation of the mammalian circadian timing system. Light is the principal entraining zeitgeber, relaying direct afferents to the SCN via the RHT. Light can also indirectly affect clock function via the IGL and GHT, and possibly via a retinal–raphe–hypothalamic tract (broken arrow). The GHT and RaHT are also thought to relay nonphotic inputs to the SCN, modulating the effects of light on clock function. The SCN coordinates clock inputs, projecting to effector regions to produce overt rhythmicity. GHT, geniculohypothalamic tract; IGL, intergeniculate leaflet; RaHT, raphe-hypothalamic tract; RHT, retinohypothalamic tract.

Table 2.2 Topographical organisation of SCN efferent projections

VL-SCN	DM-SCN	VL- and DM-SCN
DM-SCN	BNST	Paratenial thalamic nucleus
Lateral septum	DMN	
Lateral SPV	Hypothalamic PVN	
Peri-SC region	Medial POA	
Ventral tuberal area	Medial SPV	
	Thalamic PVN	
	Zona incerta	

From Leak *et al.* (1999); Leak and Moore (2001).
BNST, bed nucleus of the stria terminalis; DMN, dorsomedial hypothalamic nuclei; DM-SCN, dorsal-medial SCN; POA, preoptic area; PVN, paraventricular nuclei; SC, suprachiasmatic; SPV, subparaventricular zone; VL-SCN, ventral-lateral SCN.

pathways informing of environmental and behavioural stimuli would become redundant.

However, these output pathways shed precious little light on how they can influence all manner of circadian responses, and little is known about the functional organisation of SCN projections mediating control of effector systems. It has been hypothesised that circadian patterns of neuronal activity generated in the SCN are relayed to effector circuitry in the subparaventricular zone (SPV), a relatively cell-sparse area dorsal to the SCN that serves as an amplification stage, thereby increasing the influence of the SCN (Watts and Swanson, 1987; Watts et al., 1987). In addition, a subset of neurons projecting to the SPV are light-responsive, and may represent a direct channel through which photic inputs to the SCN can influence neural outputs (De la Inglesia and Schwartz, 2002).

Regulation of circadian functions may also be mediated by discrete sets of SCN projections. Activity–rest cycles are possibly the most extensively studied clock output, because of the availability of noninvasive monitoring techniques involving infrared movement detectors or infrared wheel-running apparatus. They are mediated through caudal projections from the SCN to the SPV and lateral hypothalamic areas (Moore and Danchenko, 2002) and are dependent upon VIP action on the VPAC(2) receptor (Harmar et al., 2002) and the production of prokineticin-2 protein molecules (Cheng et al., 2002).

Dorsal SCN projections to the paraventricular nucleus of the hypothalamus relay signals to the pineal gland to stimulate the nocturnal rise in melatonin secretion (Hastings and Herbert, 1986; Moore and Danchenko, 2002). Melatonin, the 'hormone of darkness', has also been investigated extensively and translates the dark phase into a measurable signal, particularly in seasonally breeding animals (e.g. the hamster or sheep), such that the longer the dark phase, the longer the duration of melatonin secretion. On the other hand, rostral SCN projections to the anterior hypothalamic/preoptic areas mediate core body temperature rhythms (Moore and Danchenko, 2002).

Communication between SCN neurons is key to generating and maintaining circadian rhythmicity in animals. While neural projections are important mediators of synaptic transmission between SCN cells, there is evidence that the chain of events leading to rhythm generation in the SCN also involves the release of a diffusible agent, or agents (Silver et al., 1996). However, the exact mechanisms behind the organisation of divergent clock output signals have not been fully elucidated and are beyond the scope of this review.

2.5 Comment

Much progress has been made in recent years in elucidating the functional neuroanatomy and neurochemistry of the mammalian CTS. The subset of neurons in the anterior hypothalamus that comprises the SCN, although small and compact, have an enormous impact on all manner of physiological, pharmacological and behavioural responses. The SCN efficiently coordinate and interpret converging inputs to their VL 'core' region, relaying neuronal projections to effector regions, which in turn interpret signals to regulate overt day–night oscillations in the internal milieu.

References

Abrahamson E E, Moore R Y (2001). Suprachiasmatic nucleus in the mouse: retinal innervation, intrinsic organization and efferent projections. *Brain Res* 916: 172–191.

Amir S (1992). Blocking NMDA receptors or nitric oxide production disrupts light transmission to the suprachiasmatic nucleus. *Brain Res* 586: 336–339.

Amir S, Edelstein K (1997). A blocker of nitric oxide synthase, N^G-nitro-L-arginine methyl ester, attenuates light-induced Fos protein expression in rat suprachiasmatic nucleus. *Neurosci Lett* 224: 29–32.

Artinian L R, Ding J M, Gillette M U (2001). Carbon monoxide and nitric oxide: interacting messengers in muscarinic signalling to the brain's circadian clock. *Exp Neurol* 17: 293–300.

Azmitia E C, Segal M (1978). An autoradiographic analysis of the differential ascending projections of the dorsal and median raphe nuclei in the rat. *J Comp Neurol* 179: 641–668.

Barbacka-Surowiak G, Gut M (2001). The effect of dorsal raphe nucleus (DRN) lesions on the locomotor activity rhythm in mice. *Folia Biol (Krakow)* 49: 77–84.

Belenky M A, Pickard G E (2001). Subcellular distribution of 5-HT(1B) and 5-HT(7) receptors in the mouse suprachiasmatic nucleus. *J Comp Neurol* 432: 371–388.

Cahill G M, Menaker M (1989). Effects of excitatory amino acid receptor antagonists and agonists on suprachiasmatic nucleus responses to retinohypothalamic tract volleys. *Brain Res* 479: 76–82.

Card J P, Moore R Y (1984). The suprachiasmatic nucleus of the golden hamster: immunohistochemical analysis of cell and fiber distribution. *Neuroscience* 13: 415–431.

Challet E, Naylor E, Metzger J M, *et al.* (1998). An NK_1 receptor antagonist affects the circadian regulation of locomotor activity on golden hamsters. *Brain Res* 800: 32–39.

Challet E, Dugovic C, Turek F W, van Reeth O (2001). The selective neurokinin 1 receptor antagonists R116301 modulates photic responses of the hamster circadian system. *Neuropharmacology* 40: 408–415.

Chen D, Buchanan G F, Ding J M, et al. (1999). Pituitary adenylate cyclase-activating peptide: a pivotal modulator of glutamatergic regulation of the suprachiasmatic circadian clock. *Proc Natl Acad Sci USA* 96: 13468–13473.

Cheng M Y, Bullock C M, Li C (2002). Prokineticin 2 transmits the behavioural circadian rhythm of the suprachiasmatic nucleus. *Nature* 417: 405–410.

Cutera R A, Ouarour A, Pevet P (1994). Effects of the 5-HT$_{1A}$ receptor agonist 8-OH-DPAT and other non-photic stimuli on the circadian rhythm of wheel-running activity in hamsters under different constant conditions. *Neurosci Lett* 172: 27–30.

Cutera R A, Saboureau M, Pevet P (1997). Phase-shifting effect of 8-OH-DPAT, a 5-HT$_{1A}$/5-HT$_7$ receptor agonist, on locomotor activity in golden hamster in constant darkness. *Neurosci Lett* 210: 1–4.

De la Inglesia H O, Schwartz W J (2002). A subpopulation of efferent neurons in the mouse suprachiasmatic nucleus is also light responsive. *Neuroreport* 13: 857–860.

de Vries M J, Cardozo B N, van der Want J, et al. (1993). Glutamate immunoreactivity in terminals of the retinohypothalamic tract of the brown Norwegian rat. *Brain Res* 612: 231–237.

de Vries M J, Treep J A, De Pauw E S D, Meijer J H (1994). The effects of electrical stimulation of the optic nerves and anterior optic chiasm on the circadian activity rhythm of the Syrian hamster: involvement of excitatory amino acids. *Brain Res* 642: 206–212.

Ding J M, Chen D, Weber E T, et al. (1994). Resetting the biological clock: mediation of nocturnal circadian shifts by glutamate and NO. *Science* 266: 1713–1717.

Dudley T E, Dinardo L A, Glass D (1999). *In vivo* assessment of the midbrain raphe nuclear regulation of serotonin release in the hamster suprachiasmatic nucleus. *J Neurophysiol* 81: 1469–1477.

Ebling F J P, Maywood E S, Staley K, et al. (1991). The role of N-methyl-D-aspartate-type glutamatergic neurotransmission in the photic induction of immediate early gene expression in the suprachiasmatic nuclei of the Syrian hamster. *J Neuroendocrinol* 3: 641–652.

Edelstein K, Amir S (1999). The role of the intergeniculate leaflet in entrainment of circadian rhythms to a skelten photoperiod. *J Neurosci* 19: 372–380.

Fite K V, Janusonis S (2001). Retinal projection to the dorsal raphe nucleus in the Chilean degus (*Octodon degus*). *Brain Res* 895: 139–145.

Foote W E, Taber-Pierce E, Edwards L (1978). Evidence for a retinal projection to the midbrain raphe of the cat. *Brain Res* 156: 135–140.

Garabette M L (1998). Autoreceptor and glutamatergic regulation of serotonin release from the suprachiasmatic nuclei. PhD thesis, University of Bath, Department of Pharmacy and Pharmacology.

Ghosh P K, Baskaran N, van den Pol A N (1997). Developmentally regulated gene expression of all eight metabotropic glutamate receptors in hypothalamic suprachiasmatic and arcuate nuclei – a PCR study. *Dev Brain Res* 102: 1–12.

Hannibal J, Ding J M, Chen D, et al. (1998). Pituitary adenylate cyclase-activating peptide (PACAP) in the retinohypothalamic tract: a daytime regulator of the biological clock. *Ann NY Acad Sci* 865: 197–206.

Hannibal J, Moller M, Ottersen O, Fahrenkrug J (2000). PACAP and glutamate are co-stored in the retinohypothalamic tract. *J Comp Neurol* 418: 147–155.

Harmar A J, Marston H M, Shen S, *et al.* (2002). The VPAC(2) receptor is essential for circadian function in the mouse suprachiasmatic nuclei. *Cell* 109: 497–508.

Harrington M E (1997). The ventral lateral geniculate nucleus and the intergeniculate leaflet: interrelated structures in the visual and circadian systems. *Neurosci Behav Rev* 21: 705–727.

Harrington M E, Nance D M, Rusak B (1987). Double-labeling of neuropeptide Y-immunoreactive neurons which project from the geniculate to the suprachiasmatic nuclei. *Brain Res* 410: 275–282.

Hastings M H, Herbert J (1986). Neurotoxic lesions of the paraventriculo-spinal projection block the nocturnal rise in pineal melatonin synthesis in the Syrian hamster. *Neurosci Lett* 69: 1–6.

Janik D, Mrosovsky N (1994). Intergeniculate leaflet lesions and behaviorally-induced shifts of circadian rhythms. *Brain Res* 651: 174–182.

Johnson R, Morin L, Moore R (1988a). Retinohypothalamic projections in the hamster and rat demonstrated using cholera toxin. *Brain Res* 462: 301–312.

Johnson R F, Smale L, Moore R Y, Morin L P (1988b). Lateral geniculate lesions block circadian phase-shift responses to a benzodiazepine. *Proc Natl Acad Sci USA* 85: 5301–5304.

Johnson R F, Moore R Y, Morin L P (1988c). Loss of entrainment and anatomical plasticity after lesions of the hamster retinohypothalamic tract. *Brain Res* 460: 297–313.

Jovanovska A, Prosser R A (2002). Translational and transcriptional inhibitors block serotonergic phase advances of the suprachiasmatic nucleus circadian pacemaker *in vitro*. *J Biol Rhythms* 17: 137–146.

Kalkowski A, Wollnik F (1999). Local effects of the serotonin agonist quipazine on the suprachiasmatic nucleus of rats. *Neuroreport* 10: 3241–3246.

Kawano H, Decker K, Reuss S (1996). Is there a direct retina-raphe-suprachiasmatic nucleus pathway in the rat? *Neurosci Lett* 212: 143–146.

Kim Y I, Dudek F E (1991). Intracellular electrophysiological study of suprachiasmatic nucleus neurons in rodents: excitatory synaptic mechanisms. *J Physiol* 444: 269–287.

Kim D Y, Kang H C, Shin H C, *et al.* (2001). Substance P plays a critical role in photic resetting of the circadian pacemaker in the rat hypothalamus. *J Neurosci* 21: 4026–4031.

Knowles R G, Palacious M, Palmer R M J, Moncada S (1989). Formation of nitric oxide from L-arginine in the central nervous system: a transduction mechanism for stimulation of the soluble guanylate cyclase. *Proc Natl Acad Sci USA* 86: 5159–5162.

Kriegsfeld L J, Demas G E, Lee S E Jr, *et al.* (1999). Circadian locomotor analysis of male mice lacking the gene for neuronal nitric oxide synthase (nNOS–/–). *J Biol Rhythms* 14: 20–27.

Kriegsfeld L J, Drazen D L, Nelson R J (2001). Circadian locomotor analysis of male mice lacking the gene for endothelial nitric oxide synthase (eNOS–/–). *J Biol Rhythms* 16: 142–148.

Leak R K, Moore R Y (2001). Topographical organization of suprachiasmatic nucleus projection neurons. *J Comp Neurol* 433: 312–334.

Leak R K, Card J P, Moore R Y (1999). Suprachiasmatic pacemaker organization analysed by viral transynaptic transport. *Brain Res* 819: 23–32.

Liou S Y, Shibata S, Iwasaki K, Ueki S (1986). Optic nerve stimulation-induced increase of release of ^{3}H-glutamate and ^{3}H-aspartate but not ^{3}H-GABA from the suprachiasmatic nucleus in slices of rat hypothalamus. *Brain Res Bull* 16: 527–531.

Maywood E S, Smith E, Hall S J, Hastings M H (1997). A thalamic contribution to arousal-induced, non-photic entrainment of the circadian clock of the Syrian hamster. *Eur J Neurosci* 9: 1739–1747.

Maywood E S, Okamura H, Hastings M H (2002). Opposing actions of neuropeptide Y and light on the expression of circadian clock genes in the mouse suprachiasmatic nuclei. *Eur J Neurosci* 15: 216–220.

Medanic M, Gillette M U (1992). Serotonin regulates the phase of the rat suprachiasmatic circadian pacemaker *in vitro* only during the subjective day. *J Physiol* 450: 629–642.

Meyer-Bernstein E L, Morin L P (1996). Differential serotonergic innervation of the suprachiasmatic nucleus and the intergeniculate leaflet and its role in circadian rhythm modulation. *J Neurosci* 16: 2097–2111.

Meyer-Bernstein E L, Morin L P (1999). Electrical stimulation of the median or dorsal raphe nuclei reduces light-induced FOS protein in the suprachiasmatic nucleus and causes circadian activity rhythm phase shifts. *Neuroscience* 92: 267–279.

Mick G, Yoshimura R, Ohno K, *et al.* (1995). The messenger RNAs encoding metabotropic glutamate receptor subtypes are expressed in different neuronal subpopulations of the rat suprachiasmatic nucleus. *Neuroscience* 66: 161–173.

Moga M M, Moore R Y (1997). Organization of neural inputs to the suprachiasmatic nucleus in the rat. *J Comp Neurol* 389: 508–534.

Moore R Y (1996). Entrainment pathways and the functional organization of the circadian system. *Prog Brain Res* 111: 103–119.

Moore R Y, Card J P (1990). Neuropeptide-Y in the circadian timing system. *Ann NY Acad Sci* 611: 247–257.

Moore R Y, Card J P (1994). The intergeniculate leaflet: an anatomically and functionally distinct subdivision of the lateral geniculate complex. *J Comp Neurol* 344: 403–430.

Moore R Y, Danchenko R L (2002). Paraventricular-subparaventricular hypothalamic lesions selectively affect circadian function. *Chronobiol Int* 19: 345–360.

Moore R Y, Lenn J (1972). A retinohypothalamic projection in the rat. *J Comp Neurol* 146: 1–14.

Moore R Y, Halaris A E, Jones B E (1978). Serotonin neurons of the midbrain raphe: ascending projections. *J Comp Neurol* 180: 417–438.

Moore R Y, Speh J C, Card J P (1995). The retinohypothalamic tract originates from a distinct subset of retinal ganglion cells. *J. Comp. Neurol* 352: 351–366.

Morin L P, Blanchard J (1991). Depletion of brain serotonin by 5,7-DHT modifies hamster circadian rhythm response to light. *Brain Res* 566: 173–185.

Morin L P, Blanchard J H (2001). Neuromodulator content of hamster intergeniculate leaflet neurons and their projection to the suprachiasmatic nucleus or visual midbrain. *J Comp Neurol* 437: 79–90.

Morin L P, Meyer-Bernstein E L (1999). The ascending serotonergic system in the hamster: comparison with projections of the dorsal and median raphe nuclei. *Neuroscience* 91: 81–105.

Morin L P, Blanchard L P, Moore R Y (1992). Intergeniculate leaflet and suprachiasmatic nucleus organization and connections in the golden hamster. *Vis Neurosci* 8: 219–230.

Moriya T, Yamanouchi S, Fukushima T, et al. (1996). Involvement of 5-HT_{1A} receptor mechanisms in the inhibitory effects of metamphetamine on photic responses in the rodent suprachiasmatic nucleus. *Brain Res* 740: 261–267.

Moyer R W, Kennaway D J (2000). Serotonin depletion decreases light induced c-*fos* in the rat suprachiasmatic nucleus. *Neuroreport* 11: 1021–1024.

Moyer R W, Kennaway D J, Ferguson S A, Dijstelbloem Y P (1997). Quipazine and light have similar effects on c-*fos* induction in the rat suprachiasmatic nucleus. *Brain Res* 765: 337–342.

O'Hara B F, Andretic R, Heller H C, et al. (1995). $GABA_A$, $GABA_C$, and NMDA receptor subunit expression in the suprachiasmatic nucleus and other brain regions. *Mol Brain Res* 28: 239–250.

Ohi K, Takashima M, Nishikawa T, Takahasji K (1991). N-methyl-D-aspartate receptor participates in neuronal transmission of photic information through the retinohypothalamic tract. *Neuroendocrinology* 53: 344–348.

Park H T, Baek S Y, Kin B S, et al. (1993). Profile of Fos-like immunoreactivity induction by light stimuli in the intergeniculate leaflet is different from that of the suprachiasmatic nucleus. *Brain Res* 610: 334–339.

Penev P D, Turek F W, Zee P C (1993). Monoamine depletion alters the entrainment and the responses to light of the circadian activity rhythm in hamsters. *Brain Res* 156: 164.

Pickard G E (1982). The afferent connections of the suprachiasmatic nucleus of the golden hamster with emphasis on the retinohypothalamic projection. *J Comp Neurol* 211: 65–83.

Pickard G E, Rea M A (1997). TFMPP, a 5-HT_{1B} receptor agonist, inhibits light-induced phase shifts of the circadian activity rhythm and c-*fos* expression in the mouse suprachiasmatic nucleus. *Neurosci Lett* 231: 95–98.

Pickard G E, Weber E T, Scott P A, et al. (1996). 5-HT_{1B} receptor agonists inhibit light-induced phase shifts of behavioural circadian rhythms and expression of the immediate early gene c-*fos* in the suprachiasmatic nucleus. *J Neurosci* 16: 8208–8220.

Piggins H D, Marchant E G, Goguen D, Rusak B (2001a). Phase-shifting effects of pituitary adenylate cyclase activating polypeptide on hamster wheel-running rhythms. *Neurosci Lett* 305: 25–28.

Piggins H D, Samuels R E, Coogan A N, Cutler D J (2001b). Distribution of substance P and neurokinin-1 receptor immunoreactivity in the suprachiasmatic nuclei and intergeniculate leaflet of hamster, mouse and rat. *J Comp Neurol* 438: 50–65.

Prosser R A (2001). Glutamate blocks serotonergic phase advances of the mammalian circadian pacemaker through AMPA and NMDA receptors. *J Neurosci* 21: 7815–7822.

Prosser R A, Miller J D, Heller H C (1990). A serotonin agonist phase-shifts the circadian clock in the suprachiasmatic nuclei *in vitro*. *Brain Res* 534: 336–339.

Prosser R A, MacDonald E S, Heller H C (1994a). c-*fos* mRNA in the suprachiasmatic nuclei *in vitro* shows a circadian rhythm and responds to a serotonergic agonist. *Mol Brain Res* 25: 151–156.

Prosser R A, Heller H C, Miller J D (1994b). Serotonergic phase advances of the mammalian circadian clock involve protein kinase A and K^+ channel opening. *Brain Res* 644: 67–73.

Ralph M R, Menaker M (1988). A mutation of the circadian system in golden hamsters. *Science* 241: 1225–1227.

Ralph M R, Foster R G, Davis F C, Menaker M (1990). Transplanted suprachiasmatic nucleus determines circadian period. *Science* 247: 975–978.

Rea M A (1989). Light increases Fos-related protein immunoreactivity in the rat suprachiasmatic nuclei. *Brain Res Bull* 23: 577–581.

Reuss S, Fuchs E (2000). Anterograde tracing of retinal afferents to the tree shrew hypothalamus and raphe. *Brain Res* 874: 66–74.

Romijn H J, Sluiter A A, Wortel J, et al. (1998). Immunocytochemical evidence for a diurnal rhythm of neurons showing colocalization of VIP and GRP in the rat suprachiasmatic nucleus *J Comp Neurol* 391: 397–405.

Scott G, Rusak B (1996). Activation of hamster suprachiasmatic neurons *in vitro* via metabotropic glutamate receptors. *Neuroscience* 71: 533–541.

Selim M, Glass J D, Hauser U E, Rea M A (1993). Serotonergic inhibition of light-induced Fos protein expression and extracellular glutamate in the suprachiasmatic nuclei. *Brain Res* 621: 181–188.

Shen H, Semba K (1994). A direct retinal projection to the dorsal raphe nucleus in the rat. *Brain Res* 635: 159–168.

Shibata S, Tsuneyoshi A, Hamada T, et al. (1992a). Effect of substance P on circadian rhythms of firing activity and the 2-deoxyglucose uptake in the rat suprachiasmatic nucleus *in vitro*. *Brain Res* 597: 257–263.

Shibata S, Tsuneyoshi A, Hamada T, et al. (1992b). Phase-resetting effect of 8-OH-DPAT, a serotonin$_{1A}$ receptor agonist, on the circadian rhythm of firing rate in the rat suprachiasmatic nuclei *in vitro*. *Brain Res* 582: 353–356.

Shibata S, Watanabe A, Hamada T, et al. (1994). N-methyl-D-aspartate induces phase shifts in circadian rhythm of neuronal activity of rat SCN *in vitro*. *Am J Physiol* 267: R360–R364.

Shirakawa T, Moore R Y (1994a). Glutamate shifts the phase of the circadian neuronal firing rhythm in the rat suprachiasmatic nucleus *in vitro*. *Neurosci Lett* 178: 47–50.

Shirakawa T, Moore R Y (1994b). Responses of rat suprachiasmatic nucleus neurons to substance P and glutamate *in vitro*. *Brain Res* 642: 213–220.

Silver R, LeSauter J, Tresco P A, Lehman M N (1996). A diffusible coupling signal from the transplanted suprachiasmatic nucleus controlling circadian locomotor rhythms. *Nature* 382: 810–813.

Smale L, Michels K M, Moore R Y, Morin L P (1990). Destruction of the hamster serotonergic system by 5,7-DHT: effects on circadian rhythm phase, entrainment and response to triazolam. *Brain Res* 515: 9–19.

Smart C M, Biello S M (2001). WAY-100635, a specific 5-HT1A antagonist, can increase the responsiveness of the mammalian circadian pacemaker to photic stimuli. *Neurosci Lett* 305: 33–36.

Starkey S J (1996). Melatonin and 5-hydroxytryptamine phase-advance the rat circadian clock by activation of nitric oxide synthesis. *Neurosci Lett* 211: 199–202.

Starkey S J, Grant A L, Hagan R M (2001). A rapid and transient synthesis of nitric oxide (NO) by a constitutively expressed type II NO synthase in the guinea-pig suprachiasmatic nucleus. *Br J Pharmacol* 134: 1084–1092.

Stephan F K, Zucker I (1972). Circadian rhythms in drinking behaviour and locomotor activity of rats are eliminated with hypothalamic lesions. *Proc Natl Acad Sci USA* 69: 1583–1586.

Sutin E L, Kilduff T S (1992). Circadian and light-induced expression of immediate early gene mRNAs in the rat suprachiasmatic nucleus. *Brain Res Mol Brain Res* 15: 281–290.

Takatsuji K, Miguelhidalgo J J, Tohyama M (1991). Substance-P-immunoreactive innervation from the retina to the suprachiasmatic nucleus in the rat. *Brain Res* 568: 223–229.

Tominaga K, Shibata S, Ueki S, Watanabe S (1992). Effects of 5-HT$_{1A}$ receptor agonists on the circadian rhythm of wheel-running activity in hamsters. *Eur J Pharmacol* 214: 79–84.

van de Kar L D, Lorens S A (1979). Differential serotonergic innervation of individual hypothalamic nuclei and other forebrain regions by the dorsal and median midbrain raphe nuclei. *Brain Res* 162: 45–54.

van den Pol A N (1991). Glutamate and aspartate immunoreactivity in hypothalamic presynaptic axons. *J Neurosci* 11: 2087–2101.

van den Pol A N (1994). Metabotropic glutamate receptor mGluR1 distribution and ultrastructural localization in hypothalamus. *J Comp Neurol* 349: 615–632.

van den Pol A N, Tsujimoto K L (1985). Neurotransmitters of the hypothalamic suprachiasmatic nucleus: immunocytochemical analysis of 25 neuronal antigens. *Neuroscience* 15: 1049–1086.

van den Pol A N, Romano C, Ghosh P (1995). Metabotropic glutamate receptor mGluR5 subcellular distribution and developmental expression in hypothalamus. *J Comp Neurol* 362: 134–150.

Van Reeth O, Turek F W (1989). Stimulated activity mediates phase shifts in the hamster circadian clock induced by dark pulses or benzodiazepines. *Nature* 339: 49–51.

Villar M J, Vitale M L, Parisi M N (1987). Dorsal raphe serotonergic projection to the retina. A combined peroxidase tracing-neurochemical/high-performance liquid chromatography study in the rat. *Neuroscience* 22: 681–686.

Watanabe A, Hamada T, Shibata S, Watanabe S (1994). Effects of nitric oxide inhibitors on N-methyl-D-aspartate-induced phase delay of circadian rhythm of neuronal activity in the rat suprachiasmatic nucleus *in vitro*. *Brain Res* 646: 161–164.

Watanabe A, Ono M, Shibata S, Watanabe S (1995). Effect of a nitric oxide synthase inhibitor, N-nitro-L-arginine methylester, on light-induced phase delay of circadian rhythm of wheel-running activity in golden hamsters. *Neurosci Lett* 192: 25–28.

Watts A G, Swanson L W (1987). Efferent projections of the suprachiasmatic nucleus: II. studies using retrograde transport of fluorescent dyes and simultaneous peptide immunohistochemistry in the rat. *J Comp Neurol* 258: 230–252.

Watts A G, Swanson L W, Sanchez-Watts G (1987). Efferent projections of the suprachiasmatic nucleus: I. studies using anterograde transport of *Phaseolus vulgaris* leucoagglutinin in the rat. *J Comp Neurol* 258: 204–229.

Weber E T, Rea M A (1997). Neuropeptide Y blocks light-induced phase advances but not delays of the circadian activity rhythm in hamsters. *Neurosci Lett* 231: 159–162.

Weber E T, Gannon R L, Michel A M, *et al.* (1995). Nitric oxide synthase inhibitor blocks light-induced phase shifts of the circadian activity rhythm, but not c-*fos* expression in the suprachiasmatic nucleus of the Syrian hamster. *Brain Res* 692: 137–142.

Wickland C, Turek F W (1994). Lesions of the thalamic intergeniculate leaflet block activity-induced phase shifts in the circadian activity rhythm of the golden hamster. *Brain Res* 660: 293–300.

3

The molecular biology of the mammalian circadian clock

Mauro Zordan, Rodolfo Costa, Chiaki Fukuhara and Gianluca Tosini

3.1 Introduction

A *circadian oscillator* is a part (cell, tissue or organ) of a living organism that is capable, when isolated and maintained in constant conditions, of generating a self-sustained rhythm with a period close to 24 h. When an organism is held under constant conditions, its circadian rhythm will persist with a period close to, but in most cases significantly different from 24 h. This result establishes that the rhythms are self-generated, or endogenous; the period of a rhythm under constant conditions is called its *free-running period* (see Chapter 1, Section 1.3.1). The free-running period of a rhythm is directly related to the period of the biological clock that generates it, and it is therefore a useful measure of the state of the clock (e.g. how it is affected by particular genetic mutations).

The mechanism that generates the circadian oscillation clearly operates at the subcellular (molecular) level. Furthermore, in multicellular organisms, the central pacemakers are localised to particular structures primarily within the nervous system. Basically, three experimental approaches have been used successfully to identify structures containing biological clocks: lesion of the candidate structure, *in vitro* culture of the candidate tissue/cells and transplantation of the candidate tissue. Lesions can be made surgically in discrete areas of the brain or endocrine organs, and the effects of such lesions on overt rhythms can then be observed. Despite its drawbacks, this approach has been used for the preliminary localisation of central oscillators in several species. The 'master clock' controlling most of the circadian rhythms of mammals – the suprachiasmatic nuclei of the hypothalamus (SCN) – were first identified using this method (Moore and Eichler, 1972). However, the results of lesion experiments are difficult to interpret and this approach alone is not sufficient to demonstrate whether a particular structure acts as a clock.

When the results of lesion experiments are combined with data from culture and transplantation experiments, unambiguous interpretations may become possible. It is sometimes possible to remove the candidate tissue from the organism and to culture it *in vitro* under constant conditions for several days, while recording potentially rhythmic outputs (e.g. electrical activity, hormone release, peptide synthesis). If, under such circumstances, one or more of the measured outputs persist rhythmically, this would provide a direct demonstration that the cultured tissue contains a biological clock. This approach has been used to demonstrate the presence of biological clocks in the SCN and retina of mammals (Welsh *et al.*, 1995; Tosini and Menaker, 1996).

Although experiments of the above kind are able to confirm the presence of a circadian clock in the cultured tissue, they are not suitable to establish the clock's function in the intact organism. In this respect, transplantation experiments can sometimes produce dramatic results leading to important information regarding the role of the candidate tissue or organ. For example, lesions of the SCN abolish most circadian rhythms in rodents, while transplantation of fetal SCN tissue into the lesion site can restore circadian rhythmicity of locomotor activity (Silver *et al.*, 1990) but does not restore any of the hormonal rhythms (Meyer-Bernstein *et al.*, 1999). Furthermore, using SCN tissue from mutant donors, it has been possible to show that the restored rhythm has the characteristics of the donor's rhythm rather than those of the host (Ralph *et al.*, 1990).

3.2 Molecular aspects of the circadian clock

So far, significant progress in understanding the biology of circadian rhythms has ultimately depended on research based on genetically tractable model organisms such as the fruit fly *Drosophila melanogaster*. Thus, in the early 1970s, Konopka and Benzer, working with *Drosophila*, succeeded in isolating the first clock mutants (Konopka and Benzer, 1971), hence opening the way to a systematic genetic and, more recently, molecular approach to the study of circadian biological oscillators. Much of the work currently in progress is being devoted to understanding the mechanisms underlying the biological oscillators in organisms as diverse as cyanobacteria and humans; so far, comparative analyses of the available experimental data suggest multiple independent origins for the intracellular clocks in living organisms. None the less, a theme common to all taxa can be identified in the physical mechanism governing the intracellular oscillator, which in all cases appears to be based on a combination of positive and negative transcriptional/translational feedback loops.

Forward genetic approaches to dissecting the circadian clock, starting with mutagenesis, followed by mapping and cloning the relevant loci have proved particularly successful in *Drosophila*, *Neurospora*, cyanobacteria and the mouse (see Zordan *et al.*, 2000, for a recent review). So far, the fruit fly *Drosophila melanogaster* has provided the most detailed description of the workings of the circadian clock machinery; recently it has become clear that remarkable similarities exist between the mammalian and fruit fly clocks (see Table 3.1, Figures 3.1 and 3.2). They share almost the same clock genes, although some of their regulatory relationships have been altered during the evolutionary process. The first *Drosophila* clock genes to have been identified and cloned were *period* (*per*) and *timeless* (*tim*). Both *per* and *tim* are rhythmically expressed at the RNA and protein levels. In addition, the fact that the RNA levels decline as soon as the protein levels rise suggests that the expression of both genes is under control of their protein products. Indeed, in per^0 and tim^0 mutants the RNA cycling is abolished for both genes (Rosato and Kyriacou, 2001). In order to autoregulate gene expression, the gene products PER and TIM must enter the nucleus, and they do so late at night as trimer complexes (DBT/PER/TIM) with another protein, the kinase DOUBLETIME (DBT) (Saez and Young, 1996; Kloss *et al.*, 2001). In the nucleus, the complex DBT/PER/TIM is gradually converted to DBT/PER, which in turn acts as a negative regulator by physically associating with dimers constituted by the CLOCK (CLK) and CYCLE (CYC), basic helix–loop–helix (bHLH-PAS) transcription factors, forming a complex that prevents CLK–CYC from binding to the DNA target (Lee *et al.*, 1999; Rothenfluh *et al.*, 2000; Kloss *et al.*, 2001). It is not necessary for a gene to be rhythmically expressed in order to encode a clock component, and *doubletime* (*dbt*) is such an example (Price *et al.*, 1998). The DBT protein is a homologue of human casein kinase Iε, and phosphorylates PER, when PER is free of TIM, targeting it for rapid degradation (Kloss *et al.*, 1998). The formation of a complex with TIM protects PER, allowing protein levels to rise; but after nuclear entry the conversion of DBT/PER/TIM into DBT/PER and TIM leads to the progressive phosphorylation and degradation of PER and consequently to termination/reinitiation of the molecular circadian clock cycle (Price *et al.*, 1998; Kloss *et al.*, 2001). Another kinase, Shaggy/GSK-3, which plays a major role in development at the level of the Wnt signal transduction pathway (Peifer and Polakis, 2000), seems to be involved in TIM phosphorylation, which also affects the timing of PER/TIM nuclear translocation (Martinek *et al.*, 2001). The components responsible for the cycling of TIM levels in constant conditions are not known, but under light–dark (LD) cycles, light triggers proteasome-mediated degradation of TIM (Naidoo *et al.*, 1999), probably through the action of CRYPTOCHROME (Stanewsky *et al.*, 1998). During the light phase, CRYPTOCHROME (CRY), a blue-light photoreceptor, physically interacts with TIM, inhibiting the formation of the DBT/PER/TIM complex and thereby ultimately releasing CLK/CYC to reinitiate *per* and *tim* transcription. Cryptochromes are pterin-/flavin-containing proteins presumably derived from photolyases and first identified in plants. The pterin is proposed to act as the antenna and receives light energy, which it passes to the second flavin chromophore. This in turn transfers a free excited electron to a redox signalling partner interacting with the COOH-terminal of the protein. Moreover, a study based on the yeast two-hybrid system, performed to examine the interaction of CRY with other known clock proteins, has shown that a large fragment of PER (residues 233–685) interacts strongly with full-length CRY under light conditions but not under dark (Rosato *et al.*, 2001).

The last *Drosophila* clock component to have been described, *vrille*, (*vri*) was discovered in a screen for clock-controlled genes. *Vrille*, a transcription factor essential for embryonic development, is expressed rhythmically in circadian pacemaker cells in larval and adult brains. Eliminating the normal *vri* cycle suppresses *per* and *tim* expression and causes long-period behavioural rhythms and arrhythmicity, indicating that cycling *vri* is required for a functional *Drosophila* clock. In addition, CLOCK and VRI independently regulate levels of a neuropeptide, pigment-dispersing factor, which appears to regulate overt behaviour (Blau and Young, 1999).

The first mutation affecting the biological clock of mammals was discovered only in the late 1980s. The identification of this mutation, which was called *tau*, was a fortuitous event, since the mutant turned up in a shipment of hamsters sent to Menaker's laboratory (Ralph and Menaker, 1988). *Tau* has since been shown to be a semidominant autosomal mutation, characterised by a short period in the rhythm of circadian wheel-running activity (22 h in the heterozygous mutant and 20 h in the homozygote). Recently the *tau* gene has been cloned, and it has been shown that the *tau* locus encodes for casein kinase Iε (*ckIε*) and that the *tau* gene is homologous to the *Drosophila dbt* gene.

The second mutation affecting the circadian system in mammals was discovered a few years later, and the corresponding gene was named *clock*. *Clock* is also an autosomal, semidominant mutation. It

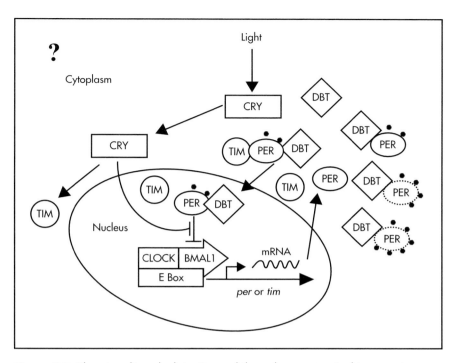

Figure 3.1 The circadian clock in *Drosophila melanogaster*. In this organism transcription of the clock genes *per* and *tim*, is positively regulated by the heterodimer formed by the protein products of the *clock* and *bmal1* (d*cycle*) genes. PER and TIM (together with DBT) form complexes that can then translocate into the nucleus, where they are converted into DBT/PER and TIM, thereby negatively regulating their own transcription by interfering with the positively acting CLOCK–BMAL1 dimers. Moreover, the inhibiting action of the PER/DBT complexes is contrasted by the light-activated product of the *cry* gene. The action of a constitutive homologue of human casein kinase type 1ε, encoded by the *dbt* gene, introduces delays in the formation of DBT/PER/TIM complexes by subtracting PER monomers, which are degraded, following targeting via DBT-dependent phosphorylation (depicted by the black circles on PER), until TIM reaches a threshold concentration in the cytoplasm and binds and stabilises PER. Clock resetting by light is effected by light-dependent degradation of TIM protein, although the precise mechanism by which light leads to this degradation has still to be described.

was isolated in Takahashi's laboratory in a deliberate mutagenesis screen for mutations affecting the circadian rhythmicity phenotype (Vitaterna *et al.*, 1994). Mice homozygous for the *clock* mutation do not show any circadian pattern in locomotor activity rhythms when the animals are kept in constant darkness (Vitaterna *et al.*, 1994). The mouse *clock* gene was later cloned (Antoch *et al.*, 1997; King *et al.*, 1997), and shown to be a rather large gene (100 kb) located on chromosome 5. *Clock* is a member of the bHLH-PAS family of transcription factors (i.e. it contains a PAS protein–protein interaction domain, along with a bHLH DNA-binding domain). Interestingly, *clock* transcription does not appear to be regulated in a circadian manner (i.e. it is not expressed in an oscillatory fashion).

Shortly after *clock* was cloned, a mammalian homologue of the *Drosophila period* gene was identified (Sun *et al.*, 1997a,b; Tei *et al.*, 1997). Like the fruit fly *period* gene, the mammalian gene possesses a PAS domain but not a functional bHLH domain, or any other known DNA-binding sites. In the mouse SCN, the transcript levels of this gene begin to rise in the late night before dawn and peak around mid-day, while in other tissues its expression appears to lag 4 h behind that observed in the SCN. Additional studies (Zylka *et al.*, 1998) showed that, in mammals, in addition to the first *period* gene just described (*per1*), there are two other genes encoding *period* products (*per2* and *per3*); these also contain a PAS domain, but no functional DNA-binding domain. The importance of the *period* genes in the generation and control of circadian rhythmicity has also been demonstrated. Zheng *et al.* (1999) reported that mice in which *per2* has been mutated in its PAS domain displayed (in constant darkness) a circadian period shorter than that of the wild-type that is followed by the loss of circadian rhythmicity. Several recent studies have reported the effects on circadian rhythmicity of targeted disruption of *per1*, *per2* and *per3*. Animals with disruption of *per1* show a shorter (about 1 h) and less stable free-running period (Bae *et al.*, 2001; Cermakian *et al.*, 2001; Zheng *et al.*, 2001). Mice with a disruption of *per1* and *per 2* became arrhythmic after a few days in constant darkness (DD) (Bae *et al.*, 2001; Zheng *et al.*, 2001). Disruption of *per3* determined only a small change (about 0.5 h) in the free-running period of the circadian locomotor activity rhythms (Shearman *et al.*, 2000a).

Since the *clock* gene possesses a functional PAS domain (i.e. a protein interaction domain) several laboratories have attempted to identify the partner of the *clock* protein (CLOCK). Using yeast two-hybrid assays, Gekakis *et al.* (1998) were the first to identify BMAL1 (also known as MOP3) as the partner of mammalian CLOCK. In the SCN, *bmal1* mRNA levels vary rhythmically and in antiphase with the *per* genes. In mice, loss of *bmal1* results in reduced levels of locomotor activity during LD and, more importantly, an immediate and complete loss of the circadian rhythm in DD (Bunger *et al.*, 2000). Recent work has led to the identification of *bmal2* (which is also capable of binding to CLOCK proteins), but its role in the generation of circadian rhythmicity is not yet clear (Okano *et al.*, 2001).

Additional studies, using a reverse genetic approach, indicated that two other genes (*cryptochrome 1* and *2*) are necessary for the generation of circadian rhythmicity in mammals. In fact, mice in which both *cryptochrome* genes (*cry1* and *cry2*) have been removed (i.e. a double knockout), are completely arrhythmic when placed in DD (van der Horst *et al.*, 1999). *Cryptochrome* mRNA levels vary rhythmically in the SCN, with a peak at light-to-dark transition (Miyamoto and Sancar, 1998).

The accepted model describing the molecular mechanisms leading to the generation of the circadian rhythm-associated phenotypes in mammals is similar to that already described for *Drosophila*, albeit with some important differences. First, in mammals the system may embody considerable redundancy, as is suggested by the presence of three *period* genes, and two *cryptochrome* genes. A further important difference in the mammalian mechanism lies in the role played by *timeless*. Specifically, mammalian *timeless* shows no evidence of rhythmic circadian regulation (Sangoran *et al.*, 1998), TIMELESS protein is not degraded on exposure to light (as in *Drosophila*) and TIMELESS does not dimerise with PERIOD as is known to occur in *Drosophila*. In addition, database searches of the *Drosophila* genome have indicated that mammalian *timeless* is not the true orthologue of *Drosophila timeless* but is likely to be the orthologue of a newly described *Drosophila* gene named *timeout* (Benna *et al.*, 2000; Gotter *et al.*, 2000). Analysis of the full human genome sequence did not indicate any other gene that could be considered a homologue of *Drosophila timeless* (Clayton *et al.*, 2001).

The current knowledge regarding the workings of the mammalian system is summarised in Figure 3.2. At the centre of the circadian clock there are interacting positive and negative transcriptional and

Table 3.1 List of the genes implicated in circadian timing system in *Drosophila* and in the mouse

Drosophila	*Mouse*
period	period 1, period 2 and period 3
timeless	No homologue identified
cryptochrome	cryptochrome 1 and cryptochrome 2
clock	clock
cycle	bmal1, bmal2
doubletime	ckIε
vrille	No homologue identified

posttranscriptional feedback loops (Shearman *et al.*, 2000b). The positive autoregulatory feedback loop begins with the heterodimer CLOCK/BMAL1 binding to the E-box (a short DNA sequence to which transcription factors of the bHLH family of proteins can bind) of the *per1* gene promoter, initiating the transcription of this gene. This results in the rise in the levels of the PER1 protein in the cytoplasm. Soon after that, PER3, and later PER2 also start to accumulate, with each of these proteins reaching peak levels at different circadian times. The difference in the phases of the peak levels of the three PER proteins suggests the existence of other (yet to be discovered) transcriptional activators besides the CLOCK/BMAL1 heterodimer. CLOCK/BMAL1 also acts as a transcriptional activator of the two mammalian *cryptochrome* genes. Once the PERIOD proteins (PER1, PER2 and PER3) and the CRY proteins (CRY1 and CRY2) have reached determined levels, they form heterodimers with each of the PER proteins, and the heterodimers then translocate to the nucleus. Two possible roles have been proposed for CKIε in the control of the molecular oscillation. First, it regulates the phosphorylation state of the PER proteins; secondly, since CKIε is able to interact with PER proteins, CKIε could play an additional role in the control of the circadian oscillator either by retaining PERs in the cytoplasm or by promoting their nuclear localisation (Lowrey *et al.*, 2000). Once they have reached the nucleus, the PER/CRY heterodimers inhibit the transcription of their own genes through inactivation of the positive effectors (the CLOCK/BMAL1 complex; Kume *et al.*, 1999). At this point, the level of the PER1, PER2, PER3, CRY1 and CRY2 proteins in the cytoplasm begins to decrease, leading to a parallel decrease in the formation of the inhibiting heterodimers. Ultimately, the level of heterodimers will be insufficient to inhibit the transcription of the *per* and *cry* genes, which will once again fall under the positive control of the CLOCK/BMAL1 complex. A recent study using mice carrying a null mutation in the *per1* gene has demonstrated that in these mice *per1* mRNA is still cycling, thus demonstrating that PER1 is not necessary for the rhythmic expression of its own gene (Bae *et al.*, 2001; Zheng *et al.*, 2001). The expression of *per2* in the *per1* mutants was not significantly different from that of wild-type mice, demonstrating that *per1* is not necessary for rhythmic expression of *per2*. In contrast, *per1* and *per2* mRNA expression is significantly reduced in *per2* mutant mice (Zheng *et al.*, 1999). In the double mutants (*per1* and *per2*), the expression of both genes was not rhythmic (Zheng *et al.*, 2001). In *per1* mutant mice, the expression of *per1* and *per2* in the peripheral tissues such as heart, kidney and muscle is delayed (Cermakian *et al.*, 2001).

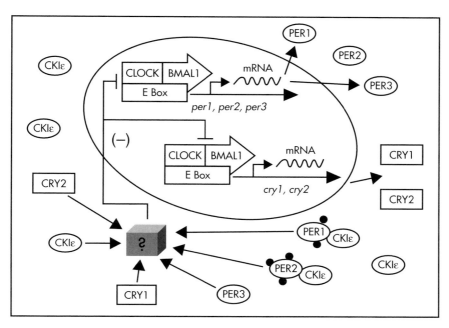

Figure 3.2 The circadian clock in mammals. In this model, transcription of the three *per* genes and *cry1* and *cry2* is positively regulated by the heterodimer formed by the protein products of the *clock* and *bmal1* genes. From *in vitro* experiments it has been shown that the products of the three *per* genes and CKIε, CRY1 and CRY2 are all able to participate in the formation of complexes in various ways. However, it is not known which complexes are actually formed *in vivo* (indicated by '?'). In addition, some or all of these proteins may participate as negative regulators of the transcription of the three *per* genes and the *cry* genes, although recent evidence implies that PER–CRY dimers may be the most important in this respect. In the figure the 'minus' sign indicates an inhibiting effect. The role of *bmal2* is unknown, and thus this gene is not represented in the model.

Table 3.2 shows the effects that mutations in the circadian clock genes produce on the circadian system. Summarising, we can say that there are three different phenotypes: a change in the free-running period; gradual loss of rhythmicity in constant conditions; and finally, immediate loss of rhythmicity. The picture that emerges from the analysis of the effects of these mutations suggests that not all the clock genes are equally important for the generation of circadian rhythmicity. In addition, the fact that some of these genes are necessary for the maintenance of rhythmicity, but are not necessary within each circadian cycle, indicates a redundancy of function among some of these genes.

Several investigations have demonstrated induction of *per1* and *per2* following a pulse of light, thus suggesting that these genes may be involved in the light entrainment pathway of the circadian clock (Albrecht *et al.*, 1997; Shearman *et al*, 1997; Shigeyoshi *et al.*, 1997; Wakamatsu *et al.*, 2001) and, indeed, the experimental evidence so far gathered seems to support this view. Akiyama *et al.* (1999) reported that *per1* antisense oligonucleotides injected intracerebroventricularly can reduce (or even block) light-induced phase shifts. Albrecht *et al.* (2001) demonstrated that mice lacking *per1* do not show phase advances in response to a light pulse, while mice lacking *per2* are deficient in phase delays. Support for a minor role of *per1* in mediating phase delays also comes from data obtained in *cry2*-deficient mice. These mice display larger phase delays than wild-type animals despite the fact that light inducibility of *per1* appears to be reduced (Thresher *et al.*, 1998). Furthermore, it has been shown that phase shift of locomotor activity in response to a pulse of light is not altered in *per1* mutant mice (Cermakian *et al.*, 2001).

Table 3.2 Effect on the circadian rhythm of locomotor activity (CRL) in constant darkness of mutations affecting one or more circadian clock genes (τ = free-running period)

Mutated gene(s)	Effect of CRL	Onset of the effect	Reference
per1	Shorter τ (23.0 h)	Immediate	Bae *et al.* (2001); Zheng *et al.* (2001); Cermakian *et al.* (2001)
per2	Arrhythmic	Delayed (<5 days)	Zheng *et al.* (1999); Bae *et al.* (2001)
per3	Shorter τ (23.5 h)	Immediate	Shearman *et al.* (2000a)
per1/per2	Arrhythmic	Immediate	Bae *et al.* (2001); Zheng *et al.* (2001)
per1/per3	Arrhythmic	Delayed (>20 days)	Bae *et al.* (2001)
per2/per3	Arrhythmic	Delayed (>10 days)	Bae *et al.* (2001)
cry1	Shorter τ (22.5 h)	Immediate	van der Horst *et al.* (1999)
cry2	Longer τ (24.6 h)	Immediate	Thresher *et al.* (1998); van der Horst (1999)
cry1/cry2	Arrhythmic	Immediate	van der Horst (1999)
bmal1	Arrhythmic	Immediate	Bunger *et al.* (2000)
clock	Arrhythmic	Delayed (<10 days)	Vitaterna *et al.* (1994)
ckIε (tau)	Shorter τ (20.0 h)	Immediate	Ralph and Menaker (1988)

Per1 and *per2* are also thought to be involved in mediating non-photic phase shifts, since stimuli that would produce phase shifts when administered during the day decrease the levels of expression of these genes (Maywood *et al.*, 1999; Horikawa *et al.*, 2000).

3.3 Which photopigment is mediating circadian photoreception?

In mammals, light information from the eye reaches the SCN, the primary circadian pacemaker, via a distinct neural projection called the retinohypothalamic tract (see Chapter 2). This tract arises from a small population of retinal ganglion cells and constitutes a relatively small percentage of the fibres of the optic nerve. Until a few years ago it was believed that the rods and cones, in conjunction with their photopigments, were responsible for collecting the light information necessary to the SCN for synchronisation to light. However, experiments conducted in the early 1990s on mice with genetically determined impairments of the eye (Foster *et al.*, 1991) cast doubt on this assumption. The most recent experiments have compared the effects of loss of photoreceptors on circadian photic entrainment in mouse models that lacked all rods and cones with respect to the congenic wild-type controls. In these mice, the absence of rods and cones did not influence photoentrainment or pineal melatonin suppression (Freedman *et al.*, 1999; Lucas *et al.*, 1999). Several studies provide support for the existence of functionally distinct visual and circadian photoreceptors in humans. These studies show that a significant number of subjects who have eyes but have lost conscious light perception, as result of retinal disease, are still able to suppress melatonin (Czeisler *et al.*, 1995) and entrain their circadian rhythms (Lockley *et al.*, 1997). Collectively these results suggest that photic entrainment in mammals must involve some as yet unidentified ocular photoreceptor, possibly localised at the level of the inner retina.

Most of the experimental evidence suggests that the circadian system of mammals is entrained by photoreceptors located within the eye. An exception is provided by a study that suggested that bright light of 13 000 lux applied to the skin at the back of the knee could shift circadian rhythms of body temperature and melatonin (Campbell and Murphy, 1998). These results led to the hypothesis that humans, and possibly other mammals, possess dermal photoreceptors that are used for photoentrainment. However, other findings show that eye loss (enucleation) in humans (Czeisler *et al.*, 1995; Lockley *et al.*, 1997) and other mammals (Foster *et al.*, 1991; Yamazaki *et al.*, 1999) always prevents photoentrainment. Application of light to the back of the knee does not produce any effect on the suppression of nocturnal levels of pineal melatonin (Lockley et al., 1998). Furthermore, it has been shown that exposure of the chest and abdomen to 13 000 lux broad-spectrum light did not produce phase shifts in the circadian rhythms of melatonin, cortisol or thyrotropin (Lindblom et al., 2000).

From what we have just described, it is evident that circadian photoreception is a phenomenon that could happen independently of classical photoreceptor cells and therefore without the intervention of classical photopigments (i.e. rhodopsin and/or cone-opsin). In the last few years, several investigations have focused on the identification of novel photopigments involved in determining the effects of light on the circadian system, and several candidates have been proposed based on genetic and molecular approaches. These candidates fall into two broad classes: (1) photopigments of the opsin:retinaldehyde family or (2) unknown photoreceptive processes in animals.

Several novel opsins in nonmammalian vertebrates have successfully been isolated using molecular biology techniques. These opsins are sufficiently different from the rod and cone opsins to warrant their assignment to new opsin gene families. Following the isolation from the chicken pineal of cDNA encoding the photopigment pinopsin, orthologues of this gene have been isolated from the pineal of several different bird, lizard, and amphibian species (for review see Foster and Soni, 1998).

In mammals, action spectra for circadian entrainment or for melatonin suppression have implicated opsin-based pigments in circadian responses to light (Provencio and Foster, 1995; Takahashi et al., 1984). The finding that these responses are not removed by the absence of the rod and cone photoreceptors (Freedman et al., 1999; Lucas et al., 1999) suggests that these novel photopigments may be localised to the inner layers of the retina, and a number of opsin photopigment candidates, including RGR (retinal-binding G protein-coupled receptor), peropsin and melanopsin, have recently been identified in mammals. RGR is not present in rods and cones and has been localised to the retinal pigment epithelium (RPE) and Müller cells (Jiang et al., 1993). In contrast to rod and cone opsins, its chromophore is all-*trans*-retinaldehyde rather than 11-*cis*-retinal. Upon exposure to light, RGR photoisomerises the all-*trans* chromophore to the 11-*cis* configuration. Its likely function, therefore, is not that of a sensory photopigment but that of a photoisomerase (Hao and Fong, 1999). Another candidate, peropsin, is also localised to the RPE, but its function is still unknown (Sun et al., 1997b). Since phylogenetic analysis places peropsin in the same group as RGR, this suggests that it may function as a photoisomerase. The most recently discovered candidate is a mammalian homologue of *Xenopus* melanopsin (Provencio *et al.*, 1998). Mammalian melanopsin is expressed within the inner retina in a small number of ganglion cells, and in an even smaller number of cells within the amacrine cell layer in the mouse retina. The function of melanopsin is unknown, but its homology to other opsins and its localisation to the inner retina suggest it as a good candidate for nonvisual photoreception (Provencio *et al.*, 2000).

As already mentioned, cryptochromes are important components of the mechanism that generates the circadian oscillation but, although their role in circadian entrainment of mammals cannot be completely excluded, the vast majority of the experimental data fail to support the idea that mammalian *cry1* and *cry2* contribute to photoentrainment. Using *in vitro* cellular assays, Griffin *et al* (1999) were unable to identify any effect of light on the activity of these proteins. In addition, the disruption of the cryptochrome genes in mice with a double *cry* knockout does not block the light-induced expression of the two clock genes *per1* and *per2* in the SCN (Okamura *et al.*, 1999). However, it is still argued by some that the mammalian CRYs have a dual function both as components of the oscillator and as photopigments. This view is more related to the originally proposed function of the mammalian CRYs as photopigments (Miyamoto and Sancar, 1998; Thresher *et al.*, 1998) and does not appear to be based on any positive photobiological results.

3.4 Role of clock genes in peripheral tissues

The experimental data gathered during the last 20 years suggested that mammals also possess circadian oscillators outside the SCN. The most convincing evidence of this comes from studies that demonstrated the persistence of some circadian rhythms in animals with SCN lesions. One of the best examples of such a phenomenon is represented by the observation that the circadian rhythm in photoreceptor disc shedding persists in SCN-lesioned animals even after optic nerve resection, demonstrating that this retinal rhythm is independent of humoral or neural SCN control (for review, see Tosini, 2000). Additional support for the idea that mammals have multiple circadian oscillators arose from the cloning of the circadian clock genes. As soon as these genes were cloned, it became evident that 'clock' genes were not only expressed in the SCN but that their expression was widespread within the body and that in several tissues/organs the expression was rhythmic (Shearman *et al.*, 1997; Zylka *et al.*, 1998; Fukuhara *et al.*, 2000) even if the rhythmicity in the peripheral tissues seems to disappear after SCN lesioning, suggesting, once again, that these rhythms are governed by the SCN (Sakamoto *et al.*, 1998). Of course, the only way ultimately to demonstrate that a specific organ and/or tissue has the capability to oscillate autonomously is to maintain it *in vitro* and to record a rhythmic output from the isolated preparation. In the last few years, circadian rhythms have been recorded from a variety of isolated extra-SCN tissues *in vitro*.

The first direct demonstration of an extra-SCN oscillator in a mammal was obtained in the retina, where a circadian rhythm of melatonin release was recorded (Tosini and Menaker, 1996, 1998). In *tau* mutant hamsters, the *tau* gene (*ckIε*) influences the retinal circadian rhythm of melatonin release in the same way that this gene influences the period of circadian oscillation in the SCN (Tosini and Menaker, 1996). Although this rhythm is generated within the neural retina – most likely in the photoreceptors – it is not yet clear which retinal cell type or types contain the clock (Tosini and Menaker, 1998). Just one year after the demonstration of the clock in the mammalian retina, the foundations of received wisdom in the circadian field were shaken by the work of Schibler's laboratory, which demonstrated that rat-1 fibroblasts (an immortalised and nonneural cell line) also contain a biological clock (Balsalobre *et al.*, 1998). In a series of very elegant experiments, Schibler and co-workers showed that a circadian oscillation in several genes could be initiated in rat-1 fibroblasts merely by exposing the cultured cells to a 'serum shock'. Analysis of the pathway involved in eliciting the circadian oscillation in these cells indicated that various signalling pathways involving PKC, cAMP-dependent protein kinases, Ca^{2+} and the glucocorticoid receptor can induce this circadian oscillation (Balsalobre *et al.*, 2000a; Yagita and Okamura, 2000).

The field was still coming to terms with this finding when Yamazaki *et al.* reported that many isolated tissues/organs of mammals can also express a circadian rhythm when cultured *in vitro* (Yamazaki *et al.*, 2000). This research team, using a transgenic rat in which luciferase expression was placed under the control of the *per1* promoter, demonstrated that *in vitro* cultures of lungs, liver, and skeletal muscle showed circadian rhythms in bioluminescence, although the rhythmicity in these organs/tissues tended to dampen out after few (2–7) cycles, while the bioluminescence rhythms in the SCN continued to oscillate for up to 32 days. This result suggested that there must be important differences between the 'central' and the 'peripheral' oscillators; and that the mechanism of the circadian clock located in the SCN must somehow be different from the mechanism of the peripheral tissues. In addition, it was also shown that different tissues/organs respond in a different way to a phase shift, since the peripheral oscillators entrain to LD cycles much less promptly than the SCN.

One of the most intriguing aspects of the mammalian circadian system remains the unveiling of how each part of the system communicates with the others and thus how the system is kept synchronised. Experimental evidence suggests that both neural and humoral factors are involved in this process. A recent study has indicated that the SCN projects directly to several peripheral areas, suggesting an active role for neural connections in controlling the expression of circadian rhythms in the periphery (Ueyama *et al.*, 1999). At the same time, it must also be mentioned that restoration of the circadian rhythms in locomotor activity in SCN-lesioned animals has also been achieved following the transplantation of SCN that has been encapsulated in a semipermeable polymeric capsule, preventing neural outgrowth but allowing the diffusion of humoral signals (Silver *et al.*, 1996). Indeed, a recent paper has shown that *per1* expression may be controlled by several pathways and different kinds of signals can affect the expression of this gene (Balsalobre *et al.*, 2000a). Among these signals, glucocorticoids seem to play an important role. Balsalobre *et al.* (2000b) have demonstrated that the glucocorticoid dexamethasone can induce circadian gene expression in cultured rat-1 fibroblasts, and injection of this glucocorticoid can affect the phase of circadian gene expression, at any phase of the cycle, of liver, kidney and heart, but it does not affect the expression of these genes in SCN neurons. However, the rhythmic expression of several genes (clock or simply clock-controlled genes) does not appear to be altered in the liver of mice with a specific genetic defect that inactivates the glucocorticoid receptor gene in hepatocytes ($GR^{alfpcre}$ mouse). This suggests that, although glucocorticoids may play a role in the entrainment of peripheral oscillators, they cannot be the only signal involved. The possibility of entraining animals to feeding regimes has been reported by several studies, and two recent papers have demonstrated that in animals subjected to a regime of restricted feeding (i.e. with food available only at particular times of the day) the circadian oscillators in the liver can be entrained independently from the SCN and the LD cycle (Damiola *et al.*, 2000; Stokkan *et al.*, 2001). Abrupt changes in the feeding time lead to a gradual resetting of the rhythmic gene expression, indicating that phase resetting is mediated by clock-dependent mechanisms. Interestingly, food-induced re-synchronisation proceeds faster in the liver than in the other peripheral tissues, suggesting that, as is the case of the resetting action of light

3.5 Relevance of the molecular clock mechanism to human biology

With the identification of the mammalian circadian clock genes, many research groups have begun to study the role that these genes may play in humans. Katzenberg *et al.* (1998) were the first to investigate whether a polymorphism in the human *clock* gene could be used as a predictor of diurnal/nocturnal preference. They found that individuals carrying one of the polymorphic alleles showed a slight difference (10–44 min) in the preferred timing for activity of sleep episodes. In a second study, the same group reported that polymorphisms in *per1* were not associated with a preference for diurnal or nocturnal pattern of activity (Katzenberg *et al.*, 1999). Another research team (Desan *et al.*, 2000) investigated whether polymorphisms in the *clock* gene were in any way linked to the occurrence of depression, but their data did not support such an association. Finally, a connection was made between an alteration in the daily activity pattern of human subjects and a mutation in a clock gene. Jones *et al.* (1999) reported the first human circadian disorder that showed a tendency for familiarity (i.e. it was hereditary). The disorder was named familial advanced sleep-phase syndrome (FASPS). The FASPS trait segregates as an autosomal dominant with high genetic penetrance. Affected individuals have a severely altered sleep–wake pattern since they experience sleepiness at an unusual time of the day (around 19.30) and early morning awakening (04.30). In addition, laboratory studies on FASPS-affected individuals demonstrated that such subjects present a shorter (by about 1 h) circadian period. In a further study, Ptacek and collaborators demonstrated that this syndrome is the consequence of a mutation affecting the *per2* gene (Toh *et al.*, 2001). FASPS-affected individuals have a serine (S662) to glycine mutation within the CKIε- binding region of the *per2* gene. Serine 662 is part of a consensus CKIε phosphorylation domain and the substitution with glycine renders the mutant protein less readily phosphorylated by CKIε. As we have already mentioned, phosphorylation of PERIOD proteins by CKIε is an important step in determining the length of the circadian period.

Interestingly, it must be noted that not all the families affected by FASPS show a mutation in the coding region of the corresponding *per2* genes. This observation raises the possibility that a further genetic analysis of these families could identify yet other alleles of the *per2* gene or perhaps even alterations in other genes.

3.6 Conclusion

The work carried out in the last few years has provided a completely new vision of how the circadian system of mammals is organised. These studies have indicated that mammals possess multiple circadian oscillators and that circadian oscillation properties are not exclusive to neural tissue (or cells). This work has also demonstrated that the mechanisms underlying the basic functioning of biological clocks show substantial conservation at various levels in the phylogeny of living organisms. This has proved to be basically true, insofar as all known circadian oscillators use intracellular feedback loops that are based on positive and negative elements. Such elements act in oscillators, in which transcription of clock genes yields proteins (negative elements) that act to block the action of positive element(s) whose role is to activate the clock genes encoding the negative element(s).

Acknowledgements

R. Costa and M. Zordan are supported by grants from Agenzia Spaziale Italiana (ASI) and from Ministero dell'Universita' e della Ricerca Scientifica (MURST). G. Tosini's laboratory is supported by grants from National Institute of Neurological Disorders (NINDS) and by the National Space Biomedical Research Institute (NSBRI).

References

Akiyama M, Kouzu Y, Takahashi S, *et al.* (1999). Inhibition of light- or glutamate-induced *m*Per1 expression represses the phase shifts into the mouse circadian locomotor and suprachiasmatic firing rhythms. *J Neurosci* 19: 1115–1121.

Albrecht U, Sun Z S, Eichele G, *et al.* (1997). A differential response of two putative mammalian circadian regulators, *mper1* and *mper2*, to light. *Cell* 91: 1055–1064.

Albrecht U, Zheng B, Larkin D, *et al.* (2001). *mPer1* and *mPer2* are essential for normal resetting of the circadian clock. *J Biol Rhythms* 16: 100–104.

Antoch M P, Song E J, Chang A M, *et al.* (1997). Functional identification of the mouse circadian *Clock* gene by transgenic BAC rescue. *Cell* 89: 655–667.

Bae K, Jin X, Maywood E S, *et al.* (2001). Differential functions of *mPer1*, *mPer2*, and *mPer3* in the SCN circadian clock. *Neuron* 30: 525–536.

Balsalobre A, Damiola F, Schibler U (1998). A serum shock induces circadian gene expression in mammalian tissue culture cells. *Cell* 93: 929–937.

Balsalobre A, Marcacci L, Schibler U (2000a). Multiple signaling pathways elicit circadian gene expression in cultured Rat-1 fibroblasts. *Curr Biol* 10: 1291–1294.

Balsalobre A, Brown S A, Marcacci L, *et al.* (2000b). Resetting of circadian time in peripheral tissues by glucocorticoid signaling. *Science* 289: 2344–2347.

Benna C, Scannapieco P, Piccin A, *et al.* (2000). A second *timeless* gene in *Drosophila* shares greater sequence similarity with mammalian *tim*. *Curr Biol* 10: R512–R513.

Blau J, Young M W (1999). Cycling *vrille* expression is required for a functional *Drosophila* clock. *Cell* 99: 661–671.

Bunger M K, Wilsbacher L D, Moran S M, *et al.* (2000). Mop3 is an essential component of the master circadian pacemaker in mammals. *Cell* 103: 1009–1017.

Campbell S S, Murphy P (1998). Extraocular circadian phototransduction in humans. *Science* 279: 396–399.

Cermakian N, Monaco L, Pando M P, *et al.* (2001). Altered behavioral rhythms and clock gene expression in mice with targeted mutation in the *Period1* gene. *EMBO J* 20: 3967–3974.

Clayton J D, Kyriacou C P, Reppert S M (2001). Keeping time with the human genome. *Nature* 409: 829–831.

Czeisler C A, Shanahan T L, Klerman E B, *et al.* (1995). Suppression of melatonin secretion in some blind patients by exposure to bright light. *N Engl J Med* 332: 6–11.

Damiola F, Le Minh N, Preitner N, *et al.* (2000). Restricted feeding uncouples circadian oscillators in peripheral tissues from the central pacemaker in the suprachiasmatic nucleus. *Genes Dev* 14: 2950–2961.

Desan P H, Oren D A, Malison R, *et al.* (2000). Genetic polymorphism at the *clock* gene locus and major depression. *Am J Med Genet* 12: 418–421.

Foster R G, Soni B G (1998). Extraretinal photoreceptors and their regulation of temporal physiology. *Rev Reprod* 3: 145–150.

Foster R G, Provencio I, Hudson D, *et al.* (1991). Circadian photoreception in the retinally degenerate mouse (*rd/rd*). *J Comp Physiol [A]* 169: 39–50.

Freedman M S, Lucas R J, Soni B, *et al.* (1999). Regulation of mammalian circadian behavior by non-rod, non-cone, ocular photoreceptors. *Science* 284: 502–504.

Fukuhara C, Dirden J C, Tosini G (2000). Circadian expression of Period 1, Period 2, and arylalkylamine N-acetyltransferase mRNA in the rat pineal gland under different light conditions. *Neurosci Lett* 286: 167–170.

Gekakis N, Staknis D, Nguyen H B, *et al.* (1998). Role of the CLOCK protein in the mammalian circadian mechanism. *Science* 280: 1564–1569.

Griffin E, Staknis D, Weitz C (1999). Light-independent role of CRY1 and CRY2 in the mammalian circadian clock. *Science* 286: 768–771.

Gotter A L, Manganaro T, Weaver D R, *et al.* (2000). A time-less function for mouse *timeless*. *Nat Neurosci* 3: 755–756.

Hao W, Fong H K W (1999). The endogenous chromophore of retinal G protein-coupled receptor opsin from the pigment epithelium. *J Biol Chem* 274: 6085–6090.

Horikawa K, Yokota S, Fuji K, *et al.* (2000). Nonphotic entrainment by $5\text{-HT}_{1A\text{-}7}$ receptor agonists accompanied by reduced *Per1* and *Per2* mRNA levels in the suprachiasmatic nuclei. *J Neurosci* 20: 5867–5873.

Jiang M, Pandey S, Fong H K W (1993). An opsin homologue in the retina and pigment epithelium. *Invest Ophthalmol Vis Sci* 34: 3669–3679.

Jones C R, Campbell S S, Zone S E, *et al.* (1999). Familial advanced sleep-phase syndrome: a short-period circadian rhythm variant in human. *Nat Med* 5: 1062–1065.

Katzenberg D, Young T, Finn L, *et al.* (1998). A CLOCK polymorphism associated with human diurnal preference. *Sleep* 21: 569–576.

Katzenberg D, Young T, Lin L, *et al.* (1999). A human period gene (HPER1) polymorphism is not associated with diurnal preference in normal adults. *Psychiatr Genet* 9: 107–109.

King D P, Zhao Y, Sangoram A M, *et al.* (1997). Positional cloning of the mouse circadian *Clock* gene. *Cell* 89: 641–653.

Kloss B, Price J L, Saez L, *et al.* (1998). The *Drosophila* clock gene *double-time* encodes a protein closely related to human casein kinase I epsilon. *Cell* 94: 97–107.

Kloss B, Rothenfluh A, Young M W *et al.* (2001). Phosphorylation of period is influenced by cycling physical associations of double-time, period, and timeless in the *Drosophila* clock. *Neuron* 30: 699–706.

Konopka R J, Benzer S (1971). Clock mutants of *Drosophila melanogaster*. *Proc Natl Acad Sci USA* 68: 2112–2116.

Kume K, Zylka M J, Sriram S, *et al.* (1999). mCRY1 and mCRY2 are essential components of the negative limb of the circadian clock feedback loop. *Cell* 98: 193–205.

Lee C, Bae K, Edery I (1999). PER and TIM inhibit the DNA binding activity of a *Drosophila* CLOCK-CYC/dBMAL1 heterodimer without disrupting formation of the heterodimer: a basis for circadian transcription. *Mol Cell Biol* 19: 5316–5325.

Lindblom N, Hejskala H, Hatonen T, *et al.* (2000). No evidence for extraocular light induced phase shifting of human melatonin, cortisol and thyrotropin rhythms. *Neuroreport* 11: 713–717.

Lockley S W, Skene D J Arendt J, *et al.* (1997). Relationship between melatonin rhythms and visual loss in the blind. *J Clin Endocrinol Metab* 82: 3763–3770.

Lockley S W, Skene D J, Thapan K, *et al.* (1998). Extraocular light exposure does not suppress plasma melatonin in humans. *J Clin Endocrinol Metab* 83: 3369–3372.

Lowrey P L, Shimomura K, Antoch M P, *et al.* (2000). Positional syntenic cloning and functional characterization of the mammalian circadian mutation *tau*. *Science* 288: 483–491.

Lucas R J, Freedman M S, Munoz M, *et al.* (1999). Regulation of the mammalian pineal by non-rod, non-cone, ocular photoreceptors. *Science* 284: 505–507.

Martinek S, Inonog S, Manoukian A S, *et al.* (2001). A role for the segment polarity gene *shaggy*/GSK-3 in the *Drosophila* circadian clock. *Cell* 105: 769–779.

Maywood E S, Mrosovsky N, Field M D, *et al.* (1999). Rapid down-regulation of mammalian *Period* genes during behavioral resetting of the circadian clock. *Proc Natl Acad Sci USA* 96: 15211–15216.

Miyamoto Y, Sancar A (1998). Vitamin B_2-based blue-light photoreceptors in the retinohypothalamic tract as the photoactive pigments for setting the circadian clock in mammals. *Proc Natl Acad Sci USA* 95: 6097–6102.

Moore R Y, Eichler V B (1972). Loss of a circadian adrenal corticosterone rhythm following suprachiasmatic lesion in the rat. *Brain Res* 42: 201–206.

Meyer-Bernstein E L, Jetton A E, Matsumoto S I, *et al.* (1999). Effects of suprachiasmatic transplants on circadian rhythms of neuroendocrine function in golden hamsters. *Endocrinology* 140: 207–218.

Naidoo N, Song W, Hunter-Ensor M, *et al.* (1999). A role for the proteasome in the light response of the timeless clock protein. *Science* 285: 1737–1741.

Okamura H, Miyake S, Sumi Y, *et al.* (1999). Photic induction of *mPer1* and *mPer2* in *Cry*-deficient mice lacking a biological clock. *Science* 286: 2531–2534.

Okano T, Sasaki M, Fukada Y (2001). Cloning of mouse *Bmal2* and its daily expression profile in the suprachiasmatic nucleus: a remarkable acceleration of *Bmal2* sequence divergence after *Bmal1* duplication. *Neurosci Lett* 300: 111–114.

Peifer M, Polakis P (2000). Wnt signaling in oncogenesis and embryogenesis – a look outside the nucleus. *Science* 287: 1606–1609.

Price J L, Blau J, Rothenfluh A, *et al.* (1998). *Double-time* is a novel *Drosophila* clock gene that regulates PERIOD protein accumulation. *Cell* 94: 83–95.

Provencio I, Foster R G (1995). Circadian rhythms in mice can be regulated by photoreceptors with cone-like characteristics. *Brain Res* 694: 183–190.

Provencio I, Jiang G, DeGrip W J, *et al.* (1998). Melanopsin: an opsin in melanophores, brain and eye. *Proc Natl Acad Sci USA* 95: 340–345.

Provencio I, Rodriguez I R, Jiang G, *et al.* (2000). A novel human opsin in the inner retina. *J Neurosci* 20: 600–605.

Ralph M R, Menaker M (1988). A mutation of the circadian system in golden hamsters. *Science* 241: 1225–1227.

Ralph M R, Foster R G, Davis F C, *et al.* (1990). Transplanted suprachiasmatic nucleus determines circadian period. *Science* 247: 975–978.

Rosato E, Kyriacou C P (2001). Flies, clocks and evolution. *Proc R Soc Lond B Biol Sci* 29; 356: 1769–1778.

Rosato E, Codd V, Mazzotta C, *et al.* (2001). Light-dependence of *Drosophila* cryptochrome is mediated by the C-terminus. *Curr Biol* 11: 909–917.

Rothenfluh A, Young M W, Saez L (2000). A TIMELESS-independent function for PERIOD proteins in the *Drosophila* clock. *Neuron* 26: 505–514.

Saez L, Young M W (1996). Regulation of nuclear entry of the *Drosophila* clock proteins PERIOD and TIMELESS. *Neuron* 17: 911–920.

Sakamoto K, Nagase T, Fukui H, *et al.* (1998). Multitissue circadian expression of rat period homolog (*rPer2*) mRNA is governed by the mammalian circadian clock, the suprachiasmatic nucleus in the brain. *J Biol Chem* 273: 27039–27042.

Sangoram A M, Saez L, Antoch M P, *et al.* (1998). Mammalian circadian autoregulatory loop: a *timeless* ortholog and *mPer1* interact and negatively regulate CLOCK-BMAL1-induced transcription. *Neuron* 5: 1101–1113.

Shearman L P, Zylka M J, Weaver D R, *et al.* (1997). Two period homologs: circadian expression and photic regulation in the suprachiasmatic nuclei. *Neuron* 19: 1261–1269.

Shearman, L P, Jin X, Lee C, *et al.* (2000a). Targeted disruption of the *mPer3* gene: subtle effects on circadian clock function. *Mol Cell Biol* 20: 6269–6275.

Shearman L P, Sriram S, Weaver D R, *et al.* (2000b). Interacting molecular loops in the mammalian circadian clock. *Science* 288: 1013–1019.

Shigeyoshi Y, Taguchi K, Yamamoto S, *et al.* (1997). Light-induced resetting of a mammalian circadian clock is associated with rapid induction of the *mPer1* transcript. *Cell* 91: 1043–1053.

Silver R, Lehman M N, Gibson M, *et al.* (1990). Dispersed cell suspensions of fetal SCN restore circadian rhythmicity in SCN-lesioned adult hamsters. *Brain Res* 525: 45–58.

Silver R., Le Sauter J, Tresco P A, *et al.* (1996). A diffusible coupling signal from the transplanted suprachiasmatic nucleus controlling circadian locomotor rhythms. *Nature* 382: 810–813.

Stanewsky R, Kaneko M, Emery P, *et al.* (1998). The *cry^b* mutation identifies cryptochrome as a circadian photoreceptor in *Drosophila*. *Cell* 95: 681–692.

Stokkan K A, Yamazaki S, Tei H, *et al.* (2001). Entrainment of the circadian clock in the liver by feeding. *Science* 291: 490–493.

Sun Z S, Albrecht U, Zhuchenko O, *et al.* (1997a). *Rigui*, a putative mammalian ortholog of the *Drosophila period* gene. *Cell* 90: 1003–1111.

Sun H, Gilbert D J, Copeland N G, *et al.* (1997b). Peropsin, a novel visual pigment-like protein located in the apical microvilli of the retinal pigment epithelium. *Proc Natl Acad Sci USA* 94: 9893–9898.

Takahashi J, DeCoursey P, Bauman L, *et al.* (1984). Spectral sensitivity of a novel photoreceptive system mediating entrainment of mammalian circadian rhythms. *Nature* 308: 186–188.

Tei H, Okamura H, Shigeyoshi Y, *et al.* (1997). Circadian oscillation of a mammalian homologue of the *Drosophila period* gene. *Nature* 389: 512–516.

Thresher R J, Hotz Vitaterna M, *et al.* (1998). Role of mouse cryptochrome blue-light photoreceptor in circadian responses. *Science* 282: 1490–1494.

Toh K L, Jones C R, He Y, *et al.* (2001). An h*Per2* phosphorylation site mutation in familial advanced sleep syndrome. *Science* 291: 1040–1043.

Tosini G (2000). Melatonin circadian rhythms in the retina of mammals. *Chronobiol Int* 17: 599–612.

Tosini G, Menaker M (1996). Circadian rhythms in cultured mammalian retina. *Science* 272: 419–421.

Tosini G, Menaker M (1998). The clock in the mouse retina: melatonin synthesis and photoreceptor degeneration. *Brain Res* 789: 221–228.

Ueyama T, Krout K E, Nguyen X V, *et al.* (1999). Suprachiasmatic nucleus: a central autonomic clock. *Nat Neurosci* 2: 1051–1053.

van der Horst G T J, Muijtiens M, Kobayashi K, *et al.* (1999). Mammalian *Cry1* and *Cry2* are essential for maintenance of circadian rhythms. *Nature* 398: 627–630.

Vitaterna M H, King D P, Chang A M, *et al.* (1994). Mutagenesis and mapping of a mouse gene, *Clock*, essential for circadian behavior. *Science* 264: 719–725.

Wakamatsu H, Takahashi S, Moriya T, *et al.* (2001). Additive effect of mPer1 and mPer2 antisense oligonucleotides on light-induced phase shift. *Neuroreport* 12: 127–131.

Welsh D K, Logothetis D E, Meister M, *et al.* (1995). Individual neurons dissociated from rat suprachiasmatic nucleus express independently phased circadian firing rhythms. *Neuron* 14: 697–706.

Yagita K, Okamura H (2000). Forskolin induces circadian gene expression of rPer1, rPer2 and dbp in mammalian rat-1 fibroblasts. *FEBS Lett* 465: 79–82.

Yamazaki S, Goto M, Menaker M (1999). No evidence for extraocular photoreceptors in the circadian system of the Syrian hamster. *J Biol Rhythms* 14: 197–201.

Yamazaki S, Numano R, Abe M, *et al.* (2000). Resetting central and peripheral circadian oscillators in transgenic rats. *Science* 288: 682–685.

Zheng B, Larkin D W, Albrecht U, *et al.* (1999). The *m*Per2 genes encodes a functional component of the mammalian circadian clock. *Nature* 400: 169–173.

Zheng B, Albrecht U, Kaasik K, *et al.* (2001). Nonreduntant roles of *m*Per1 and *m*Per2 genes in the mammalian circadian clock. *Cell* 105: 683–694.

Zordan M, Costa R, Macino G, *et al.* (2000). Circadian clocks: what makes them tick? *Chronobiol Int* 17: 433–451.

Zylka M J, Shearman L P, Weaver D R, *et al.* (1998). Three *period* homologs in mammals: differential light responses in the suprachiasmatic circadian clock and oscillating transcripts outside of brain. *Neuron* 20: 1–20.

4

Rhythms and pharmacokinetics

Gaston Labrecque and Denis Beauchamp

4.1 Introduction and objectives

Many studies published in the last 20 years or more have reported that oral administration of drugs at different times of day produced large differences in the plasma levels and in the pharmacokinetic parameters calculated from the data. At first, conventionally trained pharmacokineticists were surprised by the presence of biological rhythms in pharmacokinetics. They suggested that these time-dependent variations could be explained by changes in external factors such as posture, presence or absence of food in the stomach or disease states. The data were also disturbing to them because, hitherto, conventional pharmacokinetic studies were always carried out in the morning, and the data obtained were extrapolated to other times of day. Thus, time of day was not considered an important factor in characterising the pharmacokinetics of drugs. It was assumed that the working of the human body remained constant over the 24 h span, because the feedback mechanisms were obviously maintaining the *milieu interieur* constant over time. As has already been demonstrated in Chapter 1, this hypothesis can no longer be supported. Many examples have revealed time-dependent variations in the kinetics of drugs administered orally. Furthermore, it is possible to explain the chronokinetic data by time-dependent variations in factors involved in the absorption, distribution, metabolism and excretion of drugs.

The objectives of this chapter are to present examples of circadian variations in the pharmacokinetics of drugs and to summarise the mechanisms of the circadian variations in the absorption, distribution, metabolism and excretion of drugs. Although it is possible to find a few examples of circannual variations in the kinetics of drugs, these variations will not be considered here.

4.2 Rhythms in pharmacokinetics

4.2.1 Chronopharmacokinetic studies in humans

When single oral doses of medications are given at four or more different times within a 24 h period, statistically significant differences can be demonstrated in the maximal plasma serum concentration (C_{max}), in the time after administration needed to obtain maximal plasma levels (t_{max}), in the half-life ($t_{1/2}$) and in the area under the curve (AUC) of drugs. The differences were observed when laboratory animals or human volunteers were given oral doses of medications at different hours of the day, with a washout period of 7 days or more before each administration time.

The first studies demonstrating time-dependent changes in pharmacokinetics were carried out more than 30 years ago. Reinberg *et al.* (1967) determined the duration of sodium salicylate excretion in six healthy volunteers: complete urinary excretion of the salicylate was 22% longer when the 1 g oral dose was taken at 07.00 than when it was taken at 19.00. In another early study, nine healthy university students ingested a 100 mg dose of indometacin at five different times of day (Clench *et al.*, 1981). This study showed that C_{max} at 07.00 and 11.00 was double that at the other times; t_{max} was reached in half the time after the morning dose compared to the evening dose, but no temporal variation was detected in the AUC. Figure 4.1 illustrates this chronokinetic phenomenon using the temporal changes in the plasma levels of ketoprofen and propranolol administered orally at four times of day (Ollagnier *et al.*, 1987; Langler and Lemmer, 1988).

Time-dependent changes can be found in the pharmacokinetics of many classes of drugs. For instance, Ramesch Rao and Rambhau (1992) administered 1 g sulfamethoxazole (SMT) in two hard gelatin capsules to healthy diurnally active male volunteers at 06.00, noon, 18.00 and midnight, with a washout period of 1 week between each treatment. Plasma SMT levels were determined up to 36 h following oral administration and the usual pharmacokinetic parameters were calculated. Table 4.1 presents the mean values of pharmacokinetic parameters calculated from the data: the highest SMT plasma levels were obtained during the activity period and C_{max} was 32% larger at 06.00 and noon than at midnight ($p \leq 0.05$). As indicated by the t_{max}, K_a, $t_{1/2}$ and mean residence time (MRT) values, SMT was absorbed more rapidly during the activity period. However, no statistically significant difference was observed in the mean $AUC_{0-\infty}$ and in the volume of distribution. In another study of the chronokinetics of rifampicin, the same group of

Rhythms in pharmacokinetics **77**

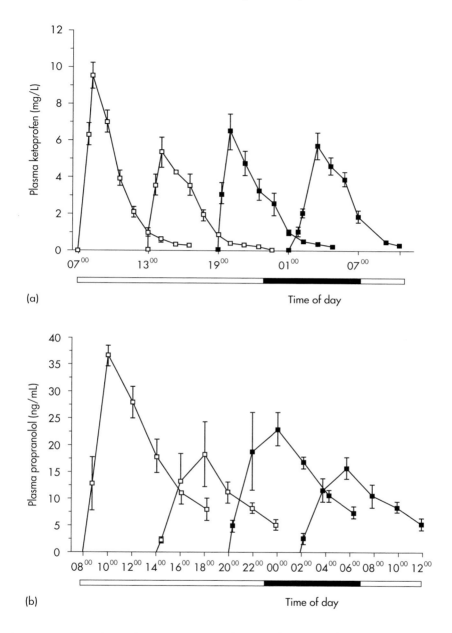

Figure 4.1 Time-dependent variation in pharmacokinetics in humans. (a) Mean plasma concentrations obtained when eight healthy volunteers took 100 mg ketoprofen orally at four times of day. (b) Mean plasma levels of propranolol obtained in four subjects when a dose of 80 mg was ingested orally at four times of day. Redrawn from Ollagnier *et al.* (1987); Langler and Lemmer (1988).

investigators found that the K_a was again significantly lower ($p \leq 0.05$) and t_{max} was longer after drug administration at midnight than at any other dosing times (Avachat et al., 1992).

Table 4.2 presents a list of drugs for which time-dependent differences in pharmacokinetic parameters have been demonstrated after oral administration: these drugs were absorbed faster in the morning than in the evening. Although most of these studies were carried out in 10 or fewer volunteers, the overall time-dependent changes were very similar from one study to the next: the rate of absorption was increased, the C_{max} was larger and the t_{max} was shorter when the drugs were taken at the beginning of the activity period.

Table 4.3 presents examples of chronokinetics found in patients or healthy volunteers receiving parenteral drugs injections at different times of day. These data emphasise the importance of time of day to the pharmacokinetics of drugs. To our knowledge, the only exceptions to this rule are the larger plasma levels found after the evening administration of dexamethasone (Lamiable et al., 1991), methotrexate (Balis et al., 1989), mercaptopurine (Balis et al., 1989; Langevin et al., 1987) and intramuscularly injected pethidine (meperidine), (Ritschell et al., 1983), and the absence of circadian changes in the pharmacokinetics of nortriptyline (Nakano and Hollister, 1978). Extensive literature reviews (including animal data) are available elsewhere (Lemmer, 1981; Reinberg and Smolensky, 1982; Bruguerolle, 1987, 1989, 1998; Bélanger et al., 1993, 1997; Labrecque and Bélanger, 1991).

Table 4.1 Mean values of pharmacokinetic parameters after oral ingestion of sulfamethoxazole (SMT) at four times of day[a]

Parameters	Time of day			
	06.00	Noon	18.00	Midnight
C_{max} (mg/L)	56.8 ± 3.8*	55.2 ± 4.5	50.7 ± 3.7	43.0 ± 4.9*
t_{max} (h)	3.0 ± 0.7	2.4 ± 0.4†	3.8 ± 0.7	5.5 ± 0.5†
K_a (h^{-1})	1.8 ± 0.4	2.1 ± 0.4†	1.8 ± 0.4	0.9 ± 0.2†
$t_{1/2}$ (h)	8.8 ± 1.1	7.9 ± 0.8	10.4 ± 0.5§	8.5 ± 0.4§
MRT (h)	13.6 ± 1.0	12.1 ± 0.8	15.2 ± 0.7	14.8 ± 0.2
AUC (mg/L/h)	772 ± 159	648 ± 105	732 ± 54	759 ± 64

Adapted from Ramesh Rao and Rambhau (1992).
[a]SMT (1 g) was administered orally to four healthy adult volunteers at four times of day.
*Significant difference ($p \leq 0.05$) from the 06.00 value.
†Significant difference ($p \leq 0.05$) from the 12.00 value.
§Significant difference ($p \leq 0.05$) from the 18.00 value.

Table 4.2 Time-dependent changes in pharmacokinetic parameters observed after oral administration of non-sustained-release drug formulations

Drug	Time of ingestion	Subjects	Parameters[a]	Results	Reference
Aspirin	06.00, 10.00, 18.00, 22.00	6 healthy male adults	C_{max} and AUC	Larger: 06.00	Markewicz and Semonowicz (1979)
Amitriptyline	09.00, 21.00	10 healthy male adults	K_a	Greater: 09.00	Nakano and Hollister (1983)
Busulfan	06.00, 12.00, 18.00, 00.00	27 acute leukaemic patients	Plasma levels	Peak: 02.00 (+29% to +47%)	Hassan et al. (1991)
		21 solid tumour patients	Peak/trough ratio Urinary excretion	Peak: 06.00 (+50%) Peak: 12.00 (+50%)	Vassall et al. (1993)
Ciclosporin	09.00, 20.00	12 healthy male adults	C_{max} and AUC Renal clearance	Peak: 09.00 Slower: 20.00	Choi and Park (1999)
Ciprofloxacin	10.00, 20.00	12 healthy male adults	Urinary excretion	Peak: 10.00; trough: 22.00	Sarveshwar et al. (1997)
Dexamethasone	08.00, 23.00	6 healthy adults	t_{max}	Longer: 23.00	Lamiable et al. (1991)
Diazepam	09.30, 21.30	10 healthy male adults	Plasma levels	Larger: 09.30	Naranjo et al. (1980)
		5 healthy male adults	C_{max} t_{max}	Larger: 09.30 Shorter: 09.30	Nakano et al. (1984) Nakano et al. (1986)
Diclofenac	07.00, 19.00	10 healthy adults	C_{max} and AUC	Greater: 07.00	Mustafa et al. (1991)
Digoxin	08.00, 20.00	9 healthy adults	C_{max} and AUC	Greater: 08.00	Mrozikiewics et al. (1988)
Dipyridamole	06.00, 14.00, 22.00	6 healthy male adults	C_{max} AUC	Larger: 06.00 Smaller: 06.00	Markiewicz et al. (1979)

Table 4.2 Continued

Drug	Time of ingestion	Subjects	Parameters[a]	Results	Reference
Ethanol	07.00, 11.00, 19.00, 23.00	6 healthy male adults	C_{max} t_{max}	Larger: 07.00 Smaller: 07.00	Reinberg et al. (1975)
	10.00, 22.00	11 healthy adults	C_{max} and AUC t_{max}	Larger: 10.00 Smaller: 10.00	Lakatua et al. (1984)
Indometacin	07.00, 11.00, 15.00, 19.00, 23.00	9 healthy adults	C_{max} t_{max}	Larger: 07.00 and 11.00 Smaller: 07.00 and 11.00	Clench et al. (1981)
Isosorbide dinitrate	08.00, 14.00, 20.00, 02.00	6 healthy male adults	AUC	Larger: 02.00	Lemmer et al. (1986)
Isosorbide mononitrate	06.30, 18.30	6 healthy adults	t_{max}	Longer: 18.30	Scheidel and Lemmer (1991)
Ketoprofen	07.00, 13.00, 19.00, 01.00	8 healthy male adults	C_{max}, K_a and AUC t_{max}	Larger: 07.00 Smaller: 07.00	Ollagnier et al. (1987)
Lithium	08.00	3 healthy male adults	Urinary excretion	08.00–20.00 > 20.00–08.00	Groth et al. (1974)
Lorazepam	07.00, 19.00	16 healthy adults	K_a Absorption half-life	Larger: 07.00 (+338%) Smaller: 07.00	Bruguerolle et al. (1985)
Nifedipine	08.00, 21.00	12 healthy male adults	C_{max} and AUC t_{max}	Larger: 07.00 Smaller: 07.00	Lemmer et al. (1991)
Nortriptyline	09.00, 21.00	10 healthy male adults	C_{max}, t_{max}, AUC	No temporal variation	Nakano and Hollister (1978)

Table 4.2 Continued

Drug	Time of ingestion	Subjects	Parameters[a]	Results	Reference
Paracetamol (acetaminophen)	08.00, 14.00, 20.00	6 male adults	C_{max}, t_{max}, AUC	No change over 24 h	Malan et al. (1985)
	07.00, 19.00	10 elderly male adults	$t_{1/2}$	Longer: 19.00 (+84%)	Bruguerolle et al. (1988)
	06.00, 14.00	18 healthy adults	Elimination half-life	Longer: 06.00 (+15%)	Shively and Vesell et al. (1975)
Prednisolone	06.00, 12.00, 18.00, 00.00	6 healthy volunteers	C_{max} and AUC t_{max}	Greatest: 12.00 Largest: 18.00	English et al. (1983)
Propranolol	08.00, 14.00, 20.00, 00.00	8 healthy adults	C_{max}	Largest: 14.00	Langler and Lemmer (1988)
	08.00, 14.00, 20.00, 02.00	4 healthy adults	C_{max}	Largest: 08.00	
Rifampicin	06.00, 12.00, 18.00, 00.00	4 healthy adults	K_a t_{max}	Largest: 06.00 (+240%) Shortest: 06.00; longest: 00.00	Avachat et al. (1992)
Sodium salicylate	07.00, 11.00, 19.00, 23.00	6 healthy adults	Urinary excretion	Longest: 07.00 (+129%)	Reinberg et al. (1967)

Table 4.2 Continued

Drug	Time of ingestion	Subjects	Parameters[a]	Results	Reference
Sulfamethoxazole	06.00, 12.00, 18.00, 00.00	4 healthy adults	C_{max}	Largest: 06.00 (+32%)	Ramesh Rao and Rambhau (1992)
			K_a	Largest: 12.00; smallest: 00.00	
			t_{max}	Largest: 12.00; smallest: 00.00	
Theophylline					
Aminophylline	08.00, 20.00	8 healthy adults	C_{max} and K_a	Larger: 08.00	St-Pierre et al. (1984)
			t_{max}	Longer: 08.00 (+50%)	
Theophylline	08.00, 20.00	8 asthmatic adults	C_{max}	Shorter: 08.00 (+20%)	Reinberg et al. (1987)
			t_{max}	Longer: 20.00 (+64%)	
Triazolam	07.00, 22.00	10 healthy adults	$t_{1/2}$	Smaller: 07.00	Smith et al. (1986)
Valproic acid	08.30, 20.30	8 male adults	C_{max} and K_a	Larger: 08.30	Nakano et al. (1986)
			t_{max}	Smaller: 08.30	
	08.30, 20.30	8 male adults	K_a	Larger: 08.30	
			t_{max}	Smaller: 08.30	

[a]C_{max}, maximal serum or plasma concentration; t_{max}, time to reach C_{max}; K_a, absorption constant; AUC, area under the curve.

Table 4.3 Time-dependent changes in pharmacokinetic parameters of drugs administered parenterally to humans

Drug	Route	Subjects	Parameters[a]	Results	Reference
Amikacin	Continuous i.v. infusion	50 neutrogenic patients	Plasma levels	Peak: 07.00; trough: 21.00	Elting et al. (1990)
		17 febrile patients	Renal clearance	Peak: 08.00; trough: 20.00	Bleyzac et al. (2000)
Carboplatin	i.v. over 30 min	7 ovarian cancer patients	Renal clearance	06.00 > 18.00 (+140%)	Kerr et al. (1990)
Cefodizime	i.v. injection	16 healthy volunteers	AUC	Peak: 00.00; trough: 18.00	Jonkman et al. (1988)
Cisplatin	i.v. over 30 min	11 ovarian or bladder cancer patients	Urine C_{max} AUC	06.00 > 18.00 (+50%) 06.00 > 18.00 (+55%)	Hrusheski et al. (1982)
Doxorubicin	i.v. over 15 min	18 breast cancer patients	Clearance β $t_{1/2}$ AUC	09.00 > 21.00 (+40%) 21.00 > 09.00 (+80%) 21.00 > 09.00 (+50%)	Canal et al. (1991)
Fluorouracil	i.v. over 15 min	28 GI cancer patients	Plasma levels	Peak: 03.00 (+40%)	Petit et al. (1988)
	Continuous i.v. infusion	4 colon cancer patients	Plasma levels	Peak: 04.00 (+260%)	Metzger et al. (1994)
dl-Folinic acid	Continuous i.v. infusion	4 colon cancer patients	Plasma levels	Peak: 07.00 (+20%)	Metzger et al. (1994)
Gentamicin	i.m. injection	9 healthy volunteers	$t_{1/2}$ Renal clearance AUC	Peak: 00.00; trough: 08.00 Peak: 08.00; trough: 00.00 Peak: 00.00; trough: 16.00	Nakano et al. (1990)
	i.v. injection	10 healthy volunteers	Renal clearance	Peak: 09.00; trough: 22.00	Choi et al. (1999)

Table 4.3 Continued

Drug	Route	Subjects	Parameters[a]	Results	Reference
Isepamicin	i.m. injection	6 healthy volunteers	$t_{1/2}$ Renal clearance AUC	Peak: 20.00; trough: 08.00 Peak: 20.00; trough: 20.00 Peak: 20.00; trough: 08.00	Yoshiyama et al. (1996)
Methotrexate	i.v. injection over 3 min	6 leukaemic patients	Clearance β Renal clearance	10.00 > 21.00 (+20%) 10.00 > 21.00 (+100%)	Koren et al. ((1992))
Netilmicin	i.v. injection i.v. injection	8 febrile patients 23 febrile patients	Plasma levels Plasma levels	Peak: 05.00; trough: 13.00 No difference over 24 h	Lucht et al. (1990) Fauvelle et al. (1994)
Pethidine (meperidine)	i.m. injection	8 sickle cell anaemia patients	Serum clearance	70% greater during the night	Ritschell et al. (1983)
Vancomycin	i.v. injection	8 febrile patients	Renal clearance	No difference over 24 h	Wang et al. (1990)
Vindesine	Continuous i.v. infusion	9 lung cancer patients	Plasma levels	Peak: 12.00 (+40%)	Focan et al. (1989)

[a] C_{max}, maximal serum or plasma concentration; t_{max}, time to reach C_{max}; AUC, area under the curve.

4.2.2 Rhythms, physicochemical properties and formulation of drugs

The temporal variations in pharmacokinetic characteristics are similar whether the drug is given in a solid form, as a capsule or a tablet, or in solution (as is the case, for instance, with ethanol), or whether the drug has a low hepatic first-pass effect, e.g. diazepam, or a high hepatic first-pass effect, e.g. propranolol. However, the physicochemical properties of drugs are important determinants of time-dependent changes in pharmacokinetics. As suggested initially by the animal study of Bélanger *et al.* (1984), lipid-soluble drugs may be more likely to show temporal variations in pharmacokinetics than are water-soluble drugs. This hypothesis is supported by human data with β-adrenoreceptor antagonists. When a lipophilic drug, such as propranolol, was given to healthy volunteers, Langler and Lemmer (1988) reported a clearly significant circadian stage-dependent variation in the absorption of this drug: peak plasma propranolol concentration was higher and t_{max} was shorter in the morning than in the evening (see Figure 4.1b). The data of Shiga *et al.* (1993) on propranolol are in agreement with those of Langler and Lemmer, but these investigators did not find any time-dependent variations when atenolol, a more water-soluble β-adrenoreceptor antagonist, was given orally to 13 hypertensive patients (see Table 4.4).

Another factor to be considered is the role of pharmaceutical formulation in the time-dependent variation in pharmacokinetics. The first studies with indometacin suggested that formulation did not influence the chronopharmacokinetic phenomenon: the time-dependent variations found were similar after oral ingestion of immediate-release or sustained-release (SR) formulations of this nonsteroidal anti-inflammatory drug

Table 4.4 Chronopharmacokinetics of oral propranolol and atenolol in 13 hypertensive patients[a]

Parameter	Propranolol		Atenolol	
	09.00	21.00	09.00	21.00
t_{max} (h)	3.2 ± 1.3	4.0 ± 1.7*	3.2 ± 1.1	4.1 ± 1.5
C_{max} (ng/mL)	17.7 ± 8.6	11.9 ± 4.0*	440 ± 148	392 ± 123
$t_{1/2}$ (h)	4.4 ± 0.9	4.2 ± 1.6	6.2 ± 0.8	6.2 ± 0.8
AUC (0–24 h)	117.4 ± 54	96.1 ± 45*	3390 ± 224	3707 ± 875

Adapted from Shiga *et al.* (1993).
[a]Patients took oral doses of atenolol (50 mg) and propranolol (20 mg) at two different times of day.
*Statistically significant difference between the 09.00 and 21.00 values.

(NSAID) (Clench et al., 1981; Guissou et al., 1983). Similar data were found in the time-dependent variation in the kinetics of theophylline ingested orally either as a nonsustained-release formulation or a once-a-day sustained-release formulation (Scott et al., 1981), but a SR theophylline tablet (Uniphyl) introduced to the Canadian market in the early 1990s completely abolished the temporal changes in the kinetics of theophylline (Rivington et al., 1985). Figure 4.2 illustrates these data.

Lemmer et al. (1991) compared the pharmacokinetics of oral immediate-release (Corticant Kapsel) and SR nifedipine (Corticant retard). Table 4.5 indicates that when nifedipine was ingested as an immediate-release formulation, significant temporal variations were found: C_{max} was larger, t_{max} was shorter and oral bioavailability was 50% greater after administration at 08.00 compared to 19.00. In contrast, no significant differences were obtained with SR nifedipine.

Finally, Scheidel and Lemmer (1991) also studied the chronokinetics of isosorbide mononitrate after morning and evening administration to healthy volunteers of a 60 mg oral dose as an immediate-release or a SR formulation. With the immediate-release formulation, the t_{max} value was obtained much faster in the morning than in the evening, but no significant diurnal variation was noted with the SR isosorbide mononitrate. To date, no explanation has been proposed to explain why some SR formulations abolish the chronopharmacokinetic phenomena while others do not. Trial and error is still the only approach that can be used to determine whether or not the time-dependent variations in the pharmacokinetics are present after oral administration of a SR formulation.

4.2.3 Rhythms and routes of drug administration

The influence on chronokinetics of the route of administration must also be considered. For instance, there are studies showing that chronopharmacokinetic variation is not found when a drug is administered rectally. This was nicely demonstrated with indometacin (Clench et al., 1981; Guissou et al., 1983) and valproic acid (Nakano et al., 1986; Yoshiyama et al., 1989). In a random crossover study, Yoshiyama et al. (1989) administered the same formulation of sodium valproate to eight male volunteers at 08.30 and 22.30 by the oral or rectal routes. Figure 4.3 illustrates that the t_{max} value was shorter in the morning when the drug was administered orally, but such a temporal variation was absent after rectal administration.

In summary, statistically significant diurnal variation can be found in the pharmacokinetics of drugs administered by the oral but not by the

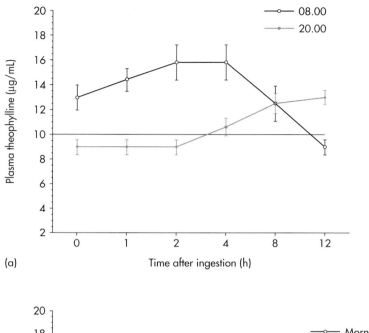

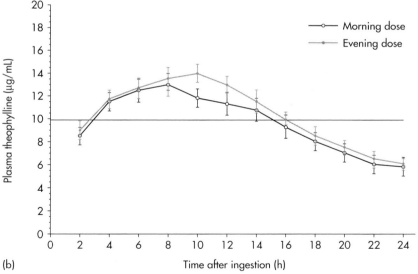

Figure 4.2 Effect of pharmaceutical formulation on theophylline chronopharmacokinetics. (a) Plasma theophylline levels obtained when 13 asthmatic patients took 300 mg of a sustained-release formulation of theophylline (Theo-Dur) for 8 days at 08.00 and 20.00. (b) Plasma theophylline levels obtained after the morning and evening ingestion of an oral sustained-release formulation of theophylline (Uniphyl) by 17 asthmatic patients. Redrawn from Scott et al. (1981); Rivington et al. (1985).

Table 4.5 Pharmacokinetic parameters calculated after oral ingestion of immediate- and sustained-release formulations of nifedipine[a,b]

Parameter	Immediate-release		Sustained-release	
	08.00	19.00	08.00	19.00
C_{max} (ng/mL)	82.0 ± 6.5	45.7 ± 6.4*	48.5 ± 5.7	50.1 ± 6.9
t_{max} (min)	22.5 ± 2.9	37.5 ± 3.9*	137 ± 18	168 ± 18
$t_{1/2}\beta$ (min)	64.6 ± 8.8	57.7 ± 6.0	228 ± 52	324 ± 150[c]
AUC (ng/mL/h)	130 ± 13	85 ± 13*	281 ± 37.3	302 ± 53.0

Adapted from Lemmer *et al.* (1991).
Values are means ± SE.
[a]Immediate-release formulation dose: 10 mg twice daily, n = 12 healthy volunteers.
[b]Sustained-release formulation dose: 20 mg twice daily, n = 10 hypertensive patients.
[c]This is the apparent $t_{1/2}$.
*Statistically significant differences ($p \leq 0.01$) in the data obtained at 08.00 and 19.00 with the immediate-release formulation.

rectal routes. It is possible to predict whether time-dependent changes will be observed because it appears that solubility in lipids is an important factor in explaining the chronokinetic data. Additional studies are needed to explain why some pharmaceutical formulations modify the chronokinetic phenomenon while others do not.

4.3 Mechanisms of chronopharmacokinetics

Circadian variation in the activity of many gastrointestinal (GI), hepatic and renal processes could explain why the absorption, distribution, metabolism and excretion of drugs change as a function of the time of administration. Thus, it is possible to predict the temporal changes in the kinetics of drugs when the chronobiology of the pathways involved in the absorption, distribution, metabolism and excretion of drugs is known. This section summarises the data presently available in this area.

4.3.1 Rhythms in drug absorption

The drug absorption process can be altered by factors such as the structure of membranes, the physicochemical properties of drugs, the pH, the rate of gastric emptying and the motility and the blood flow to the gastrointestinal tract. As far as we know, no study is available presently on temporal variations in the structure of membranes or on the effect of circadian changes of pH on drug absorption. However, circadian

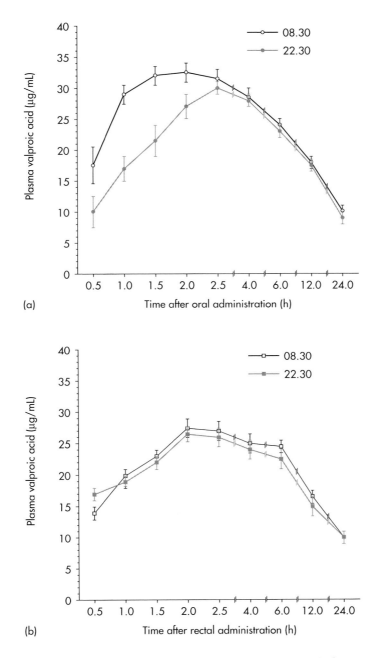

Figure 4.3 Mean plasma concentrations of valproic acid obtained after (a) a single oral dose or (b) rectal administration at 08.30 and 22.30 to eight healthy volunteers. Redrawn from Yoshiyama *et al.* (1989).

variations in gastric emptying time and blood flow could explain some chronopharmacokinetic data.

There are marked differences in the gastric emptying rates in humans. Goo *et al.* (1987) measured the gastric emptying time when an identical isotopic meal was given to 16 healthy volunteers at 08.00 and 20.00. As shown in Figure 4.4a, the gastric emptying of solids measured 60 and 80 min after meal ingestion was respectively 25% and 18% faster at 08.00 than at 20.00. The phenomenon was less marked for the emptying of liquid: a similar but statistically nonsignificant trend was found in the gastric emptying of liquid preparations. A circadian variation was also found in the propagation of the intestinal migrating myoelectric complex in humans. Figure 4.4b illustrates that GI motility in the daytime was double that in the evening or at night-time. The absence of motility at 18.00 suggests that a single meal is a powerful transient inhibitor of bowel motility (Kumar *et al.*, 1986). This subject is addressed in more detail in Chapter 6.

Circadian changes in the blood flow and cardiac output to the GI tract could also explain the time-dependent variation in the absorption of drugs. Studies carried out in adult rats receiving microspheres labelled with strontium-85 intravenously at four different times of day (Labrecque *et al.*, 1988) indicated that peak concentration of radioactivity in the intestine was maximal when the animals were active and was minimal in the sleep period. In 10 healthy male volunteers, the apparent blood flow was determined from the kinetics of indocyanine green. Figure 4.5 illustrates that the peak of estimated blood flow obtained at the beginning of the human activity period was significantly larger than at the three other times of day (Lemmer and Nold, 1991).

4.3.2 Rhythms in drug distribution

Very few investigations have been carried out on circadian variation in the process of distribution of a drug to its site of action. The factors influencing drug distribution are body size and composition, blood flow to various organs, binding of drugs to plasma proteins, and membrane permeability to drugs. This section focuses on the chronobiology of binding of drugs to plasma proteins and membrane permeability because time-dependent variation in blood flow has been alluded to above and no data are available on circadian changes in body size and composition.

In the last 15 years, time-dependent changes in human plasma proteins involved in drug binding to albumin and α_1-glycoprotein have been determined by several investigators (Reinberg *et al.*, 1977;

Mechanisms of chronopharmacokinetics 91

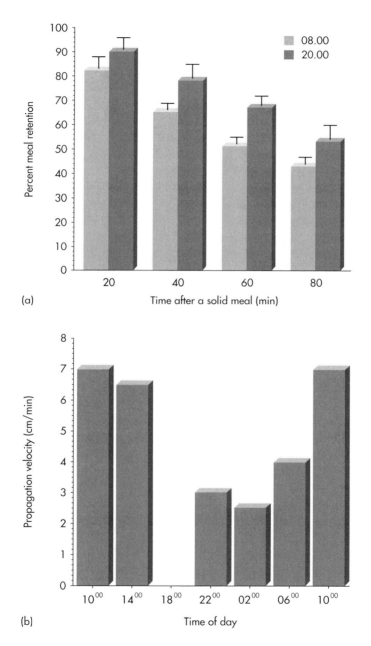

Figure 4.4 Time-dependent changes in gastrointestinal physiology pertinent to drug absorption. (a) Gastric emptying rate observed after the morning and evening ingestion of solid meals. (b) Propagation velocity of human small-bowel activity over a 24 h period. Redrawn from Goo *et al.* (1987); Kumar *et al.* (1986).

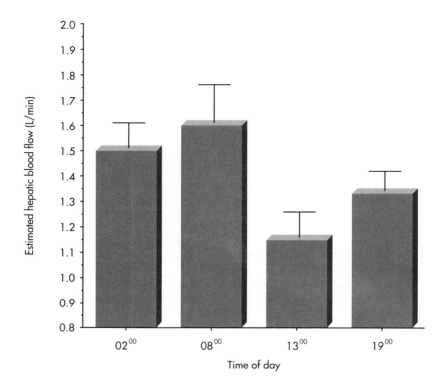

Figure 4.5 Circadian rhythm in estimated hepatic blood flow in 10 healthy volunteers. Blood flow was estimated from indocyanine green kinetics studied at four times of day. Redrawn from Lemmer and Nold (1991).

Bruguerolle et al., 1986; Focan et al., 1988; Kanabrocki et al., 1988). Peak plasma protein concentrations usually occurred early in the afternoon, while troughs were found during the night. There is a 10% difference between the peak and trough values in young adults, increasing to up to 20% in elderly volunteers (Touitou et al., 1986).

Unfortunately, circadian variation in drug binding to plasma proteins has not been studied extensively. In the first study, Naranjo et al. (1980) looked at the protein binding of diazepam in six healthy day-active volunteers. They found that the free diazepam fraction was largest during the night (i.e. 23.00–08.00) and smallest at 09.00. In six healthy volunteers, Patel et al. (1982) determined the binding of valproic acid in plasma samples collected at 2 h intervals during the day. A significant time-dependent variation was found: maximal free valproate ratio was obtained between 02.00 and 08.00 and minimal binding occurred late in the afternoon, i.e. between 16.00 and 20.00. The increase in valproate binding observed at night was correlated with an increased concentration of free fatty acids known to displace the drug from its binding sites. Highest plasma levels of free phenytoin or valproic acid were found between 02.00 and 06.00 (Patel *et al.*, 1982; Lockard *et al.*, 1985), while the lowest levels of unbound diazepam or cisplatin binding to plasma proteins occurred in the morning (Naranjo *et al.*, 1980; Hecquet *et al.*, 1984). Table 4.6 summarises the data on temporal variation in protein binding to human plasma. Bruguerolle and his co-workers seem to be the only investigators to have studied the binding of drugs in laboratory animals. They found that the highest protein binding occurred during the activity period (e.g. at 22.00 with lidocaine (lignocaine) and disopyramide (Bruguerolle *et al.*, 1982; Bruguerolle, 1984) and between 22.00 and 04.00 with carbamazepine (Bruguerolle *et al.*, 1985)) in nocturnal rodents.

These changes in free plasma levels of drugs could explain some chronopharmacokinetic data because it is generally assumed that the plasma free drug is the drug fraction absorbed and metabolised. Any increase or decrease in the free concentration could lead to parallel changes in plasma or serum levels of the drug. Much research is needed to set up guidelines in this area, but it must be realised that time-dependent variation in drug binding to plasma proteins is likely to have

Table 4.6 Temporal variations in protein binding to human plasma

Drug	Temporal variation	Reference
Carbamazepine	Lowest free fraction at 04.00	Riva *et al.* (1984)
Cisplatin	Highest protein binding at 16.00	Hecquet *et al.* (1984)
Diazepam	Highest bound fraction at 09.00	Naranjo *et al.* (1980)
Diazepam	Highest bound fraction at 09.30	Nakano *et al.* (1986)
Indometacin	No temporal change in elderly volunteers	Bruguerolle *et al.* (1986)
Phenytoin	No temporal change in volunteers	Patel *et al.* (1982)
Prednisolone	Highest binding at 06.00 and lowest binding at 18.00	English *et al.* (1983)
Prednisolone	Highest binding at midnight and lowest at 08.00	Angeli *et al.* (1978)
Valproic acid	Highest free fraction between 02.00 and 08.00	Patel *et al.* (1982)
	Unbound clearance was 15% higher in the evening in elderly volunteers.	Bauer *et al.* (1985)

Adapted from Bélanger *et al.* (1997).

clinically important consequences only for drugs that are highly protein bound (more than 90%) and have a small volume of distribution.

Bruguerolle and his co-workers used the red blood cell as a model to quantify the passage of drugs through membranes. Using lidocaine, they found that the erythrocyte to total plasma concentration ratio was 0.74 and 0.48 when the drug was administered in the activity (i.e. at 22.00) and resting (i.e. at 10.00) periods, respectively (Bruguerolle and Jadot, 1983). Further studies are needed to determine whether time-dependent variations can be detected in the membrane permeability to drugs. It would be interesting to investigate drugs with different physicochemical properties and those with high or low protein binding since the passage of drugs through membranes is dependent on these two factors.

4.3.3 Rhythms in hepatic drug metabolism

Most drugs are eliminated from the body by hepatic metabolism through the activity of many enzymatic systems with specific substrate and/or cofactor requirements. Oxidation and conjugation to endogenous substrates are the two most important reactions and time-dependent changes have been described for both. Extensive reviews on the chronobiology of drug metabolism are also available elsewhere (Bélanger et al., 1997; Labrecque and Bélanger, 1991).

4.3.3.1 The cytochrome P450 monooxygenase system

Oxidation is the most common and the most important reaction in drug metabolism. It has been studied extensively since 1967 when circadian variation was reported in the hepatic *p*-nitroanisole O-demethylase activity (Radzialowski and Bousquet, 1967, 1968) and in the aminopyrine demethylase and hexobarbital oxidase activities (Nair and Casper, 1969) in mice. These investigators noted that the oxidase activity was 40% greater in the middle of the activity period of rats than in the first half of the resting period. The work of Nair and Casper (1969) is of special interest for pharmacologists because an inverse relationship was reported between the hexobarbital oxidase activity in the liver and the sleeping time produced by the barbiturate injected at different times of day. Hexobarbital is a hypnotic drug known to be eliminated from the body by hepatic microsomal oxidation and the data indicated that the sleeping time was shortest in the activity period, i.e. when the hexobarbital oxidase activity was maximal and vice versa. This was the first time a temporal correlation was identified between the metabolism and the effect of a drug.

Bélanger and Lalande (1988) studied the temporal variation in the oxidase activity of the liver and the concentrations of microsomal components at the beginning of the activity and resting periods of rodents. These determinations were carried out in purified microsomes isolated from the same rat liver so as to rule out interindividual differences. Table 4.7 shows that the enzymatic activities of aminopyrine and p-nitroanisole demethylases and of aniline hydrolase were significantly higher in the activity than in the resting periods. On the contrary, the enzymatic activities of the biphenyl and testosterone 7α- and 6β-hydrolases were higher during the sleeping period of rodents. There was no diurnal variation in testosterone 16α-hydrolase or in the microsomal concentration of total cytochrome P450.

Miyazaki et al. (1990) also presented interesting data on hepatic P450 isoenzyme concentrations and on the circadian variation of testosterone hydroxylation. Figure 4.6a presents the temporal variations in the liver content of total P450, II CII, II C6 and II B1; Figure 4.6b illustrates the mean activity of different testosterone hydrolases. The

Table 4.7 Time-dependent variations in the activity and composition of monooxygenase of rat liver

Microsomal activity values	Time of day[a]		
	Resting period	Activity period	p
Aminopyrine N-demethylase (nmol/mg/30 min)	29.6 ± 3.4	37.4 ± 0.5	<0.02
p-Nitroanisole O-demethylase (nmol/mg/30 min)	10.2 ± 0.6	12.2 ± 0.9	<0.037
Aniline hydroxylase (nmol/mg/30 min)	6.8 ± 0.3	11.3 ± 0.4	<0.001
Biphenyl 4-hydrolase (μg/mg/30 min)	3.1 ± 0.1	1.8 ± 0.2	<0.001
Testosterone hydrolases			
7α-hydrolase	27.3 ± 5.2	18.2 ± 1.2	<0.05
6β-hydrolase	15.0 ± 2.7	10.0 ± 0.4	<0.04
16α-hydrolase	14.3 ± 2.5	12.5 ± 1.0	NS
Cytochromes P450 (nmol/mg protein)	0.47 ± 0.03	0.33 ± 0.08	NS
Microsomal protein (mg/250 mg liver)	7.3 ± 0.4	8.5 ± 0.08	NS
Cytochrome P450 reductase (nmol/mg/30 min)	44.0 ± 1.7	33.0 ± 1.5	<0.001

Adapted from Bélanger and Lalande (1988).
Results are expressed as mean values ± SE.
[a] In the resting and the activity periods, the data were obtained at 09.00 and 21.00, respectively.

temporal patterns of total P450 content differed markedly from the activity patterns of the different isoenzymes. Similarly, specific variations were obtained in the rates of various testosterone hydroxylations. The poor correlation observed between the activity of different enzymes and the P450 levels is evidently due to temporal changes in the catalytic activity of the subpopulation of isoenzymes. More research is needed to characterise further the chronobiology of the cytochrome P450 isoenzymes and their marker activities.

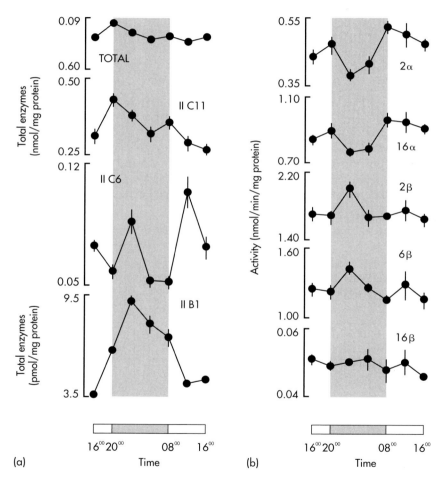

Figure 4.6 Circadian variations in the hepatic cytochrome P450. (a) The hepatic concentrations of total and isoenzymes P450 II B1, II CII. (b) The activity of five testosterone hydrolases at different times of day. Data are presented as means ± SE. Redrawn from Miyazaki et al. (1990).

4.3.3.2 Conjugation reactions

The time-dependent variations in nonoxidative pathways of drug metabolism (i.e. reduction, hydrolysis and conjugation) have not been investigated as thoroughly as the monooxygenase system. The chronobiology of the hepatic transferases involved in glucuronide and sulfate conjugation was studied in freely fed rats synchronised on a 12 h light : 12 h dark cycle (LD 12:12) (Bélanger et al., 1985). Table 4.8 indicates that the maximal rate of *p*-nitrophenol glucuronidation was significantly higher during the activity period of the animal (at 21.00) than in the resting span (at 09.00). On the contrary, the sulfation of phenol was twice as great during the activity as during the resting period. Table 4.8 indicates also that no difference was found in the apparent affinity (K_m) of the glucuronosyltransferase for its substrate at both times of day, but the apparent affinity of the sulfotransferase for its phenolic substrate was increased 4-fold during the activity period. It is interesting to note that fasting abolished these time-dependent variations of both transferases.

The temporal variation in the hepatic glucuronidation and sulfation may explain the chronopharmacokinetics of paracetamol (acetaminophen), which is known to be eliminated mainly by these two pathways in humans and rats. In human studies, the plasma $t_{1/2}$ of paracetamol was 15% longer at 06.00 than at 14.00, and the mean ratio of the glucuronide conjugate over unchanged paracetamol excreted in the first 3.5 h urine sample varied between 5.2 at 06.00 and 7.8 at 14.00 (Shively

Table 4.8 Enzyme-substrate characteristics of the transferases involved in glucuronide and sulfate conjugation

Enzymes	Substrate	Kinetic parameters[a]	Time of day	
			09.00	21.00
Glucuronosyl-transferase (microsomes)	*p*-Nitrophenol	V_{max} (nmol/g/15 min)	5.6 ± 1.2	8.8 ± 0.8*
		K_m (mmol/L)	0.20 ± 0.03	0.15 ± 0.05
Sulfo-transferase (soluble fraction)	Phenol	V_{max} (µmol/g/5 min)	3.5 ± 0.1	1.6 ± 0.2*
		K_m (mmol/L)	0.18 ± 0.01	0.04 ± 0.02*

Adapted from Bélanger et al. (1985).
[a] Data are presented as mean value ± SE obtained in 5–7 rats.
*$p < 0.05$ compared with the 09.00 values.

and Vessell, 1975). In fasted rats receiving a single dose of paracetamol, the total clearance and the total metabolism of the drug determined as the extraction ratio following administration of a single 40 mg/kg dose were both greater in the activity than in the resting periods (Bélanger *et al.*, 1987). These results obtained in humans and in rats are in good agreement if one considers that the rat is a nocturnal animal and that its biological rhythms are synchronised inversely to those of humans: in both species, paracetamol metabolism was significantly greater during the activity period.

The conjugation reaction is a pathway that is particularly important in toxicology. Indeed, reduced glutathione (GSH) has two main cellular functions: (1) it removes peroxides and other oxygen radicals produced by various enzyme systems; and (2) it forms an adduct with electrophilic compounds, with free-radical species or with other reactive intermediates of toxic drugs that may be produced during the hepatic

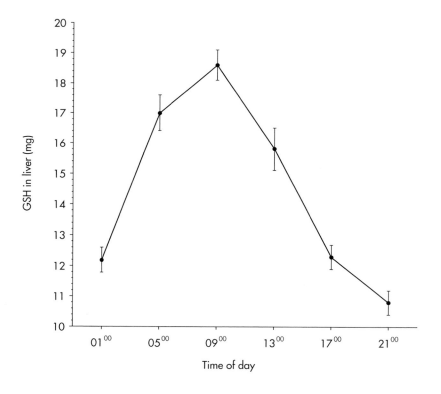

Figure 4.7 Circadian variation in the hepatic concentration of reduced glutathione in rats. Results are mean values ± SE. Redrawn from Bélanger *et al.* (1988a).

metabolism of drugs by the cytochrome P450 monooxygenase system. Many investigators have found temporal variations in the hepatic concentrations of GSH. One study carried out in our laboratory in rats synchronised under LD 12:12 (light on at 07.00) indicated (see Figure 4.7) that maximum and minimum hepatic levels of GSH were found at the beginning of the rest (09.00) and the activity (21.00) periods, respectively (Bélanger et al., 1988a).

The circadian variation in hepatic levels of GSH appears to be the major determinant in the chronohepatotoxicity of agents such as 1,1-dichloroethylene (Jaeger et al., 1973), allyl alcohol (Hanson and Anders, 1978), chloroform (Lavigne et al., 1983) and styrene (Bélanger et al., 1988b; Desgagné and Bélanger, 1986). Indeed, the liver necrosis produced by these agents is dependent on the covalent binding of reactive intermediates to proteins, nucleic acids or other macromolecules of the cell. A higher concentration of GSH in the liver leads to a greater formation of glutathione conjugate and to a decrease in the alkylation of cellular components. Conversely, lower hepatic GSH levels increased the covalent binding to macromolecules, which often caused cellular dysfunction and necrosis. Thus, the extent of liver necrosis could be related to the temporal variation in the conjugation pathway.

4.3.3.3 Factors modifying the time-dependent changes in drug metabolism

In addition to the time-dependent changes in the drug-metabolising enzymes, the extent of drug binding to plasma protein could account for circadian variations in drug metabolism. This factor will be important for drugs metabolised mainly, but slowly, by the liver (i.e. drugs with a low hepatic ratio) such as diazepam, phenytoin, theophylline and warfarin. On the other hand, circadian variation in hepatic blood flow will be very important for drugs with a high hepatic extraction ratio, such as propranolol and lidocaine.

Circadian variations in the endocrine system could also influence the rate of drug metabolism. This was shown more than 30 years ago when Radzialowski and Bousquet (1967) showed that temporal variation in the aminopyrine N-demethylase and p-nitroanisole O-demethylase activity was inversely related to plasma corticosterone in rodents. These investigators also showed that maintaining constant plasma corticosteroid levels had no effect on the activity of 4-dimethylaminoazobenzene reductase.

The effect of food on temporal variation in drug metabolism is still unclear. While Radzialowski and Bousquet (1967, 1968) reported that fasting had no effect on the time-dependent variations in enzyme activity, Bélanger et al. (1985) showed that fasting rats overnight abolished the temporal variations in the enzyme characteristics of hepatic glucuronosyltransferase and sulfotransferase activities obtained during the activity and sleep periods. Further research is needed in this area.

In summary, the results of research done in the last 30 years indicate clearly that time-dependent variation exists in the enzymatic systems involved in the hepatic metabolism of drugs. Further research is needed to identify the mechanisms of this variation.

4.3.4 Rhythms in drug excretion

4.3.4.1 Renal excretion of drugs

Drugs are eliminated from the body either unchanged or as metabolites and the kidney is the most important organ for their elimination. The processes involved in the urinary excretion of drugs and their metabolites are glomerular filtration, active tubular secretion and tubular reabsorption. Modification of urine pH during day or night can also alter the excretion rate of drugs.

Circadian rhythms have been described for glomerular filtration rate (GFR), effective renal plasma flow, tubular secretion, urine output and urinary excretion of electrolytes and many endogenous substances. The clearances of inulin and p-aminohippuric acid (PAH) were largest in the middle of the activity period (Pons et al., 1994) and smallest during the rest period of rodents. Similar data were found in humans (Vagnucci et al., 1969; Wiser and Breuer, 1981; Araki et al., 1983; Koopman et al., 1985; Waterhouse and Minors, 1989). These rhythmic patterns of normal individuals are not related to day–night differences in the activity because they persist during constant bed rest (Addis et al., 1950). It is more likely that the circadian variations found in systemic blood pressure, in the renin–angiotensin system and in renal blood flow are responsible for the time-dependent changes in renal haemodynamics (Millar-Craig et al., 1978; Redon et al., 1990). To our knowledge, tubular reabsorption has not been studied during the 24 h period.

Renal reabsorption of material from the tubules back into the blood is the third process involved in the excretion of drugs that could contribute to circadian variation in drug elimination. Active reabsorption is rare, and circadian variation in this system has not been reported. Passive reabsorption of a drug is more common and is influenced by the physicochemical properties of the drug and by the urinary pH, since the tubular cells are less permeable to the ionised form of weak acids or weak alkalis.

The data obtained with amfetamine and norephedrine are good examples of the effect of changes in urinary pH on the excretion of drugs. For example, amfetamine is more strongly ionised at lower pH (about 5.2) and it is readily excreted in the urine. When the urinary pH is more basic (about 7.0), the nonionised fraction is increased and its urinary excretion is decreased markedly (Beckett et al., 1964; Wilkinson and Beckett, 1968).

Human data indicate that urinary pH changes throughout the 24 h period; urinary pH is lower during the resting period than in the activity period (Beckett and Rowland, 1964; Jones, 1945; Elliot et al., 1959; Koike et al., 1984). These temporal changes in urinary pH could explain why the excretion of salicylate (Reinberg et al., 1967; Markiewicz and Semenowicz, 1979) and sulfonamides (Dettli and Spring, 1966) varied as a function of the time of day. Figure 4.8a illustrates that circadian changes were also observed in the urinary pH of fed rats but were absent when animals were fasted. Urinary pH was higher during the food intake period and decreased when the animals were not eating. The pH difference over the 24 h span could be very important in explaining the renal toxicity of drugs such as the aminoglycosides (AMG) (Lin et al., 1994a, b). This nephrotoxicity is due to electrostatic binding of AMG to acid phospholipids of the proximal tubules (Laurent et al., 1982). The interaction is pH-dependent; it is largest at pH 5 and it decreases as pH rises (Laurent et al., 1982). In agreement with data obtained by Nakano and colleagues (Nakano and Ogawa, 1982; Nakano et al., 1990) and Pariat et al. (1986, 1988), investigators who were using large doses of AMG, we found that nephrotoxicity of gentamicin was greatest when rats were sleeping and smallest in the middle of the activity span of the animals (see Figure 4.8b). Similar data were obtained with dibekacin (Pariat et al., 1986), amikacin (Dorian and Carbar, 1986; Dorian et al., 1987) and isepamicin (Yoshiyama et al., 1993). Thus, highest renal toxicity occurred when pH was lower and when rats were fasting; nephrotoxicity was lowest when pH was higher and when rats had free access to food (Beauchamp et al., 1996). Thus, food appears to influence the circadian variation of urinary pH, which in turn modifies the binding or the excretion of drugs. The data on the chrononephrotoxicity of cisplatin (Hrusheski et al., 1982) could also be explained by daily changes in urinary pH and possibly by the presence or absence of food at different times of day.

4.3.4.2 Biliary excretion of drugs

Biliary excretion is another elimination route for drugs, but this process appears to be more important for laboratory animals than for humans (Smith, 1973; Garrett, 1978; Levine, 1983). Bile flow and composition

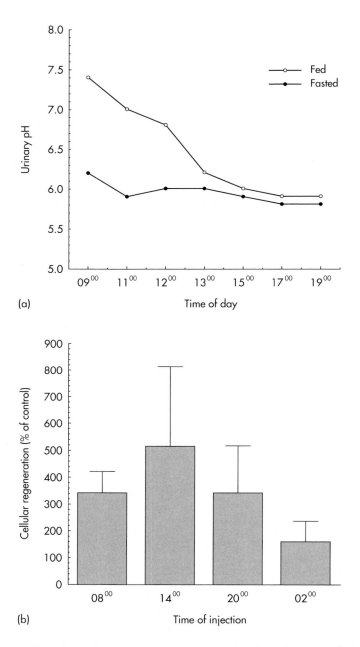

Figure 4.8 Time-dependent variations in urinary pH and renal toxicity of tobramycin and gentamicin in rats. (a) The 24 h changes in urinary pH in fasting and nonfasting rats. (b) The toxicity of gentamicin in rats at four times of day. Redrawn from Koike et al. (1984); Lin et al. (1994a).

are determinants of both the rate and extent of biliary excretion of endogenous substances such as bile acid, phospholipids and cholesterol (Smith, 1973; Levine, 1983; Vonk et al., 1978a; Okolicsanyi et al., 1986). Vonk et al. (1978b) are the only investigators to study the day–night changes in the biliary concentration and excretion rate of bromosulfophthalein. The data indicated that the biliary excretion of this dye (which is exclusively cleared into the bile) was 25% higher in the middle of the activity period of rats. A parallel increase in blood flow and modification of bile composition at different times of the day correlated well with these results (Vonk et al., 1978b; Borz and Steele, 1973; Mitropoulos et al., 1973; Ho and Drummond, 1975).

4.4 Conclusions

Investigations carried out over the last 30 years indicate clearly that it is no longer possible to support the idea that oral ingestion of immediate-release formulations of drugs at different times of day leads to constant plasma levels. On the contrary, the time of drug administration is an important factor and explains variations observed in many pharmacokinetic studies. The physiological mechanisms underlying chronopharmacokinetic data are now well defined, while the role of pharmaceutical formulation in chronopharmacokinetics is better understood. However, further research is needed before we will understand completely why the absorption, distribution, metabolism and excretion of drugs vary as a function of the hour of administration.

The presence of biological rhythms in pharmacokinetics raises the question 'When to treat?'. As it is now also known that a constant rate of drug delivery does not necessarily produce either constant plasma levels or a constant effect (Decousus et al., 1985, 1986), the way in which we consider the mode of drug administration to patients needs to be entirely revised. In our view, the timing of administration is a key factor to be considered in optimising drug effects in patients. Time-dependent variation in pharmacokinetics will influence the desired and undesired effects of drugs. Taking circadian variation in pharmacokinetics into account will help determine the optimal time for drug administration and may allow individualisation of drug treatment.

References

Addis T, Barrett E, Poo L J (1950). The relation between protein consumption and diurnal variations of the endogenous creatinine clearance in normal individuals. *J Clin Invest* 30: 206–209.

Angeli A, Frajria R, DePaoli R, et al. (1978). Diurnal variation of prednisolone binding to serum corticosteroid binding-globulin in man. *Clin Pharmacol Ther* 23: 47–53.

Araki S, Murata K, Yokoyama K, et al. (1983). Circadian rhythms in the excretion of metal and organic substances in healthy men. *Arch Environ Health* 38: 360–366.

Avachat M K, Rambhau D, Sarveswar R A O, et al. (1992). Chronokinetics of rifampicin. *Indian J Physiol Pharmacol* 36: 251–252.

Balis F M, Jeffries S L, Lange B, et al. (1989). Chronopharmacokinetics of oral methotrexate and 6-mercaptopurine: is there diurnal variation in the disposition of antileukemic therapy? *Am J Pediatr Hematol Oncol* 11: 324–326.

Bauer L A, Davis R, Wilensky A, et al. (1985). Valproic acid clearance: unbound fraction and diurnal variation in young and elderly adults. *Clin Pharmacol Ther* 37: 697–700.

Beauchamp D, Collin P, Grenier, L, et al. (1996). Effects of fasting on temporal variations in nephrotoxicity of gentamicin in rats. *Antimicrob Agents Chemother* 40: 670–676.

Beckett A H, Rowland M (1964). Rhythmic urinary excretion of amphetamine in man. *Nature* 204: 1203–1204.

Bélanger P M (1993). Chronopharmacology in drug research and therapy. *Adv Drug Res* 24: 1–79.

Bélanger P M, Lalande M (1988). Day–night variations in the activity and composition of the liver microsomal mixed-function oxidase of rat liver. *Annu Rev Chronopharmacol* 5: 219–222.

Bélanger P M, Labrecque G, Doré F (1984). Rate limiting steps in the temporal variations of the pharmacokinetics of some selected drugs. In: Haus E, Kabat H, eds. *Chronobiology 1982–1983*. Basel: Karger, 359–363.

Bélanger P M, Lalande M, Labrecque G, Doré F M (1985). Diurnal variations in the transferases and hydrolases involved in glucuronide and sulfate conjugation of rat liver. *Drug Metab Dispos* 13: 386–389.

Bélanger P M, Lalande M, Doré F, Labrecque G (1987). Time-dependent variations in the organ extraction ratios of acetaminophen in rat. *J Pharmacokinet Biopharm* 15: 133–143.

Bélanger P M, Desgagné M, Bruguerolle B (1988a). Circadian periodicity in the glutathione concentration of rat liver. *Annu Rev Chronopharmacol* 5: 215–218.

Bélanger P M, Desgagné M, Boutet M (1988b). The mechanism of the chronohepatoxicity of chloroform in rat: correlation between covalent binding to hepatic subcellular fractions and histologic changes. *Annu Rev Chronopharmacol* 5: 235–238.

Bélanger P M, Bruguerolle B, Labrecque G (1997) Rhythms in pharmacokinetics: absorption, distribution, metabolism and excretion. In: Redfern P H, Lemmer B, eds. *Physiology and Pharmacology of Biological Rhythms*. Berlin: Springer, 176–204.

Bleyzac N, Allard-Latour B, Laffont A, et al. (2000). Diurnal changes in the pharmacokinetic behavior of amikacin. *Ther Drug Monit* 22: 307–312.

Borz W M, Steele L A (1973) Synchronization of hepatic cholesterol synthesis, cholesterol and bile acid content, fatty acid synthesis and plasma free fatty acid levels in the fed and fasted rats. *Biochim Biophys Acta* 306: 85–94.

Bruguerolle, B (1984). Circadian phase dependent pharmacokinetics of disopyramide in mice. *Chronobiol Int* 1: 267–271.
Bruguerolle B (1987). Données récentes en chronopharmacocinétique. *Pathol Biol (Paris)* 35: 925–934.
Bruguerolle B (1989). Temporal aspects of drug absorption and drug distribution. In: Lemmer B, ed. *Chronopharmacology: Cellular and Biochemical Interactions*. New York: Marcel Dekker, 1–14.
Bruguerolle B (1998). Chronopharmacokinetics: current status. *Clin Pharmacokinet* 35: 83–94.
Bruguerolle B, Jadot G (1983). Influence of the hour of administration of lidocaine on its erythrocytic passage in the rat. *Chronobiologia* 10: 295–297.
Bruguerolle B, Jadot G, Valli M, *et al.* (1982). Etude chronocinétique de la lidocaine chez le rat. *J Pharmacol (Paris)* 13: 65–76.
Bruguerolle B, Bouvenot G, Bartolin R, Descottes C (1985). Temporal variations of lorazepam pharmacokinetics. *Int J Clin Pharmacol Ther Toxicol* 23: 352–354.
Bruguerolle B, Bruguerolle B, Lévi F, *et al.* (1986). Alteration of physiologic circadian time structure of six plasma proteins in patients with advanced cancer. *Annu Rev Chronopharmacol* 3: 207–210.
Bruguerolle B, Bouvenot G, Bartolin R, Pouyet A (1988). Chronokinetics of acetaminophen in elderly patients. *Annu Rev Chronopharmacol* 7: 265–268.
Canal P, Squalli A, DeForni M, *et al.* (1991). Chronopharmacokinetics of doxorubicin in patients with breast cancer. *Eur J Clin Pharmacol* 40: 287–291.
Choi J S, Park H J (1999). Circadian changes in pharmacokinetics of cyclosporin in healthy volunteers. *Res Commun Mol Pathol Pharmacol* 103: 809–812.
Choi J S, Kim C K, Lee B (1999). Administration-time differences in the pharmacokinetics of gentamicin intravenously delivered to human beings. *Chronobiol Int* 16: 821–829.
Clench J, Reinberg A, Dziewanowsak Z, *et al.* (1981). Circadian changes in the bioavailability and effects of indomethacin in healthy subjects. *Eur J Clin Pharmacol* 20: 359–369.
Decousus H, Croze M, Lévi F, *et al.* (1985). Circadian changes in anticoagulant effect of heparin infused at a constant rate. *Br Med J* 290: 341–344.
Decousus H, Ollagnier M, Cherrah Y, *et al.* (1986). Chronokinetics of ketoprofen infused intravenously at a constant rate. *Annu Rev Chronopharmacol* 3: 321–322.
Desgagné M, Bélanger P M (1986). Chronohepatoxicity of styrene in rat. *Annu Rev Chronopharmacol* 3: 103–105.
Dettli L, Spring P (1966). Diurnal variations in the elimination rate of sulfonamide in man. *Helv Med Acta* 33: 291–306.
Dorian C, Carbar J (1986). Circadian variations of amikacin nephrotoxicity in rats. *Pathol Biol* 34: 587–590.
Dorian C, Catroux P, Cambar J (1987). Chrononephrotoxicity of amikacin after 7-day chronic poisoning in rats. *Pathol Biol* 35: 735–738.
Elliot J S, Sharp R F, Lewis L (1959). Urinary pH. *J Urol* 81: 339–343.
Elting L, Bodey G P, Rosenbaum B, Fainstein V (1990). Circadian variation in serum amikacin levels. *J Clin Pharmacol* 30: 798–801.
English J, Dunne M, Marks V (1983). Diurnal variations in prednisolone kinetics. *Clin Pharmacol Ther* 33: 381–385.
Fauvelle F, Perrin P, Belfayol L, *et al.* (1994). Fever and associated changes in

glomerular filtration rate erase anticipated diurnal variations in aminoglycoside pharmacokinetics. *Antimicrob Agents Chemother* 38: 620–623.

Focan C, Bruguerolle B, Arnaud C, et al. (1988). Alteration of circadian time structure of plasma proteins in patients with inflammation. *Annu Rev Chronopharmacol* 5: 21–24.

Focan C, Doalto L, Mazy V, et al. (1989). Vindesine en perfusion continue de 48 heures (suivie de cisplatine) dans le cancer pulmonaire avancé. Données chronopharmacocinétiques et efficacité clinique. *Bull Cancer (Paris)* 76: 909–912.

Garrett E R (1978). Pharmacokinetics and clearance related to renal processes. *Int J Clin Pharmacokinet* 16: 155–172.

Goo R H, Moore J G, Greenberg E, Alazraki N P (1987). Circadian variation in gastric emptying time of meals in humans. *Gastroenterology* 91: 515–518.

Groth U, Prellwitz E., Jähnchen E (1974). Estimation of pharmacokinetic parameters of lithium from saliva and urine. *Clin Pharmacol Ther* 16: 490–498.

Guissou P, Cuisinaud G, Llorca G, et al. (1983). Chronopharmacologic study of a prolonged release form of indomethacin. *Eur J Clin Pharmacol* 24: 678–682.

Hanson S K, Anders M W (1978). The effect of diethyl maleate treatment, fasting and time of administration of allyl alcohol hepatoxicity. *Toxicol Lett* 1: 301–305.

Hassan M, Oberg G, Bakassy A N, et al. (1991). Pharmacokinetics of high-dose busulfan in relation to age and chronopharmacology. *Cancer Chemother Pharmacol* 28: 130–134.

Hecquet B, Meynardier J, Bonneterre J, Adenis L (1984). Circadian rhythm in cisplatin binding to plasma proteins. *Annu Rev Chronopharmacol* 3: 115–118.

Ho K J, Drummond J L (1975). Circadian rhythm of biliary excretion and its control mechanisms in rats with chronic biliary drainage. *Am J Physiol* 229: 1427–1437.

Hrusheski W J M, Borsch R, Lévi F (1982). Circadian time dependence of cisplatin urinary excretion. *Clin Pharmacol Ther* 32: 330–339.

Jaeger R J, Conoly R B, Murphy I D (1973). Diurnal variation of hepatic glutathione concentration and its correlation with 1,1-dichloroethylene inhalation toxicity in rats. *Res Commun Chem Pathol Pharmacol* 6: 465–471.

Jones H B (1945). On the variation of the state of acidity of the urine in the state of health. *Philos Trans R Soc Lond* 135: 335–349.

Jonkman J H G, Reinberg A, Osterhuis B, et al. (1988). Dosing time and sex-differences in the pharmacokinetics of cefodizime and in the circadian cortisol rhythm. *Chronobiologia* 15: 89–102.

Kanabrocki E L, Sothern R B, Scheving L E, et al. (1988). Ten-year replicated circadian profiles for 36 physiological, serological and urinary variables in healthy men. *Chronobiol Int* 5: 237–284.

Kerr D J, Lewis C, O'Neil B, et al. (1990). The myelotoxicity of carboplatine is influenced by its time of administration. *Hematol Oncol* 8: 59–63.

Koike M, Norimura R, Mizajiri K, Sugeno R (1984). Time-dependent elimination of cinoxacin in rats. *J Pharm Sci* 73: 1697–1700.

Koopman M G, Krediet R T, Arisz L (1985). A circadian rhythm of proteinuria in patients with a nephrotic syndrome. *Clin Sci* 69: 395–401.

Koren G, Ferrazini G, Sohl H, et al. (1992). Chronopharmacology of methotrexate pharmacokinetics in childhood leukemia. *Chronobiol Int* 9: 434–438.

Kumar D, Wingate D, Ruckebusch Y (1986). Circadian variation in the propagating velocity of the migrating motor complex. *Gastroenterology* 93: 926–930.

Labrecque G, Bélanger P M, Doré F, Lalande M (1988). 24-hour variations in the distribution of labeled microspheres to the intestine, liver and kidneys. *Annu Rev Chronopharmacol* 5: 445–449.

Labrecque G, Bélanger P M (1991). Biological rhythms in the absorption, distribution, metabolism and excretion of drugs. *Pharmacol Ther* 52: 95–107.

Lakatua D, Lesar T S, Zaska D E, et al. (1984). Observation on the pharmacokinetics of ethanol. *Annu Rev Chronopharmacol* 1: 297–300.

Lamiable D, Vistelle R, Fay R, et al. (1991). Chronopharmacocinétique de la dexaméthasone chez le sujet jeune. *Thérapie* 46: 405–407.

Langevin A M, Koren G, Soldin S J, Greenberg M (1987). Pharmacokinetic case for giving 6-mercaptopurine maintenance dose at night. *Lancet* 2: 505–506.

Langler B, Lemmer B (1988). Circadian changes in the pharmacokinetics and cardiovascular effects of oral propranolol in healthy subjects. *Eur J Clin Pharmacol* 33: 619–624.

Laurent G, Carlier M B, Rollman B, et al. (1982). Mechanism of aminoglycoside-induced lysosomal phospholipidosis: *in vitro* and *in vivo* studies with gentamicin and amikacin. *Biochem Pharmacol* 31: 3861–3870.

Lavigne J G, Bélanger P M, Doré F, Labrecque, G. (1983). Temporal variations in chloroform-induced hepatoxicity in rats. *Toxicology* 26: 267–273.

Lemmer B (1981). Chronopharmacokinetics. In: Breimer D D, Speiser P, eds. *Trends in Pharmaceutical Sciences*. Amersterdam: Elsevier, 49–68.

Lemmer B, Nold G (1991). Circadian changes in the estimated hepatic blood flow in healthy subjects. *Br J Clin Pharmacokinet* 26: 419–427.

Lemmer B, Becker H, Renczes J, et al. (1986). On the chronopharmacology of oral isosorbide dinitrate: pharmacokinetics and cardiovascular effects in healthy subjects. *Annu Rev Chronopharmacol* 3: 339–342.

Lemmer B, Nold G, Behne S, Kaiser R (1991). Chronopharmacokinetics and cardiovascular effects of nifedipine. *Chronobiol Int* 8: 485–494.

Levine W G (1983). Excretion mechanisms. In: Caldwell J, Jakoby W B, eds. *Biological Basis of Detoxification*. London: Academic Press, 251–285.

Lin L, Grenier L, Thériault G, et al. (1994a). Nephrotoxicity of low doses of tobramycin in rats: effect of the time of administration. *Life Sci* 55: 169–177.

Lin L, Grenier L, Bergeron Y, et al. (1994b). Temporal changes of pharmacokinetics, nephrotoxicity, and subcellular distribution of tobramycin in rats. *Antimicrob Agents Chemother* 38: 54–60.

Lockard J S, Viswanathan C T, Levy R H (1985). Diurnal oscillations of CSF valproate in monkey. *Life Sci* 36: 1281–1285.

Lucht F, Tigaud S, Esposito G, et al. (1990). Chronokinetic study of netilmicin in man. *Eur J Clin Pharmacol* 39: 199–201.

Malan J, Moncrieff J, Bosch E (1985). Chronopharmacology of paracetamol in normal subjects. *Br J Clin Pharmacol* 19: 843–845.

Markiewicz A, Semenowicz K (1979). Time-dependent changes in the pharmacokinetics of aspirin. *Int J Clin Pharmacol Biopharm* 17: 409–411.

Markiewicz A, Semenowicz K, Korcynska J, Pzachowska A (1979). Chronopharmacokinetics of dipyridamole. *Int J Clin Pharmacol Biopharm* 17: 222–234.

Metzger G, Massari C, Etienne M C, et al. (1994). Spontaneous or imposed circadian changes in plasma concentrations of 5-fluorouracil coadministered with folinic acid and oxiplatin: relationship with mucosal toxicity in cancer patients. *Clin Pharmacol Ther* 56: 190–201.

Millar-Craig M W, Bishop C N, Raftery E B (1978). Circadian variation in blood pressure. *Lancet* 1: 795–797.

Mitropoulos K A, Balasubramaniam S, Myran N B (1973). The effect of interruption of the enterohepatic circulation on bile acids and of cholesterol feeding on cholesterol 7-alpha-hydrolase in relation to diurnal rhythm in its activity. *Biochim Biophys Acta* 326: 428–438.

Miyazaki Y, Imaoka S, Yatagai M, *et al.* (1990). Temporal variations in hepatic cytochrome P-450 isoenzymes in rats. *Annu Rev Chronopharmacol* 7: 149–152.

Mustafa M, Suryawat S, Dwiprahasto I, Santoso B (1991). The relative bioavailability of diclofenac with respect to time of administration. *Br J Clin Pharmacol* 32: 246–247.

Mrozikiewics A, Jablecka Z, Lowicki Z, Chmara E (1988). Circadian variation in digoxin pharmacokinetics. *Annu Rev Chronopharmacol* 5: 457–459.

Nair V, Casper R (1969). The influence of light on daily rhythm in hepatic drug metabolizing enzymes in rat. *Life Sci* 8: 1291–1298.

Nakano S, Hollister L E (1978). No circadian effect on nortriptyline kinetics in man. *Clin Pharmacol Ther* 23: 199–203.

Nakano S. Hollister L E (1983). Chronopharmacology of amitriptyline. *Clin Pharmacol Ther* 33: 453–459.

Nakano S, Ogawa N (1982). Chronotoxicity of gentamicin in mice. *IRCS Med Sci* 10: 592–593.

Nakano S, Watanabe H, Nagai K, Ogawa N (1984). Circadian-stage dependent changes in diazepam kinetics. *Clin Pharmacol Ther* 36: 271–277.

Nakano S, Watanabe H, Ohdo S, Ogawa N (1986). Circadian-stage dependent changes in diazepam and valproate kinetics in man: a single and repetitive administration study. *Annu Rev Chronopharmacol* 3: 421–424.

Nakano S, Song J, Ogawa N (1990). Chronopharmacokinetics of gentamicin: comparison between man and mice. *Annu Rev Chronopharmacol* 7: 277–280.

Naranjo C A, Sellers E M, Giles H G, Abel J G (1980). Diurnal variations in plasma diazepam concentrations associated with reciprocal changes in free fraction. *Br J Clin Pharmacol* 9: 265–272.

Okolicsanyi L, Lirussi F, Strazzabosco M, *et al.* (1986). The effect of drug on bile flow and composition. An overview. *Drugs* 31: 430–488.

Ollagnier M, Decousus H, Cherrah Y, *et al.* (1987). Circadian changes in the pharmacokinetics of oral ketoprofen. *Clin Pharmacokinet* 12: 367–378.

Pariat C, Cambar J, Piriou A, Courtois P (1986). Circadian variations in the nephrotoxicity induced by high doses of gentamicin and dibekacin in rats. *Annu Rev Chronopharmacol* 3: 107–110.

Pariat C, Courtois P, Cambar J, *et al.* (1988). Circadian variations in the renal toxicology of gentamicin in rats. *Toxicol Lett* 40: 175–182.

Patel I H, Venkataramanan R, Levy R H, *et al.* (1982). Diurnal oscillations in plasma protein binding of valproic acid. *Epilepsia* 32: 282–290.

Petit E, Milano G, Lévi F, *et al.* (1988). Circadian varying plasma concentration of 5-FU during 5-day continuous venous infusion at constant rate in cancer patients. *Cancer Res* 48: 1676–1679.

Pons M, Tranchot J, L'Azou B, Cambar J (1994). Circadian rhythms of renal hemodynamics in unanesthetized, unrestrained rats. *Chronobiol Int* 11: 301–308.

Radzialowski F M, Bousquet W F (1967). Circadian rhythm in hepatic drug metabolizing activity of the rat. *Life Sci* 6: 2545–2548.

Radzialowski F M, Bousquet W F (1968). Daily rhythmic variation in the hepatic drug metabolism in the rat and mouse. *J Pharmacol Exp Ther* 163: 229–238.

Ramesh Rao B, Rambhau D (1992). Chronopharmacokinetics of sulfamethoxazole in human volunteers. *Drug Invest* 4: 199–204.

Redon P, Laval, M, Cambar J (1990). Chronobiological aspects of the renin–angiotensin system: physiology and physiopathology. *Annu Rev Chronopharmacol* 6: 183–224.

Reinberg A, Smolensky M H (1982). Circadian rhythms in drug disposition in man. *Clin Pharmacokinet* 7: 401–420.

Reinberg A, Zagula-Mally Z W, Ghata J, Halberg F (1967). Circadian rhythm in duration of sodium salicylate excretion referred to phase of excretory rhythms and routine. *Exp Biol Med* 124: 826–832.

Reinberg A, Clench J, Aymard N, *et al.* (1975). Variations circadiennes de l'éthanol et de l'éthanolémie chez l'homme adulte sain. Étude chronopharmacologique. *J Physiol (Paris)* 70: 435–456.

Reinberg A, Schuller E, Deslanerie N, *et al.* (1977). Rythmes circadiens et circannuels des leucocytes, protéines totales, immunoglobulines A, G et M. Étude chez neuf adultes jeunes et sains. *Nouv Presse Med* 6: 3819–3823.

Reinberg A, Pauchet F, Ruff F, *et al.* (1987). Comparison of once-daily evening versus morning sustained-released theophylline dosing for nocturnal asthma. *Chronobiol Int* 4: 409–419.

Ritschell W A, Bykardi G, Ford D J, *et al.* (1983). Pilot study on disposition and pain relief after i.m. administration of meperidine during the day or night. *Int J Clin Pharmacol Ther Toxicol* 21: 218–223.

Riva R, Albani F, Ambrosetto G, *et al.* (1984). Diurnal fluctuations in free and total steady state plasma levels of carbamazepine and correlation with intermittent side effects. *Epilepsia* 25: 476–481.

Rivington R N, Calcutt L, Child S, *et al.* (1985). Comparison of morning versus evening dosing with a new once-daily oral theophylline formulation. *Am J Med* 79(Suppl 6A): 67–72.

Sarveshwar Rao VD, Rambhau D, Ramesh Rao B, *et al.* (1997). Circadian variation in urinary excretion of ciprofloxacin after single-dose oral administration at 10.00 and 22.00 h in human subjects. *Antimicrob Agents Chemother* 41: 1802–1804.

Scheidel R C, Lemmer B (1991). Chronopharmacology of oral nitrates in healthy subjects. *Chronobiol Int* 8: 409–419.

Scott P H, Tabachnik E, Macleod S, *et al.* (1981). Sustained-release theophylline for childhood asthma: evidence for a circadian variation of theophylline pharmaceutics. *J Pediatr* 99: 476–479.

Shiga T, Fujimura T, Tateishi T, *et al.* (1993). Differences of chronokinetic profiles between propranolol and atenolol in hypertensive subjects. *J Clin Pharmacol* 33: 756–761.

Shively C A, Vesell E S (1975). Temporal variations in acetaminophen and phenacetin half-life in man. *Clin Pharmacol Ther* 18: 413–424.

Smith R B, Kroboth P D, Phillips J P (1986). Temporal variations in triazolam pharmacokinetics and pharmacodynamics after oral administration. *J Clin Pharmacol* 26: 120–124.

Smith R L (1973). *The Excretory Function of Bile*. London: Chapman and Hall.

St-Pierre M, Leeder S, Spino M, et al. (1984). Circadian variation in theophylline absorption. *Annu Rev Chronopharmacol* 1: 81–84.

Touitou Y, Touitou C, Bogdan A, et al. (1986). Differences between young and elderly subjects in seasonal and circadian variations of total plasma and blood volume as reflected by hemoglobin, hematocrit and erythrocyte counts. *Clin Chem* 2: 801–804.

Vagnucci A H, Shapiro A P, Mcdonald R H (1969). Effect of upright posture on renal electrolyte cycle. *Appl Physiol* 26: 720–731.

Vassal G, Challine D, Koscielny S, et al. (1993). Chronopharmacology of high-dose busulfan in children and its relationship to liver toxicity. *Cancer Res* 53: 1534–1537.

Vonk R J, Scholtens E, Strubble J H (1978a). Biliary excretion of dibromosulfophthalein in the freely moving unanesthetized rat: circadian variation and effects of deprivation of food and pentobarbital anesthesia. *Clin Sci Mol Med* 55: 399–406.

Vonk R J, Van Doorn A B C, Strubble J H (1978b). Bile secretion and bile composition in the freely moving, unanesthetized rat with a permanent biliary drainage. Influence of food intake on bile flow. *Clin Sci Mol Med* 55: 253–259.

Wang L S, Lin M S, Huang J D (1990). Lack of diurnal effect on vancomycin disposition in infected patients. *J Pharm Sci* 79: 655–656.

Waterhouse J M, Minors D S (1989). Temporal aspects of renal drug elimination. In: Lemmer B, ed. *Chronopharmacology: Cellular and Biochemical Interactions*. Basel: Marcel Dekker, 35–50.

Wilkinson G R, Beckett A H (1968). Absorption, metabolism and excretion of the ephedrines in man. I. The influence of urinary pH and urine volume output. *J Pharmacol Exp Ther* 162: 139–147.

Wiser H, Breuer H (1981). Circadian changes of clinical, chemical and endocrinological parameters. *J Clin Chem Biochem* 19: 323–327.

Yoshiyama Y, Nakano S, Ogawa N (1989). Chronopharmacokinetic study of valproic acid in man: comparison of oral and rectal administration. *J Clin Pharmacol* 29: 1048–1052.

Yoshiyama Y, Nishikawa S, Sugiyama T, et al. (1993). Influence of circadian-stage-dependent dosing schedule on nephrotoxicity and pharmacokinetics of isepamicin in rats. *Antimicrob Agents Chemother* 37: 2042–2043.

Yoshiyama Y, Kobayashi T, Ohdo S, et al. (1996). Dosing time-dependent changes of pharmacokinetics of isepamicin in man. *J Infect Chemother* 2: 106–109.

5

Rhythms and pharmacodynamics

Klaus Witte and Björn Lemmer

5.1 Principles of drug action

Specific effects of drugs can generally be attributed to an interaction between a drug and a cellular target, such as a receptor, an enzyme, a transport protein or an ion channel. The effects of a given drug depend on a variety of parameters, for example the affinity to the respective target, the amount of the target present in a given tissue, and the baseline activity of the target system. If the drug acts as a competitive receptor antagonist or a competitive enzyme inhibitor, the concentration of the endogenous agonist or substrate plays an important role. Biological rhythms can occur in all of these parameters and can therefore lead to biological time-dependent changes in the pharmacodynamics of drugs. Most studies addressing the issue of 'chronopharmacodynamics' have been conducted in animal models rather than in the clinical setting. Our present knowledge about rhythms in signal transduction therefore relies mostly on basic research. However, there is evidence from clinical studies that biological rhythms are relevant in the treatment of human diseases. In this chapter we have attempted to illustrate general principles by examining a number of specific examples and by discussing the experimental data with regard to their clinical relevance whenever possible.

5.2 Rhythms in receptors and their signalling pathways

5.2.1 β-Adrenoreceptors in the heart

Membrane-bound receptors transmit signals from the cell surface to intracellular effector molecules. Signal transduction by cardiac β-adrenoreceptors has been characterised in detail, including the molecular mechanisms of receptor regulation, their coupling to different G proteins and effectors, the involvement of different β-adrenoreceptor subtypes, and the changes occurring in different pathological states, e.g. in cardiac hypertrophy and failure (Figure 5.1). However, only a limited number

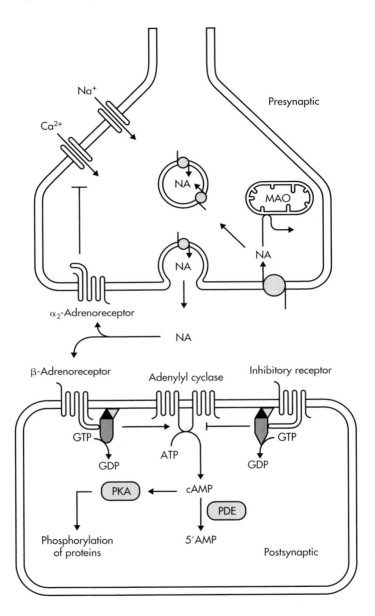

Figure 5.1 Pre- and postsynaptic components of the adrenergic signal transduction pathway. The transmitter (NA) is stored in vesicles; excitation of the presynaptic neuron leads to Ca^{2+} influx, fusion of vesicles with the presynaptic membrane, and release of the transmitter into the synaptic cleft. Transmitter release can be inhibited by presynaptic autoreceptors, e.g. α_2-adrenoreceptors. Circadian rhythms in autoreceptor function have been found in serotoninergic neurons (see Section 5.2.3).

of studies have addressed the question whether circadian rhythms in cardiac function could be due to changes in the density, affinity and function of β-adrenoreceptors, mediating the effects of the transmitter of the sympathetic nervous system, i.e. noradrenaline (norepinephrine). It has been shown in normotensive rats that the plasma concentration of noradrenaline varies considerably over 24 h, with peak values occurring at night (McCarty et al., 1981; Schiffer et al., 2001), i.e. in the activity period of the rat. Taking into account that humans are a diurnally active species, the experimental findings in rats are in line with observations in human subjects, showing significant circadian rhythms in plasma catecholamines with higher values during the day (Thurton and Deegan, 1974; Prinz et al., 1979). With regard to the regulation of blood pressure and heart rate, the tissue content and turnover of catecholamines is of greater relevance than the transmitter concentration in plasma. It was therefore logical to measure catecholamines in heart tissue at different circadian times. Surprisingly, noradrenaline content did not show significant daily variation in the hearts of normotensive Wistar rats (Lemmer and Saller, 1973, 1974). However, since most of the noradrenaline measured in tissue homogenates is accumulated in the storage granules and only a small functional pool of the transmitter is required for neuronal function, tissue concentrations of noradrenaline may not fully reflect its functional role. Moreover, tyrosine, the substrate for biosynthesis of noradrenaline and the enzymes involved (tyrosine hydroxylase, dopa decarboxylase and dopamine β-hydroxylase) are also present in the terminal nerve endings. Endogenous levels of a biogenic amine are, therefore, just the resultants of biosynthesis, storage and release. A lack of daily variation in the actual concentration of a transmitter does not exclude rhythmicity at different steps of the transmitter system and vice versa. When the turnover of noradrenaline was measured by non-radioactive methods (inhibition of its biosynthesis) as well as radioactive

---

Figure 5.1 continued
Binding of the transmitter to specific receptors on the postsynaptic membrane, e.g. β-adrenoreceptors, leads to the activation of stimulatory G proteins and, subsequently, to an increase in the formation of the second messenger cAMP. Circadian rhythms in receptor density and function have been found in cardiac tissue (see Section 5.2.1) and CNS tissue (see Section 5.2.2). The cellular consequences of an increase in cAMP are mediated by activation of protein kinases, e.g. protein kinase A (PKA), and subsequent phosphorylation of several target proteins. Finally, cAMP is degraded by phosphodiesterases (PDE), and the activity of PDE has also been found to vary with the time of day (see Section 5.2.1). NA, noradrenaline; MAO, monoamine oxidase.

methods (tracer doses of [^{3}H]noradrenaline) it turned out that the turnover of noradrenaline, being a more appropriate measure of sympathetic activity, was significantly greater in the rat's nightly activity phase than during the day (Lemmer and Saller, 1973, 1974).

Thus, there is evidence that the local concentration of the endogenous agonist at β-adrenoreceptors (i.e. noradrenaline) varies with the time of day. Consequently, competitive antagonists at receptors activated by noradrenaline can be expected to show circadian time-dependency in their efficacy. Indeed, it has been shown in healthy human volunteers that acute application of the β-adrenoreceptor antagonist propranolol led to greater reductions in heart rate in the morning than at other times of the day (Langner and Lemmer, 1988), reflecting the increase in sympathetic tone during the early morning.

Because noradrenaline exerts its cardiac effects by stimulation of receptors, a change in receptor density or function could counteract or further enhance the rhythm in agonist concentration. Because of the necessity for repeated tissue sampling, such an investigation cannot be carried out in human subjects. In rat heart tissue, Lemmer et al. (1987) observed signifcant rhythms in the basal activity of adenylyl cyclase (the enzyme mediating the cardiac effects of β-receptor activation), in the concentration of the second messenger cAMP, and in the activity of the cAMP-degrading enzyme phosphodiesterase. While cAMP concentration and phosphodiesterase activity showed 24 h rhythms, the formation of cAMP by adenylyl cyclase followed a 12 h pattern with peaks shortly after the times of light on and light off (Lemmer et al., 1987). In normotensive and spontaneously hypertensive rats, significant rhythmicity has been reported in the density of cardiac β-adrenoreceptors, with two peaks within 24 h, one in the middle of the dark span, the other in the middle of the day (Witte et al., 1995a) (Figure 5.2). Interestingly, the rhythmic pattern in β-adrenoceptors was exclusively due to changes in the $β_1$-subtype, while the density of the $β_2$-adrenoreceptor remained constant throughout the 24 h study. Surprisingly, the functional efficacy of β-adrenoreceptors – the β-adrenoreceptor stimulation of cAMP formation – showed an opposite rhythmic pattern, with troughs in the middle of the day and night spans (Witte et al., 1995a) and peaks at the times of the light–dark transition, confirming the earlier observation in normotensive Wistar rats (Lemmer et al., 1987). This finding led us to conclude that rhythms in β-adrenoreceptor receptor density are a consequence rather than the cause of changes in β-adrenoreceptor signalling. In order to characterise the mechanims responsible for the circadian time-dependent variation in stimulated cAMP formation, we additionally studied the activity of the cAMP-synthesising enzyme adenylyl cyclase after uncoupling of the G proteins by manganese ions. This experiment demonstrated that the rhythmicity in basal and stimulated cAMP formation was completely lost after uncoupling of G proteins (Witte *et al.*, 1995a), suggesting that rhythmic changes in the expression of the different G proteins and/or in their coupling efficiency to the adenylyl cyclase play a major role in the rhythmic cAMP formation. In the light of these findings, one might ask about the clinical relevance of such a rhythm in signal transduction. In fact, there is no definitive evidence that comparable changes occur in humans. However, it has been reported that circadian rhythms in blood pressure and heart rate are blunted in patients suffering from congestive heart failure (Caruana *et al.*, 1988; Giles *et al.*, 1996). This observation is of particular interest because several studies have shown a reduced efficacy of cardiac β-adrenoreceptors in failing myocardium, accompanied by an increase in the inhibitory G protein. One is tempted to speculate that the cellular changes in G protein content and β-adrenoreceptor signalling observed in heart failure could result in a loss of rhythmicity in signal transduction and, thereby, be

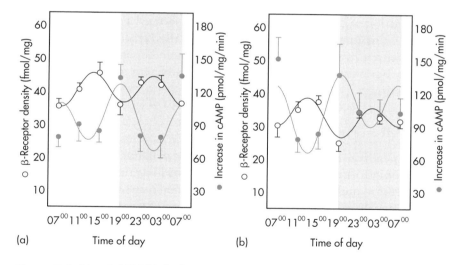

Figure 5.2 Bimodal (12 h) rhythms in β-adrenergic receptor density (open circles) and cAMP formation in response to β-adrenergic stimulation (closed circles) in cardiac tissue of (a) normotensive Wistar-Kyoto rats and (b) spontaneously hypertensive rats. In both strains of rats, rhythms in receptor density were opposite to those in receptor function. Peaks in receptor-mediated cAMP formation were found at the beginning and the end of the dark period (shaded area). Means values ± SEM, n = 42 rats per strain. Data from Lemmer *et al.* (1987); Witte *et al.* (1995a).

linked to the blunted circadian rhythms in blood pressure and heart rate. This admittedly speculative hypothesis is supported by a study in rats with experimental heart failure, showing that the severity of heart failure is correlated with desensitisation of β_1-adrenoreceptors and with the loss of circadian blood pressure and heart rate rhythms (Witte *et al.*, 2000).

5.2.2 β-Adrenoreceptors in the CNS

Circadian rhythms in receptor function have also been investigated in the central nervous system. Early studies focused on β-adrenoreceptors and their signalling in different regions of the rat brain. In whole forebrain homogenates the formation, concentration and degradation of cAMP was found to exhibit biological time-dependent variation, with either 24 h (cAMP content and degradation) or 12 h rhythms (cAMP formation), while β-adrenoreceptor density and affinity were constant throughout the 24 h study (Lemmer *et al.*, 1987). In a detailed study involving 15 different brain regions, circadian time dependency in β-adrenoreceptor binding was found in hypothalamus, cerebellum and different parts of the cortex, but was absent in hippocampus, thalamus, medulla and some cortical areas (Kafka *et al.*, 1986a). Interestingly, the concentration of cAMP in these brain regions showed a different temporal pattern from that of β-adrenoreceptors (Kafka *et al.*, 1986b), giving further support to the hypothesis that time-dependent changes in receptor binding are not ultimately responsible for rhythms in signal transduction.

5.2.3 5-HT-receptor subtypes in the CNS

Time-dependent variation has been demonstrated in the functional activity of serotonin (5-hydroxytryptamine, 5-HT) 5-HT_{1B} autoreceptors in the rat anterior hypothalamus (the region containing neurons of the suprachiasmatic nucleus), which represent the endogenous pacemaker (the clock) in mammals (Garabette *et al.*, 2000). The authors observed that stimulation of 5-HT_{1B} autoreceptors was more effective in the early light span than at the end of the day or at midnight. The differences in autoreceptor function could not be attributed to time-dependent changes in the expression of the receptor itself because the corresponding mRNA did not vary with the circadian time (Redfern *et al.*, 1999). Thus, the studies on 5-HT receptors confirm the observation that rhythms in receptor function are not necessarily linked to corresponding rhythms in receptor expression and density. The studies on 5-HT receptor function could be of clinical relevance for two reasons: (1) serotonin is known to modulate the SCN's response to photic input, and rhythmic changes in 5-HT autoreceptors may contribute to the entrainment of the endogenous clock; (2) it is known that major depression, a phasically occurring affective disorder, is frequently associated with a disturbed circadian regulation and can be treated with inhibitors of serotonin reuptake. In this regard it is of interest to note that the synthesis and release of serotonin as well as its neuronal reuptake exhibit significant circadian rhythms (for review see Martin and Redfern, 1997). Moreover, prolonged treatment of rats with the antidepressant drugs desipramine (an inhibitor of noradrenaline reuptake) and paroxetine (an inhibitor of serotonin reuptake) led to attenuation of 5-HT autoreceptor function, and this effect was biological time-dependent (Sayer *et al.*, 1999).

5.3 Rhythms in enzyme systems

5.3.1 The renin–angiotensin system

The renin–angiotensin system plays an important role in the regulation of blood pressure and volume homeostasis. The formation of the active agonist angiotensin II depends on the activity of two proteolytic enzymes: renin cleaves angiotensinogen to form angiotensin I, and the angiotensin-converting enzyme (ACE) leads to conversion of the inactive angiotensin I into angiotensin II. Rhythms in the activity of these proteases could be expected to exert an influence on the formation of angiotensin II and, hence, on the effects of inhibitors of the renin–angiotensin system. In the rat, plasma renin activity was found to show circadian variation with highest values during the day, i.e. the resting span of the rat, and a trough at midnight (Gomez-Sanchez *et al.*, 1976; Hilfenhaus 1976; Freeman *et al.*, 1982). In our own experiments in normotensive rats, we did not detect a statistically significant rhythm in plasma renin activity but, again, values tended to be higher in the resting phase (Lemmer *et al.*, 2000). The experimental data in rats are in line with several clinical studies showing that human plasma renin activity is lowest in the afternoon and increases during the nocturnal rest period (Gordon *et al.*, 1966; Portaluppi *et al.*, 1990; Brandenberger *et al.*, 1994). Thus, at the level of the first enzyme involved, there is evidence that the renin–angiotensin system becomes more active during the rest period of rats and humans and, owing to its antidiuretic action, may contribute to the decrease in urine production in the inactive period. The second enzyme, ACE, was found to show less pronounced variation over 24 h, both in rats (Stepien *et al.*, 1993; Lemmer *et al.*, 2000) and in humans (Veglio *et al.*, 1987; Witte *et al.*, 1993). Thus, biological time-dependent variation in the tone of the renin–angiotensin system is mainly due to changes in the activity of the key enzyme renin.

On the basis of the rhythm in plasma renin activity, one could expect that inhibitors of the renin–angiotensin system would be more effective during the rest period. Indeed, transgenic hypertensive TGR(mREN2)27 rats, carrying an additional mouse renin gene, showed significantly greater effects of an ACE inhibitor and an angiotensin II AT_1-receptor antagonist in the daily rest period (Schnecko *et al.*, 1995), i.e. at the time when both blood pressure (Lemmer *et al.*, 1993) and plasma renin activity (Lemmer *et al.*, 2000) are highest in this animal model of secondary hypertension. In primary hypertensive subjects, evening dosing of ACE inhibitors was associated with a marked reduction in night-time blood pressure and led to a disturbed circadian blood pressure pattern (Palatini *et al.*, 1992; Witte *et al.*, 1993). Similar findings have been reported for the effects of an angiotensin II AT_1-receptor antagonist (Pechère-Bertschi *et al.*, 1998). In summary, rhythms in plasma renin activity, leading to circadian variation in the formation of angiotensin peptides, have been documented in experimental animals as well as in human subjects, and are of clinical relevance to the effects of inhibitors of the renin–angiotensin system.

5.3.2 The nitric oxide–cyclic GMP system

Another enzymatic system that has been studied from a chronobiological point of view is the nitric oxide (NO)–soluble guanylyl cyclase pathway. Endothelium-derived NO is an important regulator of vascular tone, and a loss of endogenous NO synthesis in hypertension and diabetes has been implicated in the occurrence of cardiovascular complications associated with these diseases. In the endothelial cell, NO is synthesised by the enzyme NO synthase type III (NOS III), also called eNOS. Once synthesised, NO leaves the endothelial cell by passive diffusion, enters adjacent vascular smooth-muscle cells and, within these cells, activates the cGMP-generating enzyme soluble guanylyl cyclase (sGC). Rhythms in the activity of the NO-cGMP pathway could be due to changes in the synthesis of the first messenger, NO, or in the sensitivity of the downstream target, sGC. Because NO is a highly unstable molecule, it is difficult to measure its concentration directly in body fluids. Therefore, the concentration of the stable end products of NO oxidation, i.e. nitrite and nitrate (NOx), is usually taken as a measure of NO synthesis. Experimental studies in normotensive rats have shown that the urinary excretion of NOx is higher during the night than in the daytime (Borgonio et al., 1999; Globig et al., 1999). Interestingly, in rats, ageing was accompanied by a reduction in overall NOx excretion and by a blunted circadian rhythm (Borgonio et al., 1999). Young, spontaneously hypertensive rats showed a decrease in the 24 h excretion of NOx and its circadian amplitude, resembling the observations in aged normotensive rats (Borgonio et al., 1999). Almost identical observations have been made in a clinical study in normotensive and hypertensive subjects. In healthy, normotensive subjects, urinary excretion of NOx and its second messenger cGMP showed pronounced 24 h rhythmicity, with peak values at the end of the day and a trough in the second half of the night; the rhythmic pattern in NOx and cGMP excretion was lost in hypertensive patients (Bode-Böger et al., 2000).

Further evidence for rhythmicity in endothelial function comes from a recent study showing that the vasodilation of human forearm vasculature in response to acetylcholine, mediated by an increase in NO synthesis, varied with the circadian time: in agreement with the urinary excretion data, the smallest changes in blood flow were found at the end of the night (Elherik et al., 2000). Because the physiological rhythm in NOx excretion parallels the rhythm in blood pressure, in both rats and humans, one may assume that the increase in stress in the vessel wall associated with high blood pressure leads to activation of the vasodilating NO–cGMP pathway. This counterregulatory pathway seems to be disturbed in chronic hypertension, and its dysfunction may contribute to the loss of the nocturnal blood pressure dip frequently observed in chronic hypertensive patients.

The second step of the NO–cGMP pathway, the soluble guanylyl cyclase, has been studied in vascular tissue of normotensive Wistar-Kyoto rats. Both the basal rate of cGMP formation and the effect of exogenous NO were found to be greater in the rest period than at night and thus showed an opposite pattern to that of NOx excretion (Witte et al., 1995b). It is known that activation of NO synthesis can lead to desensitisation of soluble guanylyl cyclase (Filippov et al., 1997). Therefore, it is likely that the time-dependent pattern in guanylyl cyclase activity is due to agonist-induced desensitisation of the enzyme as a consequence of the rhythm in endogenous NO synthesis. In the light of the nocturnal increase in endogenous NO synthesis, it may seem surprising that the effects of the NOS-inhibitor L-NAME (N^G-nitro-L-arginine methyl ester) on rat blood pressure *in vivo* were greater during the day than at night (Witte et al., 1995b). However, it should be taken into account that

L-NAME acts as a competitive NOS inhibitor, competing with the enzyme's substrate L-arginine. Because the intake of food and, hence, of L-arginine occurs predominantly in the rat's nocturnal activity period, one may speculate that the effects of the competitive NOS inhibitor were counteracted by increased supply of the substrate L-arginine. Alternatively, one could assume that the rhythm in the sensitivity of the downstream target, vascular sGC, was of greater importance for the effects of the NOS inhibitor than the activity of NO synthase itself.

The experimental findings on rhythmic changes in the NO–cGMP pathway raise the question whether such changes could be of relevance in the treatment of patients receiving NO donors such as organic nitrates. Unfortunately, only a few studies have addressed this issue. A study in patients with coronary artery disease (CAD) demonstrated that the exercise-induced depression of the ECG ST segment, a marker of myocardial ischaemia, was greater in the evening than in the morning, and this pattern was preserved after nitrate treatment (Wortmann and Bachmann, 1991). In patients with Prinzmetal angina, characterised by coronary vasospasms in the absence of atherosclerotic lesions, acute application of glyceryl trinitrate (nitroglycerin) led to a greater increase in the diameter of coronary arteries when the drug was given in the morning than when it was given in the afternoon (Yasue *et al.*, 1979). Taken together with the observations that the urinary excretion of NOx drops at night (Bode-Böger *et al.*, 2000) and the endothelium-dependent vasodilation by acetylcholine is smallest in the morning (Elherik *et al.*, 2000), there is indeed some evidence that the clinical effects of NO donors could depend on the circadian time. Prospective studies will be necessary to show whether treatment of CAD patients with organic nitrates or other NO donors is able to blunt the early morning rise in the incidence of myocardial infarction (Willich *et al.*, 1989) and ischaemic episodes (Hausmann *et al.*, 1987).

5.3.3 The synthesis of cholesterol

The synthesis of cholesterol is the third enzymatic pathway presented in this chapter. The key enzyme of cholesterol synthesis is the hydroxymethylglutaryl (HMG)-CoA reductase, which is found mainly in the liver and the intestinal mucosa. In nocturnally active rats, the activity of HMG-CoA reductase shows 24 h rhythmicity, with highest values during the night (Edwards *et al.*, 1972; Szanto *et al.*, 1994). Surprisingly, a similar pattern was found in diurnal human subjects (Jones *et al.*, 1992; Cella *et al.*, 1995a), suggesting that the mechanisms of circadian regulation of cholesterol synthesis are different in the two species. There is evidence that the rhythm in cholesterol synthesis is closely related to the timing of food intake, since a shift in the timing of meals was followed by a shift in the rhythm of cholesterol synthesis in rats (Edwards et al., 1972) and in human subjects (Cella et al., 1995b). Thus, rhythms in cholesterol synthesis are influenced by food intake in both species. However, the intake of food seems to exert a stimulatory effect in nocturnally active rats but an inhibitory effect in diurnally active humans.

In the treatment of patients with elevated serum cholesterol, inhibitors of HMG-CoA reductase are now drugs of first choice. On the basis of the rhythmicity in human cholesterol synthesis, it is recommended that the lipid-lowering drugs should be given in the evening rather than in the morning. Indeed, clinical studies demonstrated that the effects of lovastatin (Illingworth, 1986) and simvastatin (Saito et al., 1991) on total cholesterol and LDL cholesterol were greater after evening dosing compared to morning dosing. Prospective clinical studies are needed to show whether evening compared to morning dosing of HMG-CoA reductase inhibitors is associated with fewer side effects and/or a better prognosis with regard to adverse cardiovascular events.

5.4 Rhythms in ion channel activity

Taking into account that rhythmic variation in the conductance of voltage-sensitive ion channels on cardiac myocytes is responsible for the spontaneous generation of the heart rate rhythm, it is plausible to assume that changes in the function of ion channels may also contribute to the generation of circadian rhythms. However, at present there is only limited information on circadian changes in the function and expression of ion channels. The most detailed study has been performed in chicken pineal cells, a cellular model of a vertebrate circadian system, containing within one cell a photoreceptive pathway, a circadian oscillator and an output pathway regulating the synthesis of melatonin. In this model system, D'Souza and Dryer (1996) detected a new type of cation channel – the so-called I_{LOT} cationic channel – which, under synchronised conditions, was active exclusively in the dark span. In the absence of a zeitgeber, i.e. in constant darkness, active I_{LOT} channels were still found predominantly during subjective night, suggesting that the function of these channels was under control of the endogenous pacemaker (D'Souza and Dryer, 1996). Rhythms in the activity of I_{LOT} channels could be involved in the circadian regulation of melatonin synthesis, which is stimulated by a nocturnal increase in intracellular cAMP and Ca^{2+}. Nocturnal opening of the I_{LOT} channels would result in membrane depolarisation, allowing Ca^{2+} influx through voltage-sensitive Ca^{2+} channels and, finally, in an increase in adenylyl cyclase activity (Takahashi, 1996).

In the mammalian circadian pacemaker, i.e. the suprachiasmatic nuclei (SCN), the intracellular Ca^{2+} concentration was shown to be higher during daytime or subjective daytime than at night, under both synchronised and free-running conditions (Colwell, 2000). Interestingly, this rhythm was abolished when the SCN slices were exposed to tetrodotoxin, a blocker of voltage-sensitive Na^+ channels, suggesting that, similar to the findings in chick pineal cells, rhythmic membrane depolarisation plays a role in the circadian regulation of intracellular events.

In the light of these experimental data, one could expect that rhythms in ion channel function have profound implications for circadian regulation in humans, and could influence the pharmacodynamics of drugs acting at Na^+ and Ca^{2+} channels. Significant 24 h variation has been found in the occurrence of late afterdepolarisations in the hearts of patients after myocardial infarction, with an increased incidence of these depolarisations in the early morning hours (Steinbigler *et al.*, 1999). Because such afterdepolarisations are due to an influx of Na^+ through a nonselective cation channel, one may speculate that rhythmic changes in the function of this channel are responsible for the incidence pattern over 24 h. Unfortunately, no detailed data are available regarding the circadian time-dependent effectiveness of antiarrhythmic drugs on different types of cardiac arrhythmias.

5.5 Summary and conclusions

The examples given in this chapter clearly demonstrate that circadian rhythms exist in several components of signal transduction cascades such as the β-adrenoreceptor pathway, the renin–angiotensin system, and the nitric oxide–cGMP signalling pathway. There is evidence that the rhythms observed on the cellular and molecular level could have clinical implications and may contribute to circadian time-dependent variation in the efficacy of drugs. Future studies will be required to show whether specific timing of drug intake is associated with a better outcome or fewer side effects than drug dosing at nonspecified times of day.

Acknowledgement

Our experimental studies were greatly supported by grants of the Deutsche Forschungsgemeinschaft (Le 318/1–Le 318/10–3).

References

Bode-Böger S M, Böger R H, Kielstein J T, *et al.* (2000). Role of endogenous nitric oxide in circadian blood pressure regulation in healthy humans and in patients with hypertension or atherosclerosis. *J Invest Med* 48: 125–132.

Borgonio A, Witte K, Stahrenberg R, *et al.* (1999). Influence of circadian time, ageing, and hypertension on the urinary excretion of nitric oxide metabolites in rats. *Mech Ageing Dev* 111: 23–37.

Brandenberger G, Follenius M, Goichot B, *et al.* (1994). Twenty-four-hour profiles of plasma renin activity in relation to the sleep–wake cycle. *J Hypertens* 12: 277–283.

Caruana M P, Lahiri A, Cashman P M M, *et al.* (1988). Effects of chronic congestive heart failure secondary to coronary artery disease on the circadian rhythm of blood pressure and heart rate. *Am J Cardiol* 62: 755–759.

Cella L K, Van Cauter E, Schoeller D A (1995a). Diurnal rhythmicity of human cholesterol synthesis: normal pattern and adaptation to simulated 'jet lag'. *Am J Physiol* 269: E489–E498.

Cella L K, Van Cauter E, Schoeller D A (1995b). Effect of meal timing on diurnal rhythm of human cholesterol synthesis. *Am J Physiol* 269: E878–E883.

Colwell C S (2000). Circadian modulation of calcium levels in cells in the suprachiasmatic nucleus. *Eur J Neurosci* 12: 571–576.

D'Souza T, Dryer S E (1996). A cationic channel regulated by a vertebrate intrinsic circadian oscillator. *Nature* 382: 165–167.

Edwards P A, Muroya H, Gould R G (1972). *In vivo* demonstration of the circadian rhythm of cholesterol biosynthesis in the liver and intestine of the rat. *J Lipid Res* 13: 396–401.

Elherik K, Khan F, McClaren M, *et al.* (2000). Circadian variation in endothelial cell function and vascular tone in healthy adults. *Circulation* 102(Suppl II): II-186 (abstract).

Filippov G, Bloch D B, Bloch K D (1997). Nitric oxide decreases stability of mRNAs encoding soluble guanylate cyclase subunits in rat pulmonary artery smooth muscle cells. *J Clin Invest* 100: 942–948.

Freeman R H, Davis J O, Williams G M, *et al.* (1982). Circadian changes in plasma renin activity and plasma aldosterone concentration in one-kidney hypertensive rats. *Proc Soc Exp Biol Med* 169: 86–89.

Garabette M L, Martin K F, Redfern P H (2000). Circadian variation in the activity of the 5-HT$_{1B}$ autoreceptor in the region of the suprachiasmatic nucleus, measured by microdialysis in the conscious freely-moving rat. *Br J Pharmacol* 131: 1569–1576.

Giles T D, Roffidal L, Quiroz A, *et al.* (1996). Circadian variation in blood pressure and heart rate in nonhypertensive congestive heart failure. *J Cardiovasc Pharmacol* 28: 733–740.

Globig S, Witte K, Lemmer B (1999). Urinary excretion of nitric oxide, cyclic GMP and catecholamines during rest and activity period in transgenic hypertensive rats. *Chronobiol Int* 16: 305–314.

Gomez-Sanchez C, Holland O B, Higgins J R, *et al.* (1976). Circadian rhythms of serum renin activity and serum corticosterone, prolactin, and aldosterone concentrations in the male rat on normal and low-sodium diets. *Endocrinology* 99: 567–572.

Gordon R D, Wolfe L K, Island D P, *et al.* (1966). A diurnal rhythm in plasma renin activity in man. *J Clin Invest* 45: 1587–1592.

Hausmann D, Nikutta P, Hartwig C-A, *et al.* (1987). ST-Strecken-Analyse im 24-h-Langzeit-EKG bei Patienten mit stabiler Angina pectoris und angiographisch nachgewiesener Koronarsklerose. *Z Kardiol* 76: 554–562.

Hilfenhaus M (1976). Circadian rhythm of plasma renin activity, plasma aldosterone and plasma corticosterone in rats. *Int J Chronobiol* 3: 213–229.

Illingworth D R (1986). Comparative efficacy of once versus twice daily mevinolin in the therapy of familial hypercholesterolemia. *Clin Pharmacol Ther* 40: 338–343.

Jones P J H, Pappu A S, Illingworth D R, *et al.* (1992). Correspondence between plasma mevalonic acid levels and deuterium uptake in measuring human cholesterol synthesis. *Eur J Clin Invest* 22: 609–613.

Kafka M S, Benedito M A, Blendy J A, *et al.* (1986a). Circadian rhythm in neurotransmitter receptors in discrete rat brain regions. *Chronobiol Int* 3: 91–100.

Kafka M S, Benedito M A, Roth R H, *et al.* (1986b). Circadian rhythms in catecholamine metabolites and cyclic nucleotide production. *Chronobiol Int* 3: 101–115.

Langner B, Lemmer B (1988). Circadian changes in the pharmacokinetics and cardiovascular effects of oral propranolol in healthy subjects. *Eur J Clin Pharmacol* 33: 619–624.

Lemmer B, Saller R (1973). Difference in the turnover of noradrenaline in rat heart during day and night. *Naunyn Schmiedebergs Arch Pharmacol* 278: 107–109.

Lemmer B, Saller R (1974). Influence of light and darkness on the turnover of noradrenaline in the rat heart. *Naunyn Schmiedebergs Arch Pharmacol* 282: 75–84.

Lemmer B, Lang P H, Schmidt S, *et al.* (1987). Evidence for circadian rhythmicity of the β-adrenoceptor–adenylate cyclase–cAMP-phosphodiesterase system in the rat. *J Cardiovasc Pharmacol* 10(Suppl 4): S138–S140.

Lemmer B, Mattes A, Böhm M, *et al.* (1993). Circadian blood pressure variation in transgenic hypertensive rats. *Hypertension* 22: 97–101.

Lemmer B, Witte K, Schänzer A, *et al.* (2000). Circadian rhythms in the renin–angiotensin-system and adrenal steroids may contribute to the inverse blood pressure rhythm in hypertensive TGR(mREN-2)27 rats. *Chronobiol Int* 17: 645–658.

Martin K F, Redfern P H (1997). 5-Hydroxytryptamine and noradrenaline synthesis, release and metabolism in the central nervous system: circadian rhythms and control mechanisms. In: Redfern P H, Lemmer B, eds. *Physiology and Pharmacology of Biological Rhythms*. Berlin: Springer-Verlag, 157–176.

McCarty R, Kvetnansky R, Kopin I J (1981). Plasma catecholamines in rats: daily variations in basal levels and increments in response to stress. *Physiol Behav* 26: 27–31.

Palatini P, Racioppa A, Raule G, et al. (1992). Effect of timing of administration on the plasma ACE inhibitory activity and the antihypertensive effect of quinapril. *Clin Pharmacol Ther* 52: 378–383.

Pechère-Bertschi A, Nussberger J, Decosterd L, et al. (1998) Renal response to the angiotensin II receptor subtype 1 antagonist irbesartan versus enalapril in hypertensive patients. *J Hypertens* 16: 385–393.

Portaluppi F, Bagni B, degli Umberti E, et al. (1990). Circadian rhythms of atrial natriuretic peptide, renin, aldosterone, cortisol, blood pressure and heart rate in normal and hypertensive subjects. *J Hypertens* 8: 85–95.

Prinz P N, Halter J, Benedetti C, et al. (1979). Circadian variation of plasma catecholamines in young and old men: relation to rapid eye movement and slow wave sleep. *J Clin Endocrinol Metab* 49: 300–304.

Redfern P H, Garabette M, Martin K F (1999). 5-HT$_{1B}$ receptor mRNA in the rat SCN does not display diurnal variation in expression as measured by RT-PCR. *Chronobiol Int* 16(Suppl 1): 88 (abstract).

Saito Y, Yoshida S, Nakaya N, et al. (1991). Comparison between morning and evening doses of simvastatin in hyperlipidemic subjects. *Arterioscler Thromb* 11: 816–826.

Sayer T J O, Hannon S D, Redfern P H, et al. (1999). Diurnal variation in 5-HT$_{1B}$ autoreceptor function in the anterior hypothalamus *in vivo*: effect of chronic antidepressant drug treatment. *Br J Pharmacol* 126: 1777–1784.

Schiffer S, Pummer S, Witte K, et al. (2001). Cardiovascular regulation in TGR(mREN2)27 rats: 24 h variation in plasma catecholamines, angiotensin peptides, and in telemetric heart rate variability. *Chronobiol Int* 18: 461–474.

Schnecko A, Witte K, Lemmer B (1995). Effects of the angiotensin II receptor antagonist losartan on 24-hour blood pressure profiles of primary and secondary hypertensive rats. *J Cardiovasc Pharmacol* 26: 214–221.

Steinbigler P, Haberl R, Jilge G, et al. (1999). Circadian variability of late potential analysis in holter electrocardiograms. *Pace* 22: 1448–1456.

Stepien M, Witte K, Lemmer B (1993). Chronobiologic evaluation of converting enzyme activity in serum and lung tissue from normotensive and spontaneously hypertensive rats. *Chronobiol Int* 10: 331–337.

Szanto A, Ruys J, Balasubramaniam S (1994). Coordinate diurnal regulation of hepatic acyl-coenzyme A: cholesterol acyltransferase and cellular levels of esterified cholesterol. *Biochim Biophys Acta* 1214: 39–42.

Takahashi J S (1996). Ion channels get the message. *Nature* 382: 117–118.

Thurton M B, Deegan T (1974). Circadian variations of plasma catecholamine, cortisol and immunoreactive insulin concentrations in supine subjects. *Clin Chim Acta* 55: 389–397.

Veglio F, Pietrandrea R, Ossola M, et al. (1987). Circadian rhythm of the angiotensin converting enzyme (ACE) activity in serum of healthy adult subjects. *Chronobiologia* 14: 21–25.

Willich S N, Linderer T, Wegscheider K, et al. (1989). Increased morning incidence of myocardial infarction in the ISAM study: absence with prior beta-adrenergic blockade. *Circulation* 80: 853–858.

Witte K, Weisser K, Neubeck M, et al. (1993). Cardiovascular effects, pharmacokinetics and converting enzyme inhibition of enalapril after morning versus evening administration. *Clin Pharmacol Ther* 54: 177–186.

Witte K, Parsa-Parsi R, Vobig M, et al. (1995a). Mechanisms of the circadian regulation of β-adrenoceptor density and adenylyl cyclase activity in cardiac tissue from normotensive and spontaneously hypertensive rats. *J Mol Cell Cardiol* 27: 1195–1202.

Witte K, Schnecko A, Zuther P, et al. (1995b). Contribution of the nitric oxide–guanylyl cyclase system to circadian regulation of blood pressure in normotensive Wistar-Kyoto rats. *Cardiovasc Res* 30: 682–688.

Witte K, Hu K, Swiatek J, et al. (2000). Experimental heart failure in rats: effects on cardiovascular circadian rhythms and on myocardial β-adrenergic signaling. *Cardiovasc Res* 47: 350–358.

Wortmann A, Bachmann K (1991). Chronotherapy in coronary heart disease: comparison of two nitrate treatments. *Chronobiol Int* 8: 339–408.

Yasue H, Omote S, Takizawa A, et al. (1979). Circadian variation of exercise capacity in patients with Prinzmetal's variant angina: role of exercise-induced coronary arterial spasm. *Circulation* 59: 938–948.

6

Rhythms and therapeutics of diseases of the gastrointestinal tract

John G Moore

6.1 Introduction

This chapter focuses on biological rhythms of gastrointestinal (GI) function insofar as they influence the pharmacokinetics, pharmacodynamic and toxicological behaviour of oral and, in some instances, parenterally administered medications. Rhythms around the circadian (~24 h) and ultradian (<22 h) time domains are emphasised; infradian rhythms (>28 h, to include monthly and yearly rhythms) are discussed only briefly. Human data are highlighted; animal data are included only if supportive of a human observation or in support of a particular theoretical view proposed by the author. In addition, rhythms of motor, secretory or absorptive function of the hollow-viscus GI organs will receive greater attention; those of the liver – important as this organ is in drug metabolism – and pancreas are beyond the scope of this chapter and are dealt with in more detail elsewhere (see Chapter 4 of this volume and also Bélanger and Labrecque, 1992; Mejean *et al*., 1992).

6.2 Gastrointestinal motility rhythms

Rhythmic gastrointestinal motor patterns influence the absorption and disposition of oral medication in several ways. Orally administered drugs must exit the stomach before absorption begins. The rate of transfer of drugs into the small bowel is thus dependent on the rate of gastric emptying, which in turn is influenced by the presence or absence of food, the physical composition (liquid or solid) of the meal, the nutrient content (lipid, carbohydrate or protein) of the meal and a host of other physiological factors (e.g. sex, posture, exercise) (Moore *et al*., 1990). In fed subjects, gastric motor activity is dominated by caudally directed peristaltic contractions that, in humans, occur at the frequency of three per minute. These muscular contractions are responsible for the grinding of digestible solid food particles to a size (less than ~1.5 mm) that permits passage from the gastric antrum into the duodenum. The antrum, or distal stomach, is the major anatomical site for this function. Emptying of liquid meals is also dependent on normal distal stomach peristalsis and, to a lesser extent, on a gastric duodenal pressure gradient generated by muscular forces in the proximal stomach.

As observed in Figure 6.1, liquid and solid phases of a mixed liquid–solid meal empty at different rates and with different patterns (Brophy *et al.*, 1986). Liquids typically empty more rapidly than digestible solids and, if ingested alone without added nutrients, obey first-order emptying kinetics. Digestible solid food emptying obeys zero-order

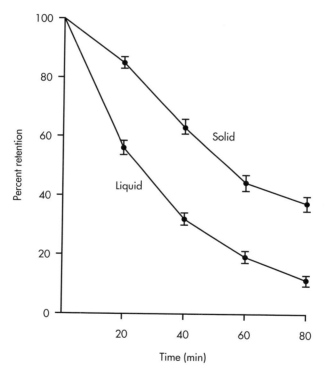

Figure 6.1 Gastric emptying of liquid and solid foods. Note the near linear and slower emptying of solids when compared to liquids. A 300 g combined solid (beef stew) and liquid (orange juice) meal, labelled with technetium-99m sulfur colloid (solid-phase marker) and indium-111 (liquid-phase marker), was ingested at 0 min. The percentage retention refers to the percentage of the radiolabelled marker remaining in the stomach over time (min). Each point represents mean ± SEM values for 32 studies in 8 male healthy subjects (4 studies each). Reproduced from Brophy *et al.* (1986).

kinetics. Indigestible solids, represented by food particles greater than ~1.5 mm after grinding and trituration, and therefore unable to pass through the pylorus, are emptied by another type of ultradian motor wave termed the migrating myoelectric complex (MMC), a particularly powerful 'housekeeping' muscular contraction that serves to sweep indigestible solids from the stomach into the small bowel. The MMC originates in the stomach and travels aborally to the ileocaecal valve. These waves occur at intervals of 90–120 min in the fasted human and are inhibited by feeding, as observed in Figure 6.2, which illustrates gastric

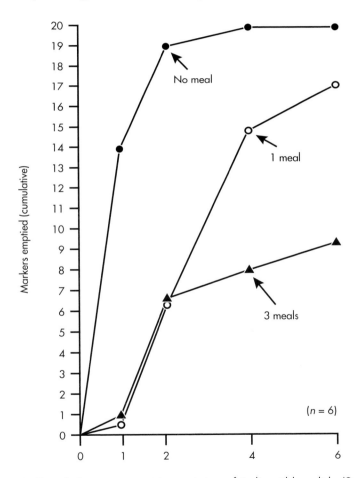

Figure 6.2 Cumulative mean gastric emptying of indigestible solids (2 mm and 10 mm inert markers) in 6 healthy subjects. Gastric emptying was most rapid under fasted conditions and was delayed in proportion to the frequency of feeding. Reproduced from Smith and Feldman (1986).

emptying rates for inert markers, representing indigestible solids, under fasted and fed conditions (Smith and Feldman, 1986). Feeding markedly inhibited emptying of the markers and in proportion to the frequency of feeding. Circadian rhythmic motor patterns superimposed on ultradian patterns are also characteristic of the healthy human GI tract. Kumar *et al.* (1986) demonstrated circadian rhythmicity in the speed of MMC propagation along the small bowel in groups of healthy and functional bowel syndrome patients intubated in the small bowel with tubes equipped with pressure-sensors. The daytime velocity (centimetres per minute) of MMC propagations was more than double the nocturnal value in both groups. Gastric emptying rates for meals also display circadian variation (Goo *et al.*, 1987). In a two-time-point study, gastric emptying rates for a meal administered at 20.00 were significantly slower than emptying rates for the same subjects at 08.00. These results are shown in Figure 4.4 in Chapter 4.

6.2.1 Pharmacological and therapeutic implications

The rate of gastric emptying, gastrointestinal transit and ultimate absorption of an oral medication is thus dependent on whether the medication is ingested as a liquid or a digestible or indigestible solid and, in turn, is strongly influenced by ultradian and circadian GI motor events. For example, circadian variation in gastric emptying results in delayed absorption of most oral medications when they are administered during the evening (Reinberg and Smolensky, 1982; Lemmer and Bruguerolle, 1994). The delay is reflected by lower maximum plasma concentrations (C_{max}) and longer times to peak drug plasma concentrations (t_{max}). With few exceptions, however, the extent of absorption (bioavailability) for most drugs, reflected by the area under the plasma concentration–time curve (AUC), does not differ with day or night administration.

However, slower evening-time absorption has therapeutic ramifications for several drugs. Slower evening-time gastric emptying and intestinal absorption rates may result in an increase in hepatic first-pass effect for high-extraction compounds such as propranolol and lead to lower bioavailability for proton pump-inhibiting antiulcer drugs that are subject to gastric acid degradation (Langner and Lemmer, 1988; Prichard *et al.*, 1985). Slower evening-time compared to morning absorption of theophylline may also be to the benefit of asthmatic patients, with increased nocturnal patterns of airway resistance (Reinberg *et al.*, 1987). Evening dosage may be more desirable for drugs with lower toxicity thresholds because acute side effects of drugs are correlated to initially high drug plasma concentrations as, for example, with isosorbide mononitrate administration. Orthostatic hypotension was more pronounced and t_{max} for the drug was significantly shorter after morning compared to evening dosage (Lemmer et al., 1991). Conversely, morning dosage may be more desirable for some drugs because of more rapid onset of action.

Drug absorption rates are also influenced by the degree to which drugs disintegrate and dissolve within the gastric lumen. Drugs that dissolve completely are, in the main, emptied as liquids, although with large combined liquid–solid meals liquid-phase and solid-phase-emptying become indistinguishable, particularly during the late meal-emptying phase (Moore et al., 1981). Drugs that dissolve completely and are entirely emptied in a non-nutrient liquid phase do not display the marked circadian variation in absorption rates shown when drugs are ingested with solids. Solid meal ingestion always delays drug absorption because of the inhibiting effect of nutrient solids on gastric emptying. This delay is exaggerated with drugs administered following or with evening meals because gastric emptying is slower in the evening compared to the morning hours (Goo et al., 1987). Gastric emptying and absorption of enteric-coated or matrix-type sustained-release medications are delayed even further by concurrent ingestion of a meal. As discussed above, food inhibits the gastric-originating MMC wave that sweeps indigestible solids, such as enteric-coated capsules or tablets, out of the stomach and into the small bowel where dissolution and absorption begin. In the absence of MMCs, gastric residence time for these drug formulations is prolonged, as can be seen for enteric-coated aspirin in Figure 6.3 (Bogentoft et al., 1978). It is thus possible that a morning-ingested enteric-coated or matrix-type sustained-release preparation will be neither emptied nor absorbed for the entire daytime hours in a frequently snacking patient. In addition, because the propagation velocity of MMCs is slower at night than in the morning, these drug formulations even when taken without meals may be expected to empty and be absorbed at slower rates at night.

6.2.2 Other rhythms influencing drug bioavailability

Ultradian and circadian motor rhythmic events thus alter the pharmacokinetic behaviour of many drug formulations. However, a number of confounding factors make predictions concerning the effects of such rhythms on drugs disposition uncertain. In the rodent model, for example, hepatic microsomal enzyme activity and blood flow, crucial in

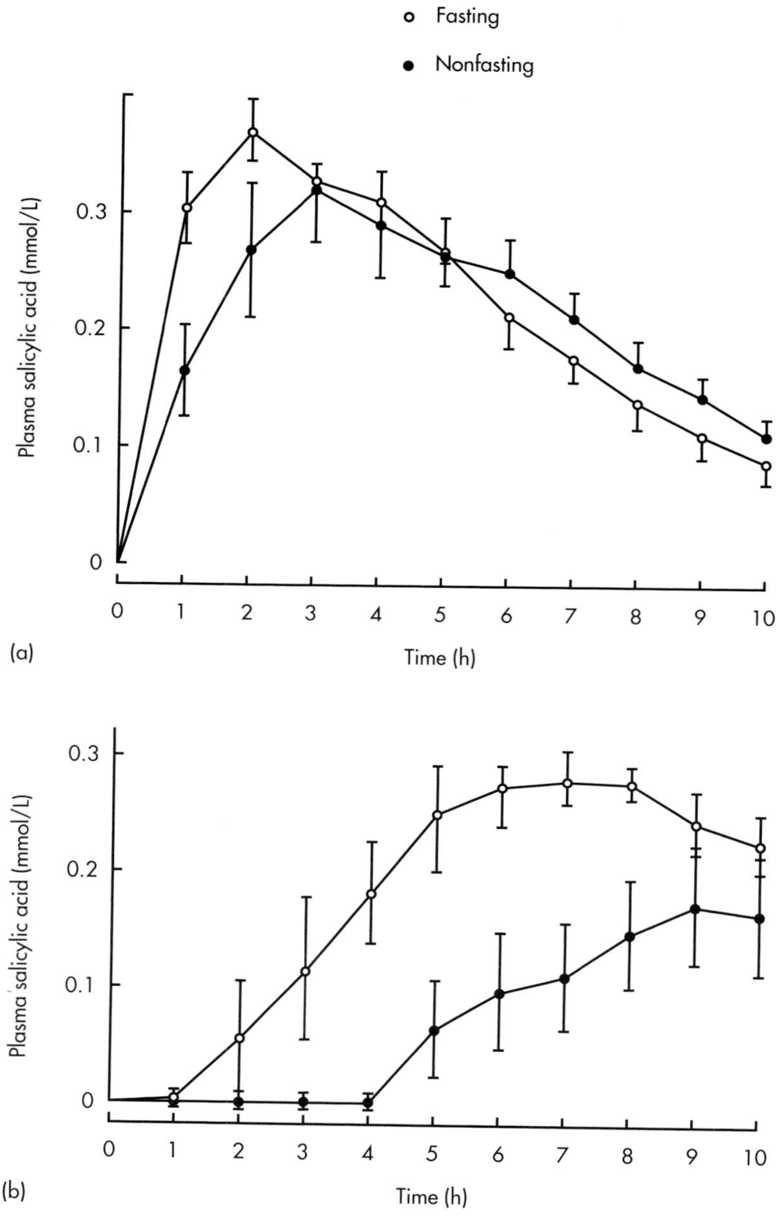

Figure 6.3 Mean ± SEM plasma concentrations of salicylic acid in 8 subjects after the administration of 1.0 g acetylsalicylic acid (aspirin; ASA) as (a) conventional or (b) enteric-coated tablets under fasting and nonfasting conditions. Reproduced from Bogentoft et al. (1978).

the metabolism and disposition of many drugs, have their own circadian rhythms (Chapter 4 of this volume; Bélanger and Labrecque, 1992). A circadian rhythm in hepatic blood flow has also been described in humans (Lemmer and Nold, 1991). In this study in supine healthy subjects, shown in Figure 6.4, estimated hepatic blood flow, as assessed by clearance of indocyanine green dye, was significantly higher at 08.00 compared to other times of the 24 h period. Assuming that hepatic and intestinal blood flow patterns run parallel, this finding could explain the higher C_{max} and/or shorter t_{max} values observed after morning compared to evening ingestion with some drugs just as readily as the difference in gastric emptying rates between the two timeframes. The liver is the major site of drug metabolism in humans, and for drugs with high extraction ratios metabolism and biotransformation depend principally

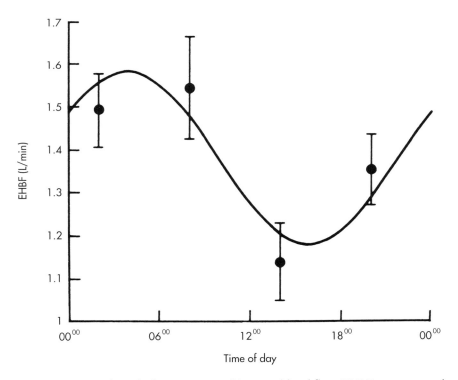

Figure 6.4 Circadian rhythm in estimated hepatic blood flow (EHBF) as measured by indocyanine green clearance in 10 healthy male subjects. Tested subjects were studied in the supine position with measurements at 02.00, 08.00, 14.00 and 20.00. Mean ± SEM values are shown. A significant ($p < 0.025$) circadian rhythm was determined by cosinor analysis with a rhythm-estimated peak time of 04.00. Reproduced from Lemmer and Nold (1991).

on the microsomal oxidising and cytosolic conjugating enzymatic activities within the hepatocyte (Bélanger and Labracque, 1992) and are, therefore, less well correlated to hepatic blood flow. Moreover, circadian alterations in circulating plasma protein levels also affect the binding, and therefore availability, of many drugs (Nakano et al., 1984).

Notwithstanding these confounding multiple influences, most orally administered drugs will be absorbed more rapidly after morning than after evening administration and this may lead to clinical consequences for a few of these drugs, as discussed above.

6.3 Rhythms in gastric acid secretion

Hydrochloric acid is secreted by oxyntic glands of the human stomach. Acid secretion may be stimulated by histamine, acetylcholine and gastrin, resulting in secretion by parietal cells of hydrogen ions into the gastric lumen against a concentration gradient greater than 2 million-fold. Histamine is released by enterochromaffin-like (ECL) cells, mast cells and histaminergic neurons after stimulation by a meal and/or by other endocrine and paracrine stimulants; acetylcholine is released from parasympathetic neurons after cephalic-vagal or gastric-vagal stimulation; and gastrin is secreted into the bloodstream from G cells of the antral and duodenal mucosa in response to cephalic-vagal stimulation, chemical reactions to food (amino acids and peptides) and gastric distension. The fact that acid secretion may be independently blocked at each of these receptor sites forms the basis for current peptic ulcer therapy. In addition, the H^+K^+-ATPase receptor site at the cell apex is blocked by proton pump inhibitors. The proton pump is believed to be the final common pathway for hydrogen ion secretion into the gastric lumen.

In the absence of any exogenous stimulation, including meals, acid is secreted in relatively low amounts to maintain a gastric pH between 0.8 and 2.5 over a 24 h period. This is termed basal acid secretion. A 24 h rhythm in basal acid secretion has been described in healthy individuals and in patients with active duodenal ulcer disease (Moore and Halberg, 1986). In the absence of food, acid output is highest in the evening (from 16.00 to midnight) and lowest during the morning hours (Figure 6.5). Higher rates of mean acid secretion were observed in duodenal ulcer patients (approximately 30% above controls) when compared to normals, but individual overlap between the two groups was considerable.

Using the method of monitoring gastric pH with glass electrodes placed intragastrically, increased acidity has been found in patients with active duodenal ulceration as compared with normals, especially in the

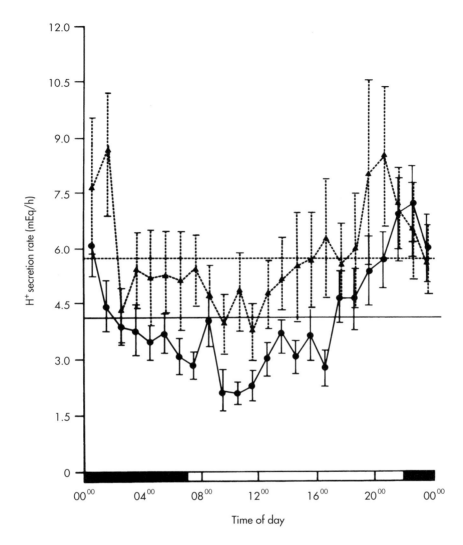

Figure 6.5 Twenty-four-hour gastric acid [H$^+$] chronogram from 14 healthy volunteers (circles) and 21 patients with active peptic ulcer (triangles). Points represent hourly mean (± SEM) acid secretory rates. Note low morning and high evening secretion in both groups. Dashed line shows the mean 24 h rate (5.76 ± 0.98 mEq/h) for the ulcer group; solid line shows mean rate (4.12 ± 0.40 mEq/h) for the healthy group. Reproduced from Moore and Halberg (1986).

evening (Merki *et al.*, 1988a) (Figure 6.6). In agreement with the study of Moore *et al.* and others, a considerable overlap of median 24 h acidity was observed between duodenal ulcer patients and nonulcer controls.

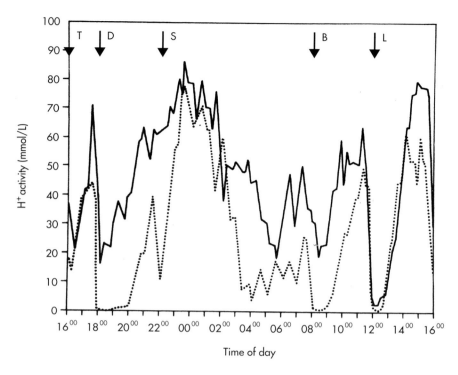

Figure 6.6 Median intragastric acidity profiles of duodenal ulcer patients (solid line) and healthy controls (broken line) over 24 h. Meals are shown above as tea (T), dinner (D), snack (S), breakfast (B) and lunch (L). Acidity is expressed as hydrogen ion activity in millimoles per litre. Reproduced from Merki et al. (1988).

The mechanisms underlying circadian rhythmicity in gastric acid secretion are not fully understood. No correlation, or even a negative correlation, was described when basal acid secretion was compared with serum gastrin concentrations (Moore and Wolfe, 1973; Gedde-dahl, 1974; Trudeau and McGuigan, 1971). Vagal stimulation may be responsible for the rhythmicity of acid secretion in the fasting state as anticholinergics effectively inhibit basal acid secretion. In addition, rhythmicity in acid secretion is lost in postvagotomy patients who continue to secrete significant amounts of basal gastric acid, suggesting that vagal nerve innervation is necessary to maintain the circadian pattern of secretion (Moore, 1973).

Under physiological conditions, the rhythmicity of gastric secretion is greatly influenced by the intake of meals and liquids. After the ingestion of a meal, intragastric hydrogen ion concentration falls immediately as a result of the buffering effect of food, despite rapidly increasing acid output (Figure 6.7). At 120–150 min after a meal, pH values drop to

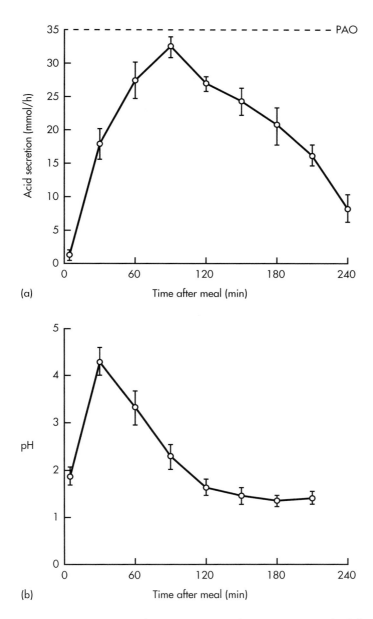

Figure 6.7 Mean (± SEM) acid secretion (a) and intragastric pH (b) following a sirloin steak in healthy subjects. (a) On day 1, after the meal was eaten, acid secretion was measured by *in vivo* intragastric titration to pH 5.5 in 6 subjects. (b) On day 2, intragastric pH in 10 subjects was allowed to seek its natural level after the meal was eaten. The mean basal secretion rate (a) and basal pH (b) prior to the meal are shown at 0 min. Peak acid output (PAO) is also indicated. Reproduced from Goldschmiedt and Feldman (1995).

low levels as food buffers become ineffective owing to saturation and because of gastric emptying. The powerful influence of meals on acid secretion, overwhelming basal acid and the circadian rhythm, is mediated via cephalic-vagal-mediated stimulation, gastric distension and chemical reactions to food involving cholinergic, histaminergic and gastrinergic mechanisms. During interdigestive periods, especially at night-time, intragastric pH levels remain low. It is believed that this nocturnal period is the time when gastric mucosa is most vulnerable to damage and also most susceptible to acid-inhibiting treatment strategies (Soll, 1989).

6.3.1 Pharmacological implications

Reduction of 24 h acid output or acidity results in improvement or healing of acid-related disease (peptic ulcer, gastro-oesophageal reflux, stress erosive gastritis) and is achieved by different pharmacological approaches. Antacids buffer acidity for approximately 60–90 min after their intake but have little influence on night-time acid secretion and, therefore, play a minor role in modern peptic ulcer therapy.

Histamine H_2-receptor antagonists (H_2RAs) are widely used antisecretory agents. There is a wide array of dosages and recommendations for the oral or intravenous administration of this class of drugs (which includes cimetidine, ranitidine, famotidine, nizatidine, roxatidine and others). Simultaneous measurements of plasma H_2RA concentrations and intragastric pH data reveal a sigmoidal dose–response relationship consistent with competitive inhibition of histamine H_2 receptors (Echizen *et al.*, 1988; Sanders *et al.*, 1989).

The influence of the circadian rhythm on basal acid secretion is reflected in the varying pH response to constant-rate intravenous infusions of histamine H_2RAs. Intragastric pH decreases during the late evening (Sanders *et al.*, 1988; Ballesteros *et al.*, 1990) (Figure 6.8), matching the time of increased basal acid secretion. This rhythmic alteration in acid secretion results in a changing dose–response relationship for H_2RAs over a 24 h period, with higher doses needed to inhibit acid output at times of peak acid secretion (Sanders *et al.*, 1988, 1991; Merki *et al.*, 1988b). This was clearly reflected in a study using individually adapted H_2RA infusion rates to achieve target pH levels over a 24 h period (Merki *et al.*, 1991 a,b). A novel computerised infusion pump was used to adjust the individual need for antisecretory drugs. Infusion rates of H_2RAs were maximal in the late afternoon and evening and decreased gradually during the night, with the lowest infusion rates during the morning (Hannan *et al.*, 1990) (Figure 6.9), clearly matching the rhythmic ebb and flow of basal acid secretion.

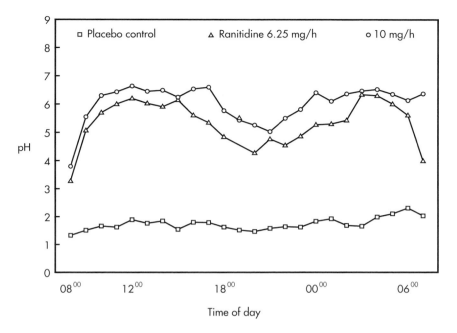

Figure 6.8 Mean gastric pH measured in 15 fasted duodenal ulcer patients during three 24 h studies. Note the decrease in pH between 18.00 and 00.00 despite continuous intravenous infusion of ranitidine at 6.25 mg/h and 10 mg/h. Reproduced from Sanders *et al.* (1988).

Meals significantly alter the rhythmicity of 24 h gastric secretion (Feldman and Richardson, 1986) (Figure 6.10) and intragastric pH levels (Merki *et al.*, 1988a) (Figure 6.7). A dramatic loss of antisecretory effects was seen after the intake of a meal, after both oral (Merki *et al.*, 1990; Johnston and Wormsley, 1988; Frislid and Berstad, 1985; Pounder *et al.*, 1977) and intravenous administration of an H_2RA (Merki *et al.*, 1991b). A full morning dose of ranitidine increases gastric pH significantly less than the same dose given at night-time (Figure 6.11), and a significant interaction between *ad lib* snacks and antisecretory effects was seen when food was taken after a full evening dose of this drug (Figure 6.12). Even high doses of intravenous famotidine were unable to suppress meal-stimulated acid secretion in a group of duodenal ulcer patients (Merki *et al.*, 1988b) (Figure 6.13). When meals were administered via nasogastric tubes, however, the interaction between meal stimulation and antisecretory effects was significantly less (Brater *et al.*, 1982). These differences may reflect the importance of the cephalic-vagal component of the meal response; modified sham-feeding, indeed, dramatically reduced the effect

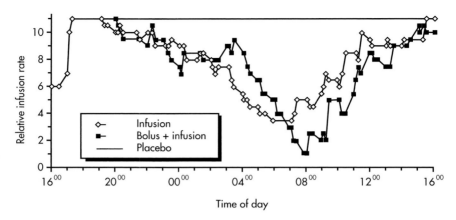

Figure 6.9 Median pump rate changes during placebo and famotidine infusion using an intragastric pH–famotidine infusion feedback system. The computer was programmed to increase or decrease famotidine infusion rates below or above an intragastric pH of 6.0. The graph shows 15 min median pump rates; a clear trend for rates to decrease in the morning is seen. Reproduced from Hannan et al. (1990).

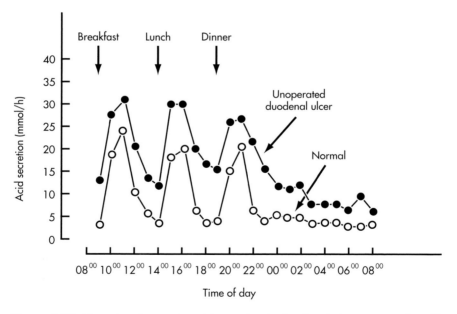

Figure 6.10 Mean hourly gastric acid secretion in duodenal ulcer patients ($n = 8$) and normal men ($n = 7$) over 24 h. Breakfast, lunch and dinner were infused into the stomach at 09.00, 14.00 and 17.00, respectively. Differences in acid secretion between groups are greatest at night. Reproduced from Feldman and Richardson (1986).

of high intravenous doses of H$_2$RAs for 2–3 h (Merki *et al.*, 1991b) (Figure 6.14). The observation that the addition of anticholinergic drugs only partially attenuated the secretory effect of sham-feeding and feeding supports the view that complete block of food-stimulated acid secretion requires simultaneous inhibition of several receptor types. The cephalic phase of gastric acid secretion is mediated by vagal mechanisms, including direct stimulation of the parietal cell, promotion of gastrin release and release of inhibitory hormones (Richardson *et al.*, 1977; Stenquist, 1979; Lam *et al.*, 1980). These pathways may be implicated in the inability of H$_2$RAs to substantially counteract the postprandial secretory drive.

H$^+$K$^+$-ATPase inhibitors, or 'proton pump inhibitors', such as omeprazole, lansoprazole and pantoprazole block the final step of acid secretion. These drugs are characterised by a long duration of action and

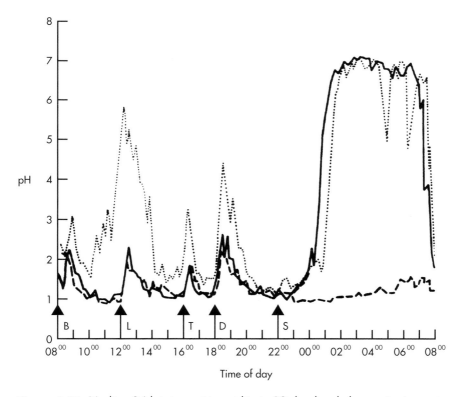

Figure 6.11 Median 24 h intragastric acidity in 12 duodenal ulcer patients receiving placebo (dashed line), ranitidine 300 mg at night (continuous line) or ranitidine 300 mg in the morning (dotted line). Meals are shown at the bottom with arrows: B, breakfast; L, lunch; T, tea; D, dinner; S, snack. Ranitidine was administered at 08.30 or 22.15. Reproduced from Merki *et al.* (1987).

little interindividual variability in response when given in high doses. Only the H^+K^+-ATPase in actively secreting membranes is inactivated by these drugs. A 40 mg dose of omeprazole administered before breakfast inhibits daytime H^+ concentration by over 90% and night-time acidity by 99% (Burget et al., 1990). Doses lower than 20 mg are inconsistent in their suppression of acidity and secretion (Howden et al., 1986); this may well be explained by the low bioavailability of the drug and high interindividual variability in drug absorption. Morning dosing of either 30 or 40 mg omeprazole was shown to be superior to night-time dosing in suppressing gastric secretion (Chiverton et al., 1992; Prichard et al., 1985), despite the lack of a difference between the two time points in the AUC of serum drug concentration. Proton pump inhibitors are the most widely used and effective acid suppressants available. However, night-time breakthrough in acid secretion, with intragastric pH levels

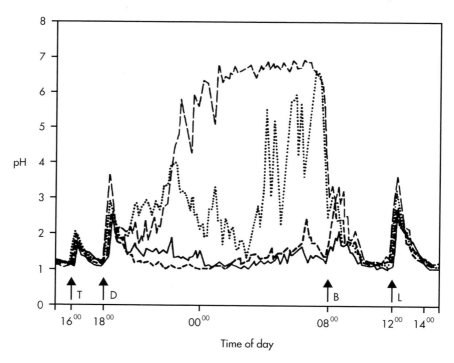

Figure 6.12 Median 24 h profiles in 20 duodenal ulcer patients during placebo at 18.30 followed by no food (dashed line, bottom); placebo at 18.30 followed by *ad lib* snacks (continuous line); ranitidine 300 mg at 18.30 with no additional food during the evening (dashed line, top); and ranitidine 300 mg at 18.30 followed by *ad lib* snacks before bedtime (dotted line). Meals are shown at bottom with arrows: T, tea; D, dinner; B, breakfast; L, lunch. Reproduced from Merki et al. (1990).

routinely falling below pH 4.0, can be demonstrated with even the highest dosage schedules (Lind *et al.*, 2000) (Figure 6.15).

6.3.2 Therapeutic implications

Ulcer healing correlates directly with the degree and duration of the inhibition of acidity as measured by intragastric pH techniques; a much smaller correlation was found between duodenal ulcer healing and reduction in total acid output (Jones *et al.*, 1987). For optimal suppression of acidity to be achieved with H_2RAs, the influence of the rhythm in basal acid secretion and the effect of meals on acid secretion need to be taken into account.

The optimal effects of oral H_2RAs on intragastric pH have been found when the full daily dose was administered after the last meal in the evening, therefore suppressing acidity during the longest interdigestive period of the day (Merki *et al.*, 1987). If intravenous administration of this class of drugs is required, patients should remain fasted throughout

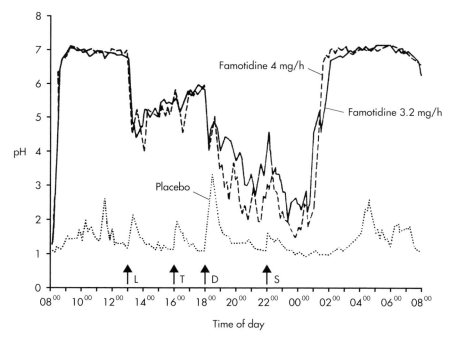

Figure 6.13 Median 24 h intragastric pH profiles during placebo (dotted line), intravenous famotidine 3.2 mg/h (solid line) and intravenous famotidine 4 mg/h (broken line). Meals are shown at the bottom with arrows: L, lunch; T, tea; D, dinner; S, snack. Reproduced from Merki *et al.* (1988b).

the treatment period, as meals interact markedly with the antisecretory effects of H_2RAs.

In contrast, the proton pump inhibitors should be taken shortly before or with a meal, which will activate the pumps in the secreting membranes, a requirement for proton pump blockade. The effect of omeprazole and lansoprazole is compromised if they are taken during interdigestive periods or administered intravenously in fasting subjects. Morning administration of H^+K^+-ATPase inhibitors has been shown to be superior to night-time dosing in suppressing 24 h acidity and gastric acid secretion.

6.4 Seasonal rhythms in peptic ulcer disease

Seasonal variation in the incidence of peptic ulcer disease was first reported in 1903 (Brunner, 1903), supported in several endoscopically

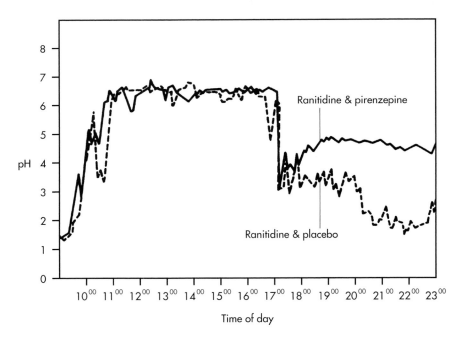

Figure 6.14 Median intragastric pH–time profiles of 10 healthy volunteers fed a standardised meal at 17.00. Ranitidine was administered intravenously by a pH–ranitidine infusion feedback system. If pH > 6.0 was not maintained after 17.00, pirenzepine or placebo was given additionally. Note that the addition of an anticholinergic only partially attenuated the acid-stimulating effect of a meal. Reproduced from Merki *et al.* (1991b).

based studies (Gibinski *et al.*, 1982; Gibinski 1987; Bretzke, 1985; Moshal, 1981; Karvonen, 1982) but challenged in other reports (Adler *et al.*, 1984; Sonnenberg *et al.*, 1992; Braverman *et al.*, 1992). Most reports document a consistently lower incidence of peptic disease in the summer months and peak occurrences during the autumn, winter or spring. Svanes *et al.* (1999) more recently reported a seasonal variation in peptic ulcer perforation in a population of 1980 patients in Norway over a 55-year period. Duodenal ulcer perforation rates showed significant 6-month rhythms with peak rates occurring in May–June–July and November–December. These observations may justify seasonal preventive treatment trials in selected patients with consistent seasonal flares of their peptic disease.

6.5 Rhythms in gastric mucosal defence

The major therapeutic aim for most peptic ulcer regimens is to reduce gastric acidity. However, it is known that more than half of all peptic ulcer patients exhibit normal or less than normal gastric acid secretion rates, and for this reason experimental efforts in recent years have been directed to possible alterations in gastric mucosal defence factors as a pathogenetic explanation. In this regard, an association has been

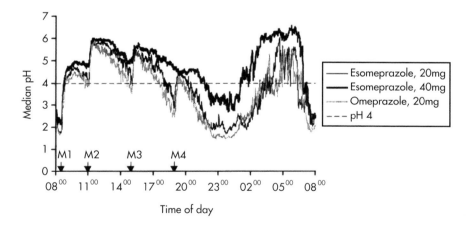

Figure 6.15 Twenty-four-hour median intragastric pH–time profiles after 5 days' dosing with esomeprazole (40 and 20 mg, once daily) and omeprazole (20 mg, once daily) in 36 patients with symptoms of gastro-oesophageal reflux disease. Arrows indicate time points at which standardised meals (M1–M4) were served. Note 'breakthrough' of pH control (i.e. pH < 4.0) between 20.00 and 02.00. Reproduced from Lind *et al.* (2000).

described between antral gastritis, peptic ulcer disease and *Helicobacter pylori*, a bacterium that resides on the surface of the antral mucosa and is known to reduce some factors of mucosal defence. Antibiotic treatment of this organism has led to a dramatic reduction in ulcer recurrence rates (Marshall *et al.*, 1988). However, it is estimated that 70% of the world's population harbours this organism and, moreover, not all peptic ulcer patients harbour *H. pylori*.

Several of these defence factors display circadian rhythmicity, as illustrated in Figure 6.16 (Moore *et al.*, 1994). This study in the rat demonstrated that gastric mucosal defence against the aggressive action of gastric acid is afforded by different mechanisms over the circadian period. During the dark phase, when this species is active and gastric acid secretion is highest, protection against the potentially damaging effect of high acidity is provided by increased gastric mucosal blood flow. During the light phase, when this species is inactive and acidity is lowest, protection against acid damage is provided by a relative increase in mucus, bicarbonate and tissue prostaglandin concentrations. In addition, a circadian rhythm in mucosal damage produced by topically applied acetylsalicylic acid (ASA; aspirin) was identified, with peak damage occurring during the dark phase. The amount of damage was correlated to the relative difference in acid and bicarbonate secretion and could not be compensated for by the relative increase in dark-phase mucosal blood flow. The relevance of this rodent study to human peptic ulcer disease is that susceptibility to gastric mucosal damage can be explained by a relative change in the ratio of gastric aggressive and defensive factors without the requirement for either an absolute increase or decrease in either factor. Similar circadian-based investigations of factors in mucosal defence have not been reported in humans, but some data exist to suggest that important day–night differences in mucosal defence do occur in humans. In a two-time-point endoscopic study performed on healthy males, an orally administered 1300 mg dose of ASA produced significantly more damage to the gastric mucosa when given at 08.00 than when given at 20.00 (Moore and Goo, 1987). However, a recent crossover study after orally administered low-dose (75 mg) and high-dose (1000 mg) ASA, in which the drug-induced lesions were rated by video-endoscopy in a blinded fashion, did not support the circadian phase dependency in ASA-induced gastric lesions (Nold *et al.*, 1995).

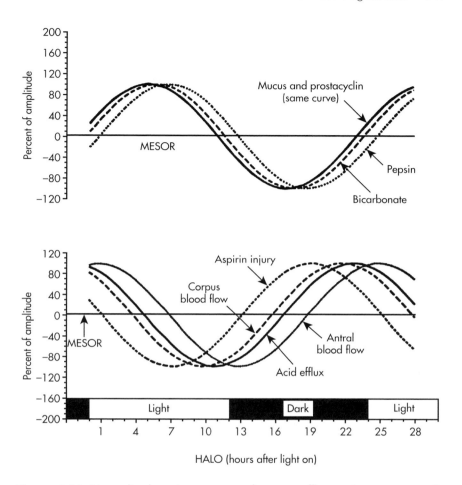

Figure 6.16 Normalised cosinor curves of mucus effluent, tissue prostacyclin activity, pepsin secretion, corpus and antral blood flow, bicarbonate and acid secretion and aspirin injury in the rat stomach over 24 h. In each plot, the MESOR (~ mean value) was set at zero and the amplitude at ±100%. All plots demonstrated significant ($p < 0.05$) circadian rhythmicity by cosinor analysis. Reproduced from Moore et al. (1994).

Acknowledgements

The author thanks Ms Sharon Henn of the VA Salt Lake City Health Care System, Research and Development, for manuscript preparation. This material is based upon work supported by the Office of Research and Development (R&D) at the VA Salt Lake City Health Care System, Salt Lake City, Utah.

References

Adler J, Ingram D, House J (1984). Perforated peptic ulcer – a seasonal disease? *Aust NZ J Surg* 54: 59–61.

Ballesteros M A, Hogan D F, Koss M A, Isenberg J I (1990). Bolus or intravenous infusion of ranitidine: effects on gastric pH and acid secretion. *Ann Intern Med* 112: 335–339.

Bélanger P M, Labrecque G (1992). Biological rhythms in the hepatic drug metabolism. In: Touitou Y, Haus E, eds. *Biological Rhythms in Clinical Laboratory Medicine.* Berlin: Springer-Verlag, 403–409.

Bogentoft C, Carlson I, Ekenved G, Magnusson A (1978). Influence of food on the absorption of acetylsalicylic acid from enteric-coated dosage forms. *Eur J Clin Pharmacol* 15: 351–355.

Brater D C, Peters M N, Eshelman F N, Richardson C T (1982). Clinical comparison of cimetidine and ranitidine. *Clin Pharmacol Ther* 32: 484–489.

Braverman D Z, Morali G A, Patz J K, Jacobsohnn W Z (1992). Is duodenal ulcer a seasonal disease? *Am J Gastroenterol* 87: 1591–1593.

Bretzke G (1985). Monate und Jahreszeitliche Verteilunge des Ulcus Ventriculi und Duodenia in Endoskopischer Uner Suchungsgruppe. *Z Gesamte Inn Med* 40: 407–408.

Brophy C M, Moore J G, Christian P E, Taylor A T (1986). Individual variability of gastric emptying measurement employing standardised radiolabeled meals. *Dig Dis Sci* 21: 799–806.

Brunner F (1903). Das acut in die freie Bauchhole perforierende Magen und Duodenal geschwuer. *Dtsh Z Chir* 69: 101–235.

Burget D W, Chiverton S G, Hunt R H (1990). Is there an optimal degree of acid suppression for healing of duodenal ulcers? A model of the relationship between ulcer healing and acid suppression. *Gastroenterology* 99: 345.

Chiverton S G, Howden C W, Burget D W, Hunt R H (1992). Omeprazole (20 mg) daily given in the morning or evening: a comparison of effects on gastric acidity, and plasma gastrin and omeprazole concentration. *Aliment Pharmacol Ther* 6: 103–111.

Echizen H, Shoda R, Umeda N, Ishizaki T (1988). Plasma famotidine concentration versus intragastric pH in patients with upper gastrointestinal bleeding and in healthy subjects. *Clin Pharmacol Ther* 44: 690–698.

Feldman M, Richardson C T (1986). Total 24-hour gastric acid secretion in patients with duodenal ulcer. Comparison with normal subjects and effects of cimetidine and parietal cell vagotomy. *Gastroenterology* 90: 540–544.

Frislid K, Berstad A (1985). Prolonged influence of a meal on the effect of ranitidine. *Scand J Gastroenterol* 20: 711–714.

Gedde-dahl D (1974). Radioimmunoassay of gastrin. Fasting serum levels in humans, with normal and high gastric acid secretion. *Scand J Gastroenterol* 9: 41–47.

Gibinski K (1987). A role of seasonal periodicity in peptic ulcer disease. *Chronobiol Int* 4: 91–98.

Gibinski K, Rybicka J, Nowak A, Czarnecka K (1982). Seasonal occurrence of abdominal pain and endoscopic findings in patients with gastroduodenal ulcer disease. *Scand J Gastroenterol* 17: 481–485.

Goldschmiedt M, Feldman M (1995). Gastric secretion in health and disease. In: Sleisenger and Fordtran, eds. *Gastrointestinal Disease/Patho-physiology/Diagnosis/Management*, 5th edn. Saunders, Philadelphia, 524–544.

Goo R H, Moore J G, Greenberg E, Alazraki N P (1987). Circadian variation in gastric emptying of meals in humans. *Gastroenterology* 93: 515–528.

Hannan A, Chesner I, Merki H S, *et al.* (1990). Use of automatic computerised pump to maintain constant intragastric pH. *Gut* 41: 1246–1249.

Howden C W, Dehodra J K, Burget D W, Hunt R H (1986). Effects of low dose omeprazole on gastric secretion and plasma gastrin in patients with healed duodenal ulcer. *Hepatogastroenterology* 33: 267–270.

Johnston D A, Wormsley K G (1988). The effect of food on ranitidine-induced inhibition of nocturnal gastric secretion. *Aliment Pharmacol Ther* 2: 507–511.

Jones D B, Howden C W, Burget D W, *et al.* (1987). Acid suppression in duodenal ulcer: a meta-analysis to define optimal dosing with antisecretory drugs. *Gut* 28: 1120–1127.

Karvonen A J (1982). Occurrence of gastric mucosal erosions in association with other upper gastrointestinal diseases, especially peptic ulcer disease, as revealed by elective gastroscopy. *Scand J Gastroenterol* 17: 977–984.

Kumar D, Wingate D, Ruckebusch Y (1986). Circadian variation in the propagating velocity of the migrating motor complex. *Gastroenterology* 91: 926–930.

Lam S K, Isenberg J S, Grossman M I, *et al.* (1980). Gastric acid secretion is abnormally sensitive to endogenous gastrin released after peptone test meals in duodenal ulcer patients. *J Clin Invest* 65: 555–562.

Langner B, Lemmer B (1988). Circadian changes in the pharmacokinetics and cardiovascular effects of oral propranolol in healthy subjects. *Eur J Clin Pharmacol* 33: 619–624.

Lemmer B, Bruguerolle B (1994). Chronopharmacokinetics – are they clinically relevant? *Clin Pharmacokinet* 26: 419–427.

Lemmer B, Nold G (1991). Circadian changes in estimated hepatic blood flow in healthy subjects. *Br J Clin Pharmacol* 32: 627–629.

Lemmer B, Scheidel B, Behne S (1991). Chronopharmacokinetics and chronopharmacodynamics of cardiovascular active drugs: propranolol, organic nitrates, nifedipine. *Ann NY Acad Sci* 618: 166–218

Lind T, Rydberg L, Kyleback A, *et al.* (2000). Esomeprazole provides improved acid control vs omeprazole in patients with symptoms of gastroesophageal reflux disease. *Aliment Pharmacol Ther* 14: 861–867.

Marshall B J, Goodwin C S, Warren J R, *et al.* (1988). Prospective double-blind trial of duodenal ulcer relapse after eradication of *Campylobacter pylori*. *Lancet* 2: 1437–1441.

Mejean L, Kolopp M, Provin P (1992). Chronobiology, nutrition and diabetes mellitus. In: Touitou Y, Haus E, eds. *Biological Rhythms in Clinical and Laboratory Medicine*. Berlin: Springer-Verlag, 375–385.

Merki H S, Witzel L, Harre K, *et al.* (1987). Single dose treatment with H_2-receptor antagonists: is bedtime administration too late? *Gut* 28: 451–454.

Merki H S, Fimmel C J, Walt R P, *et al.* (1988a). Pattern of 24 hour intragastric acidity in active duodenal ulcer disease and in healthy controls. *Gut* 29: 1583–1587.

Merki H S, Witzel L, Kaufman D, *et al.* (1988b). Continuous intravenous infusions

of famotidine maintain high intragastric pH in duodenal ulcer. *Gut* 29: 453–457.
Merki H S, Halter F, Wilder-Smith C W, *et al.* (1990). Effect of food on H$_2$-receptor blockade in normal subjects and duodenal ulcer patients. *Gut* 31: 148–150.
Merki H S, Hunt R H, Walt R P, *et al.* (1991a). A new programmable infusion pump for individual control of intragastric pH: validation and effect of ramitidine. *Eur J Gastroenterol Hepatol* 3: 9–13.
Merki H S, Wilder-Smith C W, Walt R P, Halter F (1991b). The cephalic and gastric phases of gastric secretion during H$_2$-antagonist treatment. *Gastroenterology* 101: 599–606.
Moore J G (1973). High gastric acid secretion after vagotomy and pyloroplasty in man evidence for non-vagal mediation. *Am J Dig Dis* 18: 661–669.
Moore J G, Goo R H (1987). Day and night aspirin-induced mucosal damage and protection by ranitidine in man. *Chronobiol Int* 4: 111–116.
Moore J G, Halberg F (1986). Circadian rhythm of gastric acid secretion in men with active duodenal ulcer. *Dig Dis Sci* 31: 1185–1191.
Moore J G, Wolfe G (1973). The relation of plasma gastrin to the circadian rhythm of gastric acid secretion in man. *Digestion* 9: 97–105.
Moore J G, Moody F, Datz F, Christian P (1981). Gastric emptying of liquids and solids following gastric partition in the morbidly obese. *Gastroenterology* 80: 1234.
Moore J G, Datz F L, Christian P E (1990). Exercise increases solid food gastric emptying rates in men. *Dig Dis Sci* 35: 448–452.
Moore J G, Larsen K R, Baratini P, Dayton M T (1994). Asynchrony in circadian rhythms of gastric function in the rat. A model for mucosal injury. *Dig Dis Sci* 39: 1619–1624.
Moshal M G, Spitaels J M, Tobbs J V, *et al.* (1981). Eight year experience with 3392 endoscopically proven ulcers in Durban 1972–1979. *Gut* 22: 327–331.
Nakano S, Watanabe H, Nagai K, Ogawa N (1984). Circadian stage-dependent changes in diazepam kinetics. *Clin Pharmacol Ther* 36: 271–277.
Nold G, Drossard W, Lehmann K, Lemmer B (1995). Gastric mucosal lesions after morning versus evening application of 75 mg or 1000 mg acetylsalicylic acid (ASA). *Naunyn Schmiedebergs Arch Pharmacol* 351: R17.
Pounder R E, Williams J G, Hunt R H, *et al.* (1977). The effects of oral cimetidine on food-stimulated gastric acid secretion and 2-hour intragastric acidity. In: Burland W, Simkins M, eds. *Proceedings of the Second International Symposium on Histamine H$_2$ Receptor Antagonists*. Amsterdam: Excerpta Medica, 189.
Prichard P J, Yeomans N D, Mihaly G W, *et al.* (1985). Omeprazole: a study of its inhibition of gastric pH and oral pharmacokinetics after morning and evening dosage. *Gastroenterology* 88: 64–69.
Reinberg A, Smolensky M H (1982). Circadian changes of drug disposition in man. *Clin Pharmacokinet* 7: 401–402.
Reinberg A, Pauche F, Ruff F, *et al.* (1987). Comparison of once daily evening versus morning sustained-release theophylline dosing for nocturnal asthma. *Chronobiol Int* 4: 409–420.
Richardson C T, Walsh J H, Cooper K A, *et al.* (1977). Studies on the role of cephalic-vagal stimulation in the acid secretory response to eating in normal subjects. *J Clin Invest* 60: 435–441.

Sanders S W, Moore J G, Buchi K N, Bishop A L (1988). Circadian variation in the pharmacodynamics effect of intravenous ranitidine. *Annu Rev Chronopharmacol* 5: 335–338.

Sanders S W, Buchi K N, Moore J G, Bishop A L (1989). Pharmacodynamics of intravenous ranitidine after bolus and continuous infusion in patients with healed duodenal ulcers. *Clin Pharmacol Ther* 46: 545–551.

Sanders S W, Ballesteros M A, Hogan D K, *et al.* (1991). Effect of basal gastric acid secretion on the pharmacodynamics of ranitidine. *Chronobiol Int* 8: 186–193.

Smith H J, Feldman M (1986). Influence of food and marker length on gastric emptying of indigestible radiopaque markers in healthy humans. *Gastroenterology* 91: 1452–1455.

Soll A H (1989). Duodenal ulcer and drug therapy. In: Sleisenger M H, Fordtran J S, eds. *Gastrointestinal Disease/Pathophysiology/Diagnosis/Management*, 4th edn. Philadelphia: Saunders, 814–879.

Sonnenberg A, Wasserman I H, Jacobsen J (1992). Monthly variation of hospital admission and mortality of peptic ulcer disease: a reappraisal of ulcer periodicity. *Gastroenterology* 103: 1192–1193.

Stenquist B (1979). Studies on vagal activation of gastric acid secretion in man. *Acta Physiol Scand (Suppl)* 465: 1–31.

Svanes C, Lie S A, Lie R T, *et al.* (1999). Causes of death in patients with peptic ulcer perforation: a long-term follow-up study. *Scand J Gastroenterol* 34: 18–24.

Trudeau W L, McGuigan J E (1971). Relations between serum gastrin levels and rates of gastric hydrochloric acid secretion. *N Engl J Med* 284: 408–412.

7

Rhythms and therapeutics of asthma

Esther L Langmack and Richard J Martin

7.1 Introduction

Asthma is a very prevalent respiratory disease afflicting more than 15 million people in the United States (Mannino *et al.*, 1998) and many millions more worldwide. In spite of advances in asthma treatment, morbidity and mortality related to asthma have increased dramatically over the last several decades (National Center for Health Statistics, 1991, 1993), and direct medical expenditure related to asthma exceeds 6 billion dollars annually in the United States (Weiss and Sullivan, 2001). The number of patients consulting their doctor with indications of asthma in the United States has more than doubled since 1975, owing in part to a doubling in the prevalence of the disease (Mannino *et al.*, 1998)

Asthma is a chronic inflammatory disease of the central and peripheral lung airways (Busse and Lemanske, 2001). Autopsy and bronchoscopic studies have revealed increases in inflammatory cells and mediators in asthmatic airways, even in those patients with the mildest disease (Laitinen *et al.*, 1993). Eosinophils, mast cells, T- and B-lymphocytes and other lung cells interact through a complex intercellular signalling network, resulting in epithelial cell sloughing, smooth muscle hypertrophy and hyperplasia, goblet cell hyperplasia, and sub-basement-membrane thickening, the histopathological hallmarks of the disease. Physiologically, asthma is characterised by airflow obstruction that can be partially or fully reversed by administration of bronchodilators and by the presence of airway hyperreactivity to various internal and external stimuli. Clinically, these inflammatory and physiological changes present as symptoms of intermittent wheezing, chest tightness, cough and dyspnoea.

One aspect of asthma that is not well recognised is that it is a circadian disease that worsens at night in many patients. Nocturnal asthma, defined as a significant (typically ≥12%), overnight decline in forced expiratory volume in one second (FEV_1) or peak expiratory flow rate (PEFR) is a common and potentially deadly problem. In a 1988 survey of 7600 asthmatic outpatients, 74% awakened at least one night per week with asthma symptoms, while 39% had symptoms nightly (Turner-Warwick, 1988). Significantly, asthmatics are more likely to die from their disease at night than during the day. In one study, 53% of deaths from asthma over a one-year period occurred at night, and 79% of these patients had prior complaints of asthma affecting their sleep, occurring every night in 42% (Robertson et al., 1990). Nocturnal asthma symptoms are an important defining characteristic of asthma severity in current guidelines for asthma diagnosis and treatment (National Asthma Education and Prevention Program, 1997).

Sleep disruption by nocturnal asthma adversely affects daytime life, as reflected by its effects on cognitive functioning and school performance (Bender and Annett, 1999). Adults with nocturnal asthma performed more poorly than healthy controls in psychometric tests involving focused attention and concentration (Weersink et al., 1997b) and rapid information processing (Fitzpatrick et al., 1991). When circadian variation in lung function was reduced with anti-asthma medications, test performance rose to the level of normal controls in one study (Weersink et al., 1997b). In a survey of parents of school-aged children with asthma, 45% of children awakened at least once with asthma over a 4-week period (Diette et al., 2000). Those children with as few as 1–3 nocturnal awakenings from asthma over a 4-week period were significantly more likely to miss school and have difficulty with school performance than their asthmatic peers without nocturnal awakenings.

7.2 Chronobiology of asthma

7.2.1 Circadian variation in lung function

Both normal and asthmatic subjects who are diurnally (daytime) active display circadian variation in pulmonary function, with maximal lung function occurring at around 16.00 and minimal lung function around 04.00, as assessed by measurement of FEV_1 and PEFR (Figure 7.1). However, the decline in lung function overnight is much greater in asthmatics, in whom FEV_1 falls by as much as 10–50%, compared to healthy subjects, who have only a 5–8% overnight reduction in FEV_1 (Hetzel and Clark, 1980). Circadian variation in lung function in asthmatics is not merely a medication washout effect. During the washout phase of a pharmaceutical protocol involving over 3000 patients, nocturnal asthma symptoms occurred independently of medication effects, with over 90% of dyspnoeic episodes happening between 20.00 and 07.00 (Dethlefsen and Repgas, 1985) (Figure 7.2).

Chronobiology of asthma **155**

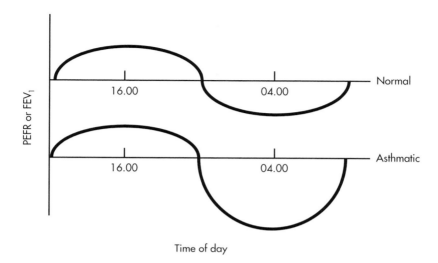

Figure 7.1 Both normal subjects and asthmatic patients have circadian alterations in lung function, with nadirs occurring at approximately 04.00. Circadian variation in lung function is increased in asthmatics compared to normal subjects. PEFR, peak expiratory flow rate; FEV_1, forced expiratory volume in one second. Reproduced with permission from Martin (1993).

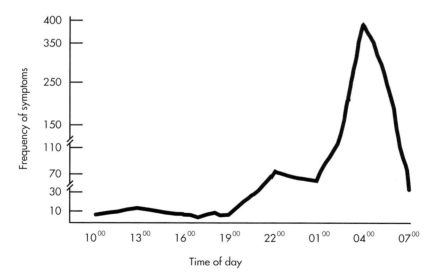

Figure 7.2 This figure illustrates the marked frequency of nocturnal asthma symptoms independent of medication in 3129 mainly asthmatic patients. Redrawn from data in Dethlefsen and Repgas (1985) and reproduced with permission from Martin (1993).

In asthmatics, a portion of the observed decrease in lung function occurs independently of sleep. Ballard *et al.* (1989) measured breath-by-breath lower airway resistance in supine nocturnal asthmatics and healthy controls from midnight to 06.00 during one night of normal sleep and during a separate night when sleep was prevented. The asthmatic subjects had progressive increases in lower airway resistance on both nights, but the rate of increase was twice as great during the night of normal sleep compared to the night of sleep-deprivation (Figure 7.3). Thus, sleep is not necessary for the overnight decrement in asthmatic lung function, but it does contribute to the observed decline. Furthermore, sleep pattern does not explain the overnight rise in lower airway resistance in asthmatics. No differences were found in sleep latency (how long it takes an individual to go to sleep), sleep efficiency (how well the patient sleeps), or sleep stage distribution between asthmatics and controls; and lower airway resistance did not differ between sleep stages in either group. The transition from an upright to supine position also does not account for nocturnal worsening of asthma. Asthmatic subjects kept supine for an entire 24 h period still demonstrated nocturnal decline in airflows (Clark and Hetzel, 1977).

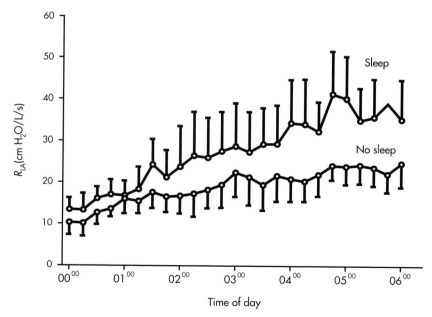

Figure 7.3 Nocturnal changes in lower airway resistance (R_{LA}) in 6 asthmatic subjects during sleep (upper line) and with prevention of sleep (lower line) between midnight and 06.00. Adapted from Ballard *et al.* (1989).

To investigate further the mechanisms of increasing lower airway resistance during the night in nocturnal asthmatics, Ballard and colleagues (Ballard *et al.*, 1990) measured functional residual capacity (FRC) in asthmatic and normal control subjects during sleep. Functional residual capacity – the lung volume achieved at the end of exhalation during normal tidal breathing – is greater in asthmatics than normal controls. The increase in FRC is thought to be protective in asthma because, as lung volume increases, airway resistance falls and airflows improve. This process is referred to as the mechanical coupling between airways and lung parenchyma. Using a horizontal body plethysmograph, Ballard *et al.* observed that FRC was higher in asthmatics before sleep onset and fell in both asthmatic and normal control groups during sleep, but to a much greater extent in the asthmatic group. The fall in FRC was felt to account, at least in part, for the increase in lower airway resistance observed in nocturnal asthmatics. Further investigation, however, revealed that increasing FRC by applying negative extrathoracic pressure, via a thoracic cuirass device, did not alter overnight changes in FEV_1 or bronchial hyper-reactivity (Martin *et al.*, 1993). More recently, Irvin *et al.* (2000), again using the cuirass technique, demonstrated that, although increasing lung volume reduced lower pulmonary resistance in nocturnal asthmatics while they were awake, restoring FRC to pre-sleep values during sleep did not cause lower pulmonary resistance to fall significantly. They concluded that during sleep there is an uncoupling of the lung parenchyma from the airway, possibly mediated by tissue inflammation and neural activity changes, that contributes to the overnight fall in lung function.

Physiological changes resulting in decreased lung function occur at the level of the peripheral airways, in addition to the central airways, in individuals with nocturnal asthma. Kraft and colleagues (Kraft *et al.*, 2001b) measured peripheral airway resistance in 10 nocturnal asthmatics, 4 non-nocturnal asthmatics and 4 normal controls by instilling warmed, humidified air through a bronchoscope into an isolated lung segment at several different flow rates. At both 04.00 and 16.00, they found that peripheral resistance in nocturnal asthmatics was at least twice as great as in non-nocturnal asthmatics and controls. Although peripheral resistance was higher in nocturnal asthmatics at 04.00 than 16.00, this difference was not statistically significant. However, plateau pressure, which reflects patency of small (diameter <2 mm), distal collateral airway channels, was significantly greater ($p = 0.0004$) at 04.00 than 16.00 in nocturnal asthmatics, but not in the other groups. Administration of subcutaneous terbutaline, a β_2-adrenoreceptor agonist, significantly reduced the plateau pressure in the nocturnal asthma group at 04.00. The authors concluded that small peripheral airways make an important contribution to the overnight decline in lung function, and that smooth muscle contraction, parenchymal inflammation, and/or oedema may compromise the patency of these distal airways.

In addition to changes in lung airflows and airway resistance, the bronchoconstrictive response to external stimuli is enhanced in asthmatics at night. One study (Martin *et al.*, 1990) demonstrated increased nocturnal bronchial reactivity to inhaled methacholine, an ACh (acetylcholine) muscarinic receptor agonist that precipitates bronchospasm at lower concentrations in asthmatics than in healthy individuals. Asthmatic subjects were challenged with inhaled methacholine at 16.00 and 04.00. Bronchial reactivity was greater in all asthmatics at 04.00 than at 16.00 (Figure 7.4). Nocturnal asthmatics, those subjects with >20% overnight fall in FEV_1, had slightly greater bronchial reactivity at 16.00 than the non-nocturnal asthmatics. However, the nocturnal asthmatics

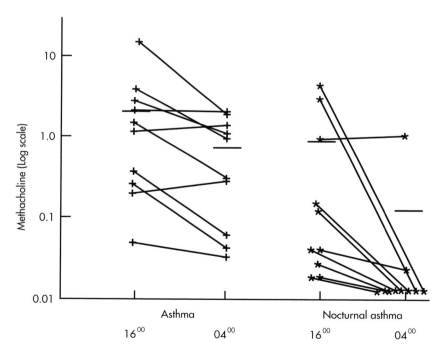

Figure 7.4 Circadian variation in bronchial reactivity to methacholine between 16.00 and 04.00 in asthmatics with minimal alteration in overnight FEV_1 (left) compared to those with marked decline in overnight FEV_1 >20%) (right). Horizontal lines represent mean values. Constructed from data in Martin *et al.* (1990) and reproduced with permission from Martin (1993).

demonstrated an 8-fold overnight increase in bronchial hyperreactivity, compared to only a 2-fold increase in non-nocturnal asthmatics. Day–night differences in bronchial reactivity have also been demonstrated using histamine (De Vries *et al.*, 1962) and house dust extract (Gervais *et al.*, 1977). Bronchial reactivity to each of these stimuli was enhanced at night, compared to daytime measurements.

7.2.2 Mechanisms of nocturnal worsening of asthma

It is highly likely that more than one biological process is responsible for the circadian variation in lung function and bronchial reactivity in nocturnal asthma (Figure 7.5). Recent investigation in this area has brought to light the important role of distal airway and alveolar inflammation in the pathogenesis of nocturnal asthma. The nocturnal decline in anti-inflammatory hormones, such as cortisol, and in endogenous bronchodilators, such as adrenaline (epinephrine), may exacerbate tissue inflammation and bronchoconstriction. Other extrinsic factors, including allergen exposure, gastro-oesophageal reflux disease, airway cooling, upper airway obstruction and sinus inflammation, may also contribute to nocturnal worsening of asthma in a minority of patients.

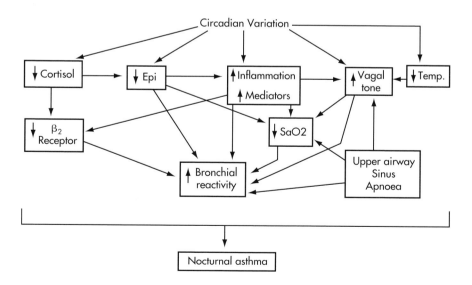

Figure 7.5 Potential mechanisms that contribute to the worsening of asthma at night. Epi, epinephrine (adrenaline); Temp., temperature; SaO_2, blood oxygen saturation. Reproduced with permission from Martin (1993).

7.2.2.1 Inflammatory cells and mediators

Martin and colleagues were the first to demonstrate that airway inflammation increases in nocturnal asthmatics during the night and is related to the overnight decrement in lung function (Martin et al., 1991). They performed bronchoscopy with bronchoalveolar lavage (BAL) on 7 nocturnal asthmatics and 7 non-nocturnal asthmatics at 16.00 and 04.00 on separate days. Nocturnal asthmatics were defined as those asthmatics who had an overnight fall in PEFR of at least 15% on 4 out of 7 nights, whereas non-nocturnal asthmatics demonstrated an overnight fall in PEFR of less than 10%. Total leukocytes, eosinophils and neutrophils increased significantly between 16.00 and 04.00 in BAL from nocturnal asthmatics, but not from non-nocturnal asthmatics. At 04.00, the nocturnal asthmatics had significantly higher total leukocytes, eosinophils, neutrophils, lymphocytes and epithelial cells in BAL than the non-nocturnal group, but no differences in BAL cell counts were observed between groups at 16.00. There were no differences in sleep latency, sleep efficiency or sleep stages between the two groups. When the groups were combined, the overnight fall in PEFR was significantly correlated with BAL total leukocytes ($r = 0.59$, $p < 0.05$), neutrophils ($r = 0.66$, $p < 0.05$) and eosinophils ($r = 0.62$, $p < 0.05$) at 04.00. In nocturnal asthmatics, the overnight decline in PEFR was significantly correlated with the change in eosinophils ($r = 0.84$, $p < 0.02$), a key effector cell in asthmatic lung inflammation.

The technique of BAL employed by Martin et al. (1991) samples cells lining both the lung airways and distal alveoli. Consequently, the exact location of the inflammatory process cannot be determined using this technique. In order to assess directly the contribution of alveolar tissue inflammation to nocturnal worsening of asthma, Kraft et al. (1996a) performed both endobronchial biopsy and transbronchial biopsy at 16.00 and 04.00 in 11 nocturnal asthmatics and 10 non-nocturnal asthmatics. Transbronchial biopsy samples distal airways and alveolar tissue, while endobronchial biopsy samples proximal airway tissue. Morphometric tissue analysis revealed that, in nocturnal asthmatics, eosinophils and macrophages increased in alveolar tissue between 16.00 and 04.00 (Figure 7.6). Compared to the non-nocturnal asthma group, nocturnal asthmatics had a significantly greater increase in alveolar tissue eosinophils between 16.00 and 04.00. In addition, when data from the two groups were combined, there was a significant correlation between percentage overnight fall in FEV_1 and number of eosinophils in alveolar tissue ($r = -0.54$, $p = 0.03$), but not proximal airway tissue, at 04.00. In a subsequent study, Kraft and colleagues characterised the presence of CD4+ lymphocytes in alveolar and proximal airway tissue from nocturnal asthmatics and non-nocturnal asthmatics at 04.00 and 16.00. CD4+ T-lymphocytes elaborate interleukin (IL)-3, IL-4, IL-5, and other cytokines, which enhance eosinophil survival and function in asthma (Kraft et al., 1999a). Using immunohistochemical staining and morphometric analysis, they discovered that at 04.00 nocturnal asthmatics had significantly more CD4+ lymphocytes in alveolar tissue than did non-nocturnal asthmatics. The number of CD4+ T-lymphocytes in alveolar tissue, but not in proximal airway tissue, correlated inversely with percent predicted FEV_1 at 04.00 ($r = -0.68$, $p = 0.0018$) and positively with the number of alveolar tissue eosinophils ($r = 0.66$, $p = 0.01$). Taken together, the results of these two studies suggest that inflammation in the distal lung, characterised by increases in eosinophils and CD4+ lymphocytes in alveolar tissue, plays a greater role in the pathogenesis of nocturnal asthma than do inflammatory changes in the proximal airways. These findings have important implications for the therapeutic approach to asthma in general, including the development of drugs and delivery devices that reduce distal lung inflammation.

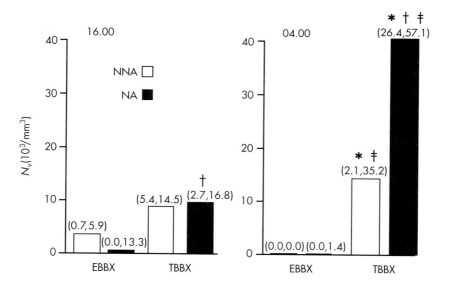

Figure 7.6 The number per volume (N_v) of eosinophils in nocturnal asthma (NA) and non-nocturnal asthma (NNA) groups in endobronchial (airway tissue, EBBX) and transbronchial (alveolar tissue, TBBX) biopsies at 04.00 and 16.00. Values are expressed as medians with the 25–75 interquartile range in parentheses above each bar. * † ‡ Means with the same symbol differ significantly, $p \leq 0.05$. Reproduced with permission from Kraft et al. (1996a).

Other features of lung inflammation are also enhanced at night in nocturnal asthmatics. Expression of the epithelial cell marker CD51 in the distal airways at 04.00 is increased in nocturnal asthmatics compared to non-nocturnal asthmatics and normal controls (Kraft et al., 1998b). CD51 corresponds to the alpha v-subunit of the integrin receptor and plays an important role in cellular adhesion and stabilisation of the extracellular matrix. Expression of CD51 was also positively correlated with the degree of airway obstruction ($r = -0.57$, $p = 0.001$), which suggests that both inflammation and tissue repair increase in the distal airways of nocturnal asthmatics at night.

The cellular mechanisms by which lung inflammation increases at night in nocturnal asthma have been the focus of two publications by Kraft and colleagues. In the first study (Kraft et al., 1999b) peripheral blood mononuclear cells (PBMCs) were isolated at 04.00 and 16.00 from 11 nocturnal asthmatics, 12 non-nocturnal asthmatics and 16 normal controls who were not taking either oral or inhaled corticosteroids. Glucocorticoid receptor (GR) binding affinity was measured in PBMCs using a radioligand binding assay, and PBMC responsiveness to hydrocortisone and dexamethasone was determined in a lymphocyte proliferation assay. The PBMCs from nocturnal asthmatics exhibited significantly lower GR binding affinity at 04.00 compared with 16.00, a circadian variation not observed in the other two groups. In nocturnal asthmatics, at 04.00, there was also a significant relationship between FEV_1 and GR binding affinity, such that a lower FEV_1 was associated with reduced GR binding affinity. Lastly, PBMCs from nocturnal asthmatics, but not the other groups, exhibited less suppression of proliferation by corticosteroids at 04.00 than at 16.00, evidence of significant circadian variability in corticosteroid responsiveness. In a second study (Kraft et al., 2001a), the expression of GRβ, an endogenous intracellular inhibitor of steroid action, was analysed in lymphocytes and mononuclear cells obtained by BAL at 04.00 and 16.00 from 10 nocturnal and 7 non-nocturnal asthmatics. They found that GRβ expression was significantly increased at night in nocturnal asthmatics, but not in non-nocturnal asthmatics, particularly in BAL macrophages. Only in the nocturnal asthmatics, at 04.00 macrophage expression of the cytokines IL-8 and tumour necrosis factor (TNF-) α was less readily suppressed by dexamethasone than at 16.00. Expression of mRNA for IL-13, a cytokine that reduces GR binding affinity, was increased at 04.00 only in nocturnal asthmatics, and addition of neutralising antibodies to IL-13 reduced GRβ expression by BAL macrophages. The results of these studies suggest that the corticosteroid responsiveness of both peripheral blood and pulmonary mononuclear cells declines at night, contributing to an overall increase in lung inflammation. The pulmonary macrophage may play an important role in decreased corticosteroid responsiveness in nocturnal asthma.

Cysteinyl leukotrienes (LTC_4, LTD_4, LTE_4) and leukotriene B_4 (LTB_4) are pro-inflammatory mediators that increase in the lungs of nocturnal asthmatics at night. These lipid-derived mediators enhance bronchoconstriction, mucus production and inflammatory cell migration in asthma (Horwitz et al., 1998). The cysteinyl leukotrienes and LTB_4 are all products of the 5-lipoxygenase pathway in the arachidonic acid cascade. Bellia and colleagues observed that night-time urinary LTE_4 from nocturnal asthmatics was 1.6 and 1.8 times greater than in non-nocturnal asthmatics and controls, respectively (Bellia et al., 1996). Urinary LTE_4 was also greater in nocturnal asthmatics during the night, compared to the morning and afternoon, a diurnal variation not observed in the other two groups. Wenzel et al. (1995) measured leukotriene concentrations in the lungs by BAL and in urine. They found that nocturnal asthmatics had significantly greater BAL fluid concentrations of cysteinyl leukotrienes and LTB_4 than normal controls at 04.00 but not at 16.00. In the nocturnal asthmatics, the overnight increase in BAL fluid LTB_4 correlated significantly with the magnitude of the overnight fall in FEV_1 ($r = -0.66$, $p < 0.0001$). Urinary LTE_4 levels also increased overnight in nocturnal asthmatics, but not in normal controls. Treatment with zileuton, a 5-lipoxygenase inhibitor, ameliorated the overnight percentage decline in FEV_1 in the nocturnal asthmatics ($-27.9\% \pm 4.2\%$ to $-19.5\% \pm 4.0\%$), with a trend towards statistical significance ($p = 0.086$). However, zileuton did significantly decrease BAL fluid LTB_4 and urinary LTE_4.

Exhaled nitric oxide, a noninvasive marker of lung inflammation, was greater at 04.00 than at 16.00 in nocturnal asthmatics but not non-nocturnal asthmatics or healthy controls in one study (ten Hacken et al., 1998). Nocturnal asthmatics also had higher exhaled nitric oxide levels at all day and night time points. A separate study (Georges et al., 1999) reported opposite results, however, with a significant fall in mean exhaled nitric oxide levels from 16.00 to 04.00 in nocturnal asthmatics, but not in non-nocturnal asthmatics. In neither study was the change in exhaled nitric oxide levels significantly correlated with the overnight decline in FEV_1. The contribution of nitric oxide to the pathogenesis of nocturnal asthma remains to be determined.

7.2.2.2 Changes in hormones and neurotransmitters

Circadian variation in levels of hormones and neurotransmitters may contribute to the overnight decrease in lung function in nocturnal asthma. Cortisol, a glucocorticoid produced by the adrenal gland, has anti-inflammatory effects on the lung including reductions in inflammatory mediator production, cellular migration, mucus production and tissue oedema. In both asthmatic and nonasthmatic individuals, peak serum cortisol levels occur upon awakening and reach their nadir between 22.00 and midnight (Barnes *et al.*, 1980). Although the cortisol nadir occurs 6 h before the time of minimal lung function at 04.00, it may well set the stage for worsening inflammation in the airways and alveolar tissue of nocturnal asthmatics.

To examine the relationship between serum cortisol and nocturnal asthma, Kraft *et al.* (1998a) measured serum cortisol concentrations in 15 nocturnal asthmatics, 15 non-nocturnal asthmatics and 15 normal controls at 08.00, 12.00, 16.00, 20.00 and 23.00, and at 04.00 and 08.00 the following day. None of the subjects took oral or inhaled corticosteroids within 6 weeks of the study. In all three groups, cortisol levels varied in a circadian fashion, with peak levels at 08.00 (all groups) and the trough occurring at either 20.00 (controls) or 23.00 (nocturnal and non-nocturnal asthmatics). Cortisol levels in the non-nocturnal asthmatics and controls were significantly lower than in the nocturnal asthmatics at a single time point, 20.00 (non-nocturnal asthmatics: 3.5 ± 0.8 mg/mL; controls: 3.4 ± 0.8 mg/mL; nocturnal asthmatics 4.9 ± 0.8 mg/mL; $p = 0.007$). The authors reasoned that the magnitude of this difference was unlikely to be clinically significant. They concluded that nocturnal asthma is not associated with substantially lower cortisol levels during daytime or night-time hours.

Nocturnal asthmatics have been infused with hydrocortisone in two studies, in attempts to abolish the circadian variation in serum cortisol levels. Soutar and colleagues administered physiological doses of hydrocortisone to 6 nocturnal asthmatics and noted at least a 20% improvement in their overnight PEFR in 5 subjects, with a range of 20% to 92% (Soutar *et al.*, 1975). Beam *et al.* (1992a) gave continuous, pharmacological doses of hydrocortisone (100 mg/mL) overnight to 11 nocturnal asthmatics. Hydrocortisone infusion significantly improved the overnight decrement in FEV_1 from $-46\% \pm 4\%$ to $-12\% \pm 3\%$ in the group as a whole. However, there was a wide spectrum of individual responses, with 2 of the 11 subjects demonstrating less than 10% change in their nocturnal drop in FEV_1, 2 of 11 showing complete abolition of nocturnal bronchospasm, and the remaining 7 demonstrating between 45% and 87% improvement over baseline values. These results suggest that there may be steroid-resistant processes contributing to nocturnal worsening of asthma in some patients, a finding supported by *in vitro* studies (Kraft *et al.*, 2001a) showing nocturnal declines in glucocorticoid responsiveness of PBMCs and BAL mononuclear cells from nocturnal asthmatics, described above.

Circadian changes in other hormones and neurotransmitters, including adrenaline, acetylcholine, and melatonin may also play a role in nocturnal worsening of asthma. Plasma concentrations of adrenaline, an endogenous α- and β-adrenoreceptor agonist with bronchodilatory action, peak in the late afternoon (16.00) and reach trough levels in early morning hours (04.00) in day-active individuals (Barnes *et al.*, 1980). Administration of physiological doses of adrenaline to nocturnal asthmatics attenuated but did not abolish the overnight decline in PEFR, however (Barnes *et al.*, 1980). Vagal or cholinergic tone is increased at night in asthmatics and healthy individuals. Asthmatics may have enhanced parasympathetic drive to the sinoatrial node, consistent with increased parasympathetic neural outflow, during both day and night, as evidenced by heart rate variation induced by deep breathing, Valsalva manoeuvre, and rising from a recumbent position (Kallenbach *et al.*, 1985). However, atropine, an ACh muscarinic receptor antagonist, ameliorated but did not prevent the decline in PEFR at 04.00 in nocturnal asthmatics, compared to placebo (Morrison *et al.*, 1988). More recent studies have focused on the role of the hormone melatonin in nocturnal asthma. Although serum melatonin levels are very low during the day in nocturnal asthmatics, non-nocturnal asthmatics and normal control subjects, the kinetics of nocturnal melatonin secretion differed between the three groups, with melatonin levels rising slightly later in the evening in nocturnal asthmatics (Silkoff *et al.*, 2001). Melatonin increased *in vitro* production of the cytokines IL-1, IL-6 and TNF-α in PBMCs from asthmatics at 04.00 compared to 16.00 (Sutherland *et al.*, 2001), raising the possibility that melatonin may modulate night-time pulmonary inflammation in asthma.

7.2.2.3 β_2-Adrenoreceptor function and gene regulation

Mechanisms of endogenous bronchodilation may be impaired in nocturnal asthmatics during the night. Not only do blood adrenaline levels fall at 04.00 (Barnes *et al.*, 1980), but β_2-adrenoreceptor function may be impaired as well. The number and function of β_2-adrenoreceptors decreased significantly from 16.00 to 04.00 in circulating blood leukocytes from nocturnal asthmatics, compared to non-nocturnal asthmatics and healthy controls (Szefler et al., 1991). β_2-Adrenoreceptor downregulation may be related to a genetic polymorphism, substitution of glycine for arginine at position 16. Interestingly, this genetic polymorphism is found with greater frequency in nocturnal asthmatics compared to non-nocturnal asthmatics (80% of nocturnal asthmatics compared to 52% of non-nocturnal asthmatics) (Turki et al., 1995), suggesting there may be a genetic predisposition to nocturnal asthma.

7.2.2.4 Allergic factors

Most asthmatics will develop an immediate (within 15–20 min) bronchoconstrictive response to inhaled allergen to which they are sensitive. Thus, night-time exposure to bedding components (e.g. feathers), pet dander or house dust may contribute to nocturnal worsening of asthma in sensitive patients. However, the fact that nocturnal asthma is also frequently observed in the hospital or laboratory environment, which is virtually free of common allergenic triggers, underscores the importance of other factors in the pathogenesis of nocturnal asthma.

It is well recognised that aeroallergen exposure can also cause a late asthmatic response, characterised by delayed bronchoconstriction, airway oedema and increased airway inflammatory changes 3–8 h after the exposure (O'Byrne, 1998). This phenomenon is observed in 40–50% of adult asthmatics. The late asthmatic response may occur during the night in some patients who are exposed to allergen during the day (Siracusa et al., 1978; Davies et al., 1976). With bronchial challenge testing, it has been shown that asthma may not only occur during the night after initial exposure but may recur for several nights thereafter, without additional allergen exposure (Taylor et al., 1979).

The timing of aeroallergen exposure influences the frequency, rapidity of onset and severity of the late asthmatic response. Mohiuddin and Martin (1990) challenged 10 mild asthmatics with inhaled allergen at 08.00 and 20.00 on separate days. They discovered that the evening challenge was more likely to provoke a late asthmatic response than morning challenge. At 20.00, the time to onset of the late asthmatic response was 3.1 ± 0.3 h, compared to 9.4 ± 2.0 h at 08.00. The maximal decrease in FEV_1 for the late asthmatic response was greater after the 20.00 challenge, $-43\% \pm 3.1\%$ compared to $-32.8\% \pm 5.6\%$ at 08.00. Bronchial hyperresponsiveness to inhaled methacholine was also significantly greater 24 h following evening challenge than after morning challenge. Evening allergen exposure may contribute to development of nocturnal asthma in allergic asthmatics.

7.2.2.5 Extrinsic factors affecting nocturnal asthma

Other processes may influence the development of nocturnal bronchoconstriction in a small subset of asthmatics. GORD (gastro-oesophageal reflux disease), with or without aspiration of gastric contents, has been linked to nocturnal asthma in some studies (Cuttitta *et al.*, 2000; Richter, 2000) but not in others (Sontag, 2000; Tan *et al.*, 1990). Medical antireflux therapy (Field and Sutherland, 1998; Kiljander *et al.*, 1999) and, in severe cases, surgical therapy (Field *et al.*, 1999) are beneficial in some symptomatic patients with GORD, although the impact on lung function appears to be far less than on symptoms. Airway cooling is a common trigger for asthma, and it has been hypothesised that the normal decrease in body temperature during sleep, approximately 1°C, could provoke nocturnal bronchospasm. However, preventing the temperature decrease by administering warmed, humidified air to nocturnal asthmatics improved, but did not eliminate, the overnight drop in lung function in one study (Chen and Chai, 1982). Upper airway obstruction, as seen in obstructive sleep apnoea, may exacerbate nocturnal asthma (Chan *et al.*, 1988), possibly by stimulating pharyngeal and laryngeal neural pathways to produce reflex bronchoconstriction. Treatment of patients who suffer from both obstructive sleep apnoea and asthma with a continuous positive airway pressure (CPAP) device has been shown to improve morning and evening lung function (Chan *et al.*, 1988). However, CPAP alone does not help patients with nocturnal asthma who do not also have obstructive sleep apnoea (Martin and Pak, 1991). Upper airway inflammation related to upper respiratory tract infection or chronic rhinosinusitis may be associated with worsening asthma symptoms, during both the day and night, possibly through reflex-mediated bronchospasm or aspiration of upper airway secretions (Brugman *et al.*, 1993). The role of these extrinsic processes requires further clarification. In the meantime, it is reasonable to conclude that remediation of these factors in selected affected asthmatics may improve their nocturnal asthma control to some degree.

7.3 Chronotherapeutic approach to asthma treatment

The chronotherapeutic approach (Table 7.1) to asthma treatment is based on a sound understanding of circadian variations in pulmonary physiology and inflammation, as well as the onset, duration and circadian variations in efficacy of each medication. This approach recognises the importance of targeting more intense pharmacological therapy at a time when the disease is worse, during the night, while the patient is asleep. Numerous clinical studies have shown that a chronopharmacological approach not only improves nocturnal symptoms and lung function but also results in better daytime asthma control. In addition to medications, it is essential to address any potentially reversible factors that may be exacerbating asthma overnight, such as obstructive sleep apnoea, rhinosinusitis, allergen or irritant exposure, or GORD.

The efficacy of therapeutic interventions should be assessed by asking patients to monitor their asthma symptoms and lung function. Circadian variation in lung function can be assessed at home using a simple and inexpensive peak flow meter. Patients record their PEFR in the late afternoon (time of best lung function), at the time of any

Table 7.1 Chronotherapeutic approach to the treatment of nocturnal asthma

Mild disease (10–20% overnight fall in PEFR and/or night-time symptoms > 2 times/month)

- Inhaled corticosteroids (first line)
- Leukotriene modifier (second line)
- If no improvement, add inhaled long-acting β_2-adrenoreceptor agonist and/or theophylline (once- or twice-daily preparation)

Moderate to severe disease (> 20% overnight fall in PEFR and/or night-time symptoms > once/week)

- Inhaled corticosteroids *plus* inhaled long-acting β_2-adrenoreceptor agonist *or* theophylline (once- or twice-daily preparation)
- If no improvement, consider oral corticosteroids daily (at 15.00) or twice daily (at 08.00 and 15.00), depending on relative severity of daytime and night-time symptoms
- Once symptoms and lung function have improved, add once-daily theophylline or inhaled long-acting β_2-adrenoreceptor agonist, then taper oral corticosteroids

Acute treatment of nocturnal awakenings

- Short-acting inhaled β_2-adrenoreceptor agonist; consider inhaled ACh mR antagonist

For all patients, regardless of severity

- Treat any co-existing obstructive sleep apnoea, gastro-oesophogeal reflux disease, sinusitis/rhinitis. Reduce allergen and irritant exposures

PEFR, peak expiratory flow rate; ACh mR, acetylcholine muscarinic receptor.

awakening during the night, and upon arising in the morning. Deterioration in asthma control is signalled by an increase in day–night variability in PEFR or an overall decline in PEFR. Because most exacerbations of asthma develop gradually over several days, these changes can alert the patient and physician to impending problems. Early intervention often averts the need for emergency care.

7.3.1 Theophylline

Treatment of nocturnal asthma with theophylline exemplifies the chronotherapeutic approach to the treatment of asthma. Successful treatment requires a working knowledge of the pharmacokinetics of different long-acting theophylline preparations in order to achieve higher serum theophylline concentrations at night than during the day (see Chapter 4).

Theophylline relaxes bronchial smooth muscle by inhibiting the enzyme phosphodiesterase, thus increasing intracellular cAMP concentrations. In addition to its mild-moderate bronchodilator activity, theophylline has also been shown to reduce some features of asthmatic airway inflammation (Sullivan *et al.*, 1994; Kidney *et al.*, 1995; Kraft *et al.*, 1996b), including inflammation at night (Kraft *et al.*, 1996c). Kraft *et al.* (1996c) treated nocturnal asthmatics with a once-daily theophylline preparation (Uniphyl tablets) at 19.00 in a randomised, double blind, placebo-controlled crossover study. Theophylline doses were titrated to achieve a serum theophylline concentration of 14–20 µg/mL at 07.00. Subjects underwent BAL at 04.00 at the end of each 2-week treatment period. Compared to placebo, theophylline significantly improved overnight lung function, reduced the percentage of neutrophils in BAL, and diminished production of the granulocyte chemoattractant LTB_4 by stimulated BAL macrophages. The reduction in BAL neutrophils was significantly and positively correlated with the serum theophylline concentration ($r = 0.77$, $p = 0.03$).

Two studies of nocturnal asthma (Martin *et al.*, 1989; D'Alonzo *et al.*, 1990) have shown that maintaining constant serum theophylline concentrations throughout the day is not as beneficial as having peaks and troughs at the appropriate times during the circadian cycle. Martin *et al.* (1989) found that administration of a once-daily theophylline preparation (Uniphyl) at 19.00 was superior to a twice-daily theophylline preparation (Theo-Dur) given at 07.00 and 19.00. Subjects in this study were treated only with theophylline and a 'rescue' β_2-adrenoreceptor agonist. The once-daily preparation resulted in peak serum theophylline concentrations of 15–16 µg/mL between 03.00 and

07.00, with a trough of 7–8 μg/mL at 19.00 (Figure 7.7). In contrast, the twice-daily preparation produced relatively constant therapeutic serum theophylline concentrations in the range 11–14 μg/mL. There was no significant difference between the preparations in FEV_1 measured every 2 h during the daytime. The major difference between the preparations was seen in their effects on nocturnal lung function. Compared to the twice-daily preparation, the once-daily preparation significantly improved FEV_1 at 07.00, reduced nocturnal asthma symptoms, and diminished the magnitude of the overnight fall in FEV_1 (−9% for the once-daily preparation, versus −28% for the twice-daily preparation). D'Alonzo et al. (1990) compared the effect of increasing serum theophylline concentrations on FEV_1 during the 4 h period from 14.00 to 18.00 with that in the period 02.00 to 06.00. They found that there was a significant positive correlation between serum theophylline concentration and the progressive increase in FEV_1 during the nocturnal period (02.00 to 06.00), but not during the afternoon (14.00 to 18.00). Lack of a significant correlation between serum theophylline concentrations and lung function has also been noted in other daytime studies (Martin et al., 1989; Rivington et al., 1985). Thus, during the day, a lower serum

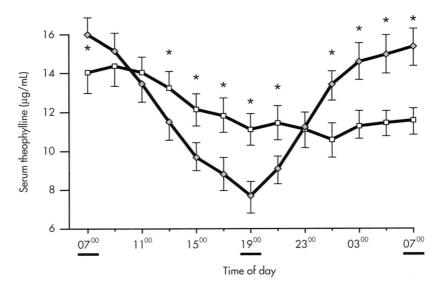

Figure 7.7 Serum theophylline concentrations for the once-daily preparation (diamonds) and the twice-daily preparation (squares). Asterisks indicate $p < 0.05$ comparing the two preparations at the marked hour. Underlining indicates dosing schedule for the preparations (see text for discussion). Adapted from Martin et al. (1989).

theophylline level is needed to achieve a given level of bronchodilation, because of the natural tendency for lung function to be maximal during the day. At night, a higher serum theophylline level provides better lung function and symptom control.

Concerns about the potential adverse effects of theophylline on sleep quality and architecture have not been borne out in clinical studies of nocturnal asthma. In the previously described study by Martin *et al.* (1989), no significant differences in polysomnographic variables were detected between the two groups with higher and lower nocturnal serum theophylline concentrations. Sleep latency, sleep efficiency and sleep stage distribution were similar between the two groups. Similarly, Zwillich *et al.* (1989) found that twice-daily sustained-release theophylline (Theo-Dur) had no significant impact on sleep architecture. In this study, theophylline was superior to the long-acting inhaled β_2-adrenoreceptor agonist bitolterol in subjects' self-ratings of sleep quality. Other studies in normal subjects (Fitzpatrick *et al.*, 1992; Kaplan *et al.*, 1993) and nocturnal asthmatics (Selby *et al.*, 1997) have shown minimal to no adverse effects of theophylline on sleep quality or cognitive functioning.

From a practical standpoint, serum theophylline concentrations should be measured periodically, but this need not be done at 04.00. For the patient taking a once-daily theophylline preparation at 18.00 (with dinner), the serum theophylline concentration is drawn at approximately 08.00, and the 04.00 concentration can be extrapolated from this measurement. Based on the foregoing clinical studies, a higher target serum concentration should be obtained during sleep (16–18 μg/mL) and the level can fall during the daytime (7–9 μg/mL). As with any other medication, the patient should be monitored for side effects and the possibility of drug–drug interactions.

7.3.2 Oral β_2-adrenoreceptor agonists

Extended-release preparations of oral β_2-adrenoreceptor agonists, like once-daily theophylline, can provide a relatively long period of bronchodilation (10–12 h). In one study (Bogin and Ballard, 1992) of nocturnal asthmatics, 8 mg of pulsed-release salbutamol (albuterol) sulfate (Proventil Repetabs) taken at bedtime significantly reduced overnight fall in FEV_1, improved morning PEFR, and decreased subjective awakenings from sleep, compared to placebo. Other studies of oral, extended release β_2-adrenoreceptor agonist preparations have also demonstrated the efficacy of these drugs in improving overnight lung function and reducing nocturnal awakenings (Milledge and Morris, 1979; Fairfax *et al.*, 1980; Moore-Gillon, 1988; Storms *et al.*, 1992).

Systemic side effects of oral β_2-adrenoreceptor agonists may limit their use in some patients. However, overall sleep quality does not appear to be significantly affected by oral β_2-adrenoreceptor agonist use. Bogin and Ballard (1992) detected no difference between pulsed-release salbutamol sulfate and placebo on multiple polysomnographic measures of sleep architecture. Pulsed-release salbutamol was associated with an increase in mean heart rate (placebo 66 ± 2 beats per minute; salbutamol 73 ± 2 beats per minute; p = 0.005), but this was not felt to be clinically significant. Tremor (Storms *et al.*, 1992) and gastrointestinal intolerance (Fairfax *et al.*, 1980; Bogin and Ballard, 1992) have been reported by subjects using oral β_2-adrenoreceptor agonist preparations in studies of nocturnal asthma, although most subjects did not stop using the drug because of side effects in these short-term studies.

7.3.3 Inhaled β_2-adrenoreceptor agonists

The long-acting inhaled β_2-adrenoreceptor agonists are a relatively recent addition to the list of pharmacological options for treating asthma. Compared to short-acting β_2-adrenoreceptor agonists, such as salbutamol, the inhaled β_2-adrenoreceptor agonist salmeterol has a longer duration of action ($\geq$12 h), making it a good candidate for relief of nocturnal bronchospasm. In addition to its bronchodilatory activity, salmeterol has been found to have some mild anti-inflammatory effects (Butchers *et al.*, 1991; Pedersen *et al.*, 1993; Proud *et al.*, 1998), although this has not been observed in other studies (Kraft *et al.*, 1997; Gardiner *et al.*, 1994). In current asthma treatment guidelines (National Asthma Education and Prevention Program, 1997) long-acting β_2-adrenoreceptor agonists are indicated for patients with moderate or severe, persistent disease who are already taking an inhaled corticosteroid. As with theophylline and oral β_2-adrenoreceptor agonists, use of long-acting inhaled β_2-adrenoreceptor agonists without concomitant anti-inflammatory therapy is not recommended (National Asthma Education and Prevention Program, 1997).

Salmeterol has been found to be an effective treatment for nocturnal asthma in short-term studies (Fitzpatrick *et al.*, 1990; Faurschou *et al.*, 1994; Wiegand *et al.*, 1999). Fitzpatrick and colleagues compared the effects of 50 μg and 100 μg of salmeterol inhaled twice daily for 2 weeks on lung function and sleep quality in 20 nocturnal asthmatics, most of whom were taking an inhaled or oral corticosteroid concomitantly

(Fitzpatrick et al., 1990). Both the 50 μg twice-daily and 100 μg twice-daily regimens significantly reduced the overnight fall in PEFR, compared to placebo, with no difference detected between the two doses. However, only salmeterol 50 μg twice daily improved sleep quality; the subjects spent less time awake or in light sleep and more time in stage 4 slow-wave sleep compared to placebo. Salmeterol 100 μg twice daily may have had a greater stimulatory effect on the central nervous system, resulting in poorer sleep quality. Utilising a different dosing regimen, Faurschou et al. (1994) compared 3 weeks of treatment with salmeterol 50 μg twice daily and salmeterol 100 μg once daily at night. Subjects had nocturnal asthma that was poorly controlled despite the use of other drugs, including oral and inhaled corticosteroids, theophylline or sodium cromoglicate. Using diary card and PEFR assessment of lung function, they found that both dosing regimens of salmeterol were superior to placebo in their ability to increase morning PEFR and reduce daytime and night-time asthma symptoms, nocturnal awakenings, and supplemental daytime rescue β_2-adrenoreceptor agonist use. Although the 50 μg twice daily regimen significantly reduced circadian variation in PEFR, compared to placebo ($p = 0.003$), 100 μg once daily proved superior to 50 μg twice daily ($p = 0.008$) (overnight change in PEFR: placebo -23.5 L/min; 50 μg twice daily -4.2 L/min; 100 μg once daily 13.71 L/min). This study did not investigate the effects of these dosing regimens on sleep quality or architecture.

In long-term (3–12 month) studies (Britton et al., 1992; Pearlman et al., 1992; Lundback et al., 1993; Greening et al., 1994; Lockey et al., 1999) of its overall efficacy for asthma, salmeterol was found to be effective in reducing circadian variation in lung function, in addition to improving daytime lung function, asthma symptoms and asthma-related quality of life. The majority of subjects in these studies were treated concomitantly with an inhaled corticosteroid. Long-term administration of salmeterol was not associated with deterioration in asthma control in any of these studies, despite previous concerns that regular, long-term use of short-acting β_2-adrenoreceptor agonists might decrease asthma control (Sears et al., 1990).

To determine whether monotherapy with salmeterol has anti-inflammatory action in the setting of nocturnal asthma, Kraft et al. (1997) performed BAL at 04.00 in moderately severe nocturnal asthmatics treated with either salmeterol (50 μg at 07.00 and 19.00) or placebo in a double blind, randomised crossover study. Salmeterol treatment was associated with significantly fewer nocturnal awakenings and less supplemental salbutamol use, compared to placebo. However, salmeterol did not significantly reduce the overnight fall in FEV_1 or improve the FEV_1 at 04.00. Regarding markers of airway inflammation, there was a trend towards a decrease in BAL eosinophils ($p = 0.08$) and LTB_4 production by stimulated BAL macrophages ($p = 0.07$) after 6 weeks of salmeterol treatment. Kraft and colleagues reasoned that salmeterol may not have improved overnight lung function because their subjects were treated with salmeterol alone, whereas subjects in other studies (Fitzpatrick et al., 1990; Britton et al., 1992; Lundback et al., 1993; Greening et al., 1994; Lockey et al., 1999) were, by and large, treated with both salmeterol and an inhaled corticosteroid. In contrast, Weersink and colleagues (Weersink et al., 1997a) found that 6 weeks of treatment with salmeterol (50 μg twice daily) alone did significantly reduce circadian variation in PEFR, but in a population with milder asthma than in the study of Kraft et al. However, the combination of the inhaled corticosteroid fluticasone (fluticasone 250 μg twice daily) and salmeterol was found to be superior to salmeterol alone for reducing bronchial hyperreactivity in Weersink's study.

Two studies have compared inhaled salmeterol with theophylline preparations for the treatment of nocturnal asthma. Selby et al. (1997) compared the effects of 14 days of treatment with either salmeterol 50 μg twice daily or a twice-daily theophylline preparation (Theodur) on sleep quality, psychometric performance and overnight lung function. Subjects had a 15–73% overnight fall in PEFR during the baseline run-in period. Daytime plasma theophylline levels were titrated to 10–20 μg/mL; night-time levels were not measured. The median trough plasma theophylline level was 11.1 μg/mL. Most subjects were also taking inhaled corticosteroids throughout the study. Subjects underwent cognitive testing and polysomnography at the end of each treatment period. For the two medications, the overnight fall in PEFR was similar (salmeterol -2.3%; theophylline -3.5%; $p = 0.4$), but there were significantly fewer nocturnal awakenings with salmeterol. Compared to theophylline, salmeterol treatment resulted in significantly fewer nocturnal microarousals, higher quality of life scores and improved performance on one of several cognitive tests (visual vigilance). No patient preference for either therapy was noted.

In a second study comparing salmeterol and theophylline, Wiegand and colleagues found salmeterol to be better than theophylline in improving overnight lung function (Wiegand et al., 1999). A significant overnight fall in FEV_1 was observed during 15 days of treatment with placebo or theophylline, but not with salmeterol 42 μg twice daily. Compared to placebo and theophylline, salmeterol significantly improved morning and evening PEFR and decreased both daytime and night-time salbutamol use. Sleep quality, as measured by subject questionnaire, improved with salmeterol and placebo, but not with theophylline. Polysomnographic analysis showed less stage 2 sleep during treatment with theophylline compared to placebo, but there were no other statistically significant differences in sleep architecture. Drug-related adverse events (headache, tremor, dizziness, anxiety, agitation, nausea, vomiting) were more frequent ($p = 0.008$) during treatment with theophylline than with placebo or salmeterol. The lack of improvement after treatment with theophylline, compared to placebo, in the study of Wiegand et al. deserves some comment. The twice-daily theophylline regimen may have been inadequate to produce maximal nocturnal bronchodilatation. Theophylline concentrations were not measured during the night and the exact times of dosing are not given. On the day prior to overnight spirometry, mean serum theophylline concentrations measured 5–9 h after the morning dose were only 12.2 μg/mL (range 6.2–18.4 μg/mL), and concentrations at the critical time of 03.00 to 07.00 were likely even lower, based on the findings of Martin et al. (1989) discussed above.

Salmeterol has also been compared with long-acting oral β_2-adrenoreceptor agonist preparations. Martin et al. (1999) compared the efficacy of extended-release salbutamol sulfate and salmeterol for the treatment of nocturnal asthma. In a double blind, double-dummy, randomised crossover design study, 46 subjects with nocturnal asthma received either 3 weeks of oral extended-release salbutamol sulfate (Volmax) or salmeterol 42 μg twice daily. Both medications were taken between 06.00 and 08.00 and between 18.00 and 20.00. A larger dose of salbutamol sulfate was taken in the evening (8 mg) than in the morning (4 mg), based on the chronotherapeutic approach of Postma and colleagues (Postma et al., 1984). Subjects were allowed to continue on a stable dose of inhaled corticosteroid, but they could not use systemic corticosteroids, theophylline, ipratropium bromide or β_2-adrenoreceptor agonists other than inhaled rescue salbutamol. Daily diary card and PEFR and FEV_1 measurements revealed that, compared to baseline measurements, both the oral salbutamol tablets and salmeterol produced significant and similar improvements in morning PEFR, overnight change in PEFR, and percentage of nights without awakenings. Both treatments significantly decreased the use of rescue salbutamol and nocturnal asthma symptoms. The overall incidence of adverse effects was not statistically different between groups, although five subjects reported tremor on extended-release salbutamol tablets, compared to no subjects on salmeterol. The authors concluded that both extended-release salbutamol tablets and inhaled salmeterol resulted in similar bronchodilation and good control of nocturnal asthma symptoms.

In a separate study, Crompton et al. (1999) compared salmeterol with bambuterol, an oral, long-acting pro-drug of the β_2-adrenoreceptor agonist terbutaline. Using electronic diary cards and PEFR recording, Crompton et al. compared the efficacy of bambuterol (20 mg once daily) with salmeterol (50µg twice daily) in asthmatics who were also taking a stable dose of an inhaled corticosteroid. Bambuterol and salmeterol were found to be similar in their ability to reduce the percent overnight fall in PEFR (median reduction for bambuterol -11%, for salmeterol -14%, $p = 0.17$) and improve morning PEFR (50 L/min increase over baseline for bambuterol, 55 L/min for salmeterol, $p = 0.53$). No significant differences were found between the drugs in their ability to reduce nocturnal rescue β_2-adrenoreceptor agonist use or nocturnal asthma symptoms. Both drugs reduced the incidence of nocturnal awakenings by at least 20%. In addition, both treatments reduced daytime rescue β_2-adrenoreceptor agonist use and asthma symptoms.

In summary, the long-acting inhaled β_2-adrenoreceptor agonist salmeterol has been shown to be effective in the treatment of nocturnal asthma in multiple clinical studies. Salmeterol is best employed as add-on therapy for patients with moderate persistent or severe persistent disease whose symptoms and lung function are not adequately improved by the use of inhaled corticosteroids alone (National Asthma Education and Prevention Program, 1997). Compared to other long-acting bronchodilators, such as oral β_2-adrenoreceptor agonists and theophylline, salmeterol appears to be at least as efficacious, if not superior, in controlling nocturnal symptoms and improving overnight lung function. Side effect profiles, patient preference and medication costs will likely influence treatment selection.

7.3.4 Oral and inhaled corticosteroids

Corticosteroid treatment is the cornerstone of anti-inflammatory therapy for asthma. Inhaled corticosteroids are preferred to oral or intravenous preparations, except for the treatment of serious acute exacerbations, in order to minimise systemic side effects, including adrenal suppression (National Asthma Education and Prevention Program, 1997). The chronotherapeutic approach to corticosteroid administration for asthma takes into account the important influence of dosage timing on both the adverse and the desirable effects of these medications. Corticosteroid therapy is adjusted to achieve the most favourable balance between time-dependent variations in both adversity and efficacy.

A common concern of both asthmatics and health care providers is the risk of adrenal gland suppression associated with corticosteroid medications, particularly those given by the oral or intravenous route. In healthy, diurnally active people, the adrenal glands exhibit a circadian rhythm in cortisol production, with peak serum cortisol levels occurring at the time of awakening and trough levels at approximately 23.00. The circadian pattern of adrenal cortisol production in asthmatics does not differ substantially from that of normal individuals (Kraft *et al.*, 1998a). Adrenal cortisol production is regulated by the pituitary hormone adrenocorticotrophic hormone (ACTH), which is in turn regulated by the hypothalamic hormone corticotrophin-releasing factor (CRF). When plasma corticosteroid levels rise, owing to either endogenous production or corticosteroid drug therapy, secretion of ACTH and CRF is inhibited via a negative feedback mechanism. However, there are periods within the circadian cycle during which this negative feedback mechanisms is less susceptible to the effects of exogenous corticosteroids, as demonstrated in several studies. Grant *et al.* (1965) and DiRaimondo and Forsham (1956) compared the effects of administering oral corticosteroids as either a single morning dose or several divided doses spread out over the day. In these studies, adrenal suppression, as reflected by plasma and urinary 17-hydroxycorticosteroid (17-OHCS) levels, was minimal with the single morning dose schedule, whereas the divided-dose schedule resulted in significant suppression of 17-OHCS production at the same total daily dose. Studies by Nichols *et al.* (1965) and Ceresa *et al.* (1969) compared the effects of a single dose of systemic corticosteroid given at different times during the day and night on plasma 17-OHCS levels in healthy volunteers. Maximal adrenal suppression occurred when exogenous corticosteroids were given at the end of the activity cycle or during sleep, in the period between midnight and 04.00. Adrenal suppression was minimised when corticosteroids were given as a single dose between 08.00 and 16.00. Two or four times as much corticosteroid was needed between 08.00 and 16.00 to produce adrenal suppression as in the period between midnight and 04.00 (Ceresa *et al.*, 1969). Thus, regardless of the time of administration, if corticosteroid doses are sufficiently high, adrenal suppression can certainly occur. However, the risk of adrenal suppression by systemic corticosteroids can be minimised by administering a single dose of the medication in the period between 08.00 and 16.00. Minimising the total corticosteroid dosage and the duration of therapy and recognising individual susceptibilities can also reduce this risk.

In the treatment of asthma, periods of maximal therapeutic efficacy for corticosteroids also exist within the circadian cycle. Reinberg *et al.* (1974) examined the impact of different dosing times (07.00 and 19.00, or 03.00 and 15.00) of 40 mg of intramuscular methylprednisolone in 12 asthmatic boys aged 7–15 years. Peak expiratory flow rates were measured every 2 h while awake and once at night. The largest change in 24 h average PEFR was seen when methylprednisolone was given at 15.00 (116 ± 2.96 L/min) or at 19.00 (112 ± 2.65 L/min) and the smallest change (104 ± 2.93 L/min) occurred at the 03.00 dosing time ($p < 0.005$ for 15.00 versus 03.00; $p < 0.05$ for 19.00 versus 03.00). Subsequent studies of longer duration (1–5 weeks) (Reinberg *et al.*, 1977, 1983) in adult asthmatics demonstrated that dosing oral corticosteroids at 08.00 and 15.00 proved more efficacious in increasing 24 h mean PEFRs than dosing at 15.00 and 20.00, without producing significant adrenal suppression. Thus, it is possible to achieve improved asthma control without significantly increasing the risk of adrenal suppression by judicious dosage timing.

As discussed previously, inflammation in the lungs of nocturnal asthmatics is not a static process but varies with the sleep–wake cycle, worsening during the early morning hours. Corticosteroid therapy can be timed to reduce this inflammation and the associated deterioration in lung function, as shown by the work of Beam and colleagues (Beam *et al.*, 1992b). In a double blind, placebo-controlled, crossover design study, 7 adult nocturnal asthmatics were given a single 50 mg dose of oral prednisone at 08.00, 15.00 and 20.00 in a random order, each separated by approximately one week. Subjects underwent spirometry and bronchoscopy with BAL at 04.00. Only the 15.00 dose of prednisone significantly attenuated the overnight fall in FEV_1, compared to placebo (Figure 7.8). Neither the 08.00 nor the 20.00 prednisone dose improved overnight lung function, compared to placebo. In addition, the 15.00 prednisone dose, but not the 08.00 or the 20.00 dose, significantly reduced the number of eosinophils, neutrophils, lymphocytes and macrophages in BAL fluid at 04.00 (Figure 7.9). The 08.00 and 20.00 prednisone doses had no effect on the number of any type of cell in BAL fluid. Lastly, only the 15.00 prednisone dose significantly reduced blood eosinophil numbers at both 20.00 and 04.00. These results suggest that only the 15.00 dose interrupted the cascade of events that produce lung inflammation and airflow limitation in nocturnal asthma.

The introduction of inhaled corticosteroids represented a significant advance in anti-inflammatory asthma therapy, particularly from the standpoint of reducing systemic side effects. Multiple studies have

Chronotherapeutic approach to asthma treatment 179

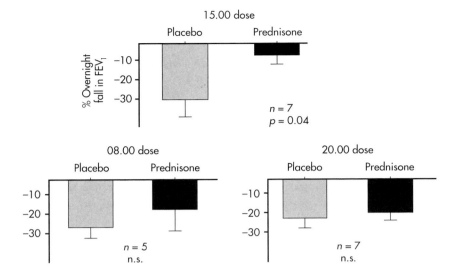

Figure 7.8 The effect of oral prednisone and placebo administration schedule on the overnight fall in FEV_1. Only the dose of prednisone at 15.00 was effective in reducing the overnight fall in FEV_1. n.s., not significant. Adapted from Beam *et al.* (1992b).

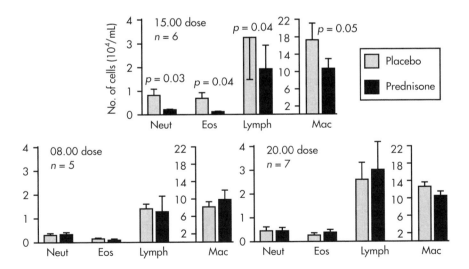

Figure 7.9 The effect of placebo (shaded bars) versus prednisone (solid bars) dosing at 15.00, 08.00 and 20.00 on inflammatory cell numbers in bronchoalveolar lavage fluid at 04.00. Only the dose of prednisone at 15.00 produced a significant reduction in all cell types. Neut, neutrophils; Eos, eosinophils; Lymph, lymphocytes; Mac, macrophages. Constructed from data in Beam *et al.* (1992b).

established the efficacy of inhaled corticosteroids in reducing airway inflammation and airway hyperresponsiveness, and improving both daytime and night-time lung function (Barnes, 1998). However, most inhaled corticosteroid preparations are dosed twice or four times during the day. Rarely is consideration given to the potential impact of these schedules on overnight lung function. To determine the impact of dosage timing of inhaled corticosteroids upon asthma control, Pincus et al. (1995) randomised 32 asthmatics to receive 4 weeks of inhaled triamcinolone acetonide, dosed either as 800 µg once daily or 200 µg four times daily. They found that the once-daily regimen and the four times daily regimen both improved daytime FEV_1, morning PEFR and evening PEFR to a similar degree and resulted in comparable reductions in rescue β_2-adrenoreceptor agonist use. Neither dosing regimen significantly reduced morning serum cortisol levels or 24 h urinary cortisol excretion. In a second study, Pincus et al. (1997) compared inhaled triamcinolone 800 µg once daily at 08.00, triamcinolone 800 µg once daily at 17.30, and triamcinolone 200 µg four times daily in subjects with mild-moderate asthma. Significant improvement in both morning and evening PEFR was seen for the 17.30 group and the four times daily group but not the 08.00 group, and the result for the 17.30 group was statistically superior to the 08.00 group. Compared to the 17.30 group, those receiving four times daily dosing had a greater improvement in median morning PEFR (four times daily: 46 L/min; dose at 17.30: 13 L/min), but this did not reach statistical significance ($p = 0.30$). Use of rescue β_2-adrenoreceptor agonist medication declined similarly in all treatment groups. No significant differences in serum cortisol levels or 24 h urinary cortisol between dosing schedules were detected. Taken together, the results of these two studies suggest that giving the entire daily dosage of triamcinolone (triamcinolone 800 µg) at a single time in the afternoon can be effective asthma therapy. Additionally, when data from the 15.00 dose group in the first study and the 17.30 dose group in the second study are compared, a greater improvement in morning and evening PEFR is seen with dosing at 15.00, suggesting that this may be the optimal time for single-dose administration of inhaled corticosteroids in asthma.

A recent study (Noonan et al., 2001) comparing once-daily and twice-daily treatment of asthma with mometasone furoate (MF), a high-potency inhaled corticosteroid, via a dry powder inhaler (DPI) device, provides additional insight into the impact of dosage timing with inhaled corticosteroids. After a 2-week open-label run-in phase, during which all subjects received MF-DPI 200 µg twice daily, Noonan et al. randomised 286 subjects to receive one of the following regimens for 3 months:

MF-DPI 200 μg twice daily; MF-DPI 200 μg once daily in the morning; MF-DPI 200 μg once daily in the evening; MF-DPI 400 μg once daily in the morning; or placebo. Owing to the complexity of the study, it was not possible to study a group receiving MF-DPI 400 μg once daily in the evening. Comparing the effects of the different regimens on mean change in FEV_1 (Figure 7.10), the 400 μg dose given only in the morning produced the same results as the 200 μg twice daily dosing regimen. This result is different from those of Pincus et al. (1997), in which a single morning dose of triamcinolone was inferior to the other dose-timing schedules. However, MF is a much more potent inhaled corticosteroid than triamcinolone, and thus administration of the entire daily dose in the morning yields the same outcome as twice-daily dosing. Of great interest was the finding that administering half the daily dose, MF-DPI 200 μg, at a single time in the evening had a statistically comparable effect on FEV_1 to giving MF-DPI 400 μg in the morning and to 200 μg twice daily. However, MF-DPI 200 μg in the evening was less effective in reducing nocturnal awakenings than the other regimens. The results of this study demonstrate an important chronotherapeutic principle,

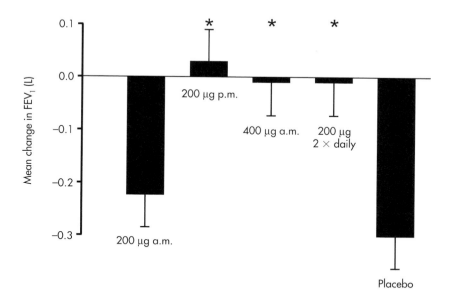

Figure 7.10 The effect of dosage timing with mometasone furoate dry powder inhaler (MF-DPI) on mean change in FEV_1 at study end point. MF-DPI 200 μg given only in the evening was as effective as MF-DPI 400 μg once in the morning and MF-DPI 200 μg twice daily at study end point and during the 12-week treatment phase. *$p < 0.01$ versus placebo. Constructed from data in Noonan et al. (2001).

namely that if the correct dose timing schedule is selected, a lower of dose of a potent medication, such as MF, can be as effective as a higher dose. The results also demonstrate that a single, higher morning dose of medication may be therapeutically equivalent to splitting that dose for twice-daily administration.

Many asthmatics with severe, persistent disease who require long periods of oral corticosteroids have nocturnal asthma. Since the nocturnal worsening of asthma is associated with an increase in inflammation and mediator release at night, it is reasonable to assume that therapy directed at these nocturnal events would be of benefit. An important fact to consider is that increasing a steroid dose 10-fold will only increase the length of action 2-fold. As a result, increasing the morning oral corticosteroid dose in these individuals leads to more steroid complications without improvement in night-time asthma control. A single afternoon dosing time should be considered for these patients. For patients with milder disease who are maintained on inhaled corticosteroids, a single daily dose of medication may be as effective as multiple doses during the day, particularly for higher-potency medications, and may result in better adherence to medication.

7.3.5 Leukotriene modifiers

Within the last 10 years, the role of leukotrienes in bronchoconstriction and airway inflammation in asthma has been established in basic and clinical research studies (Horwitz *et al.*, 1998). Two classes of leukotriene-modifier medications have become available, leukotriene-receptor antagonists (zafirlukast, montelukast, pranlukast) and 5-lipoxygenase enzyme inhibitors (zileuton). Leukotriene modifiers are considered to be second-line controller medications for mild, persistent asthma in current asthma management guidelines (National Asthma Education and Prevention Program, 1997).

Only the work by Wenzel and colleagues, described previously, evaluated the efficacy of a leukotriene-modifier drug in a well-defined population of nocturnal asthmatics in the laboratory setting (Wenzel *et al.*, 1995). Subjects were not on inhaled corticosteroids, sodium cromoglicate or nedocromil. In their study, short-term (7 days) treatment with zileuton reduced the magnitude of overnight decline in FEV_1 in nocturnal asthmatics from $-27.9\% \pm 4.2\%$ to $-19.5\% \pm 4.0\%$, compared to placebo, a difference that trended towards statistical significance ($p = 0.086$). The FEV_1 at 03.30 increased in 9 out of 12 subjects while taking zileuton, compared to placebo.

Long-term clinical studies designed to assess the overall impact of leukotriene modifiers on asthma control have shown that zafirlukast (Spector *et al.*, 1994), montelukast (Reiss *et al.*, 1998; Noonan *et al.*, 1998) and zileuton (Israel *et al.*, 1996; Liu *et al.*, 1996) reduce nocturnal asthma symptoms and improve morning PEFRs, a marker of overnight asthma control, in addition to other physiological and symptom-related outcomes. Combination therapy with montelukast plus a low-dose inhaled beclometasone (200 µg twice daily) (Laviolette *et al.*, 1999) or montelukast plus the antihistamine loratadine (Reicin *et al.*, 2000) has been shown to reduce nocturnal awakenings in mild-moderate asthma more effectively than low-dose beclometasone (Laviolette *et al.*, 1999) or montelukast alone (Laviolette *et al.*, 1999; Reicin *et al.*, 2000).

Leukotriene modifiers have been compared to inhaled corticosteroids in several studies, with nocturnal awakenings reported on diary cards as a secondary outcome measure. For example, Malmstrom and colleagues compared 12 weeks of treatment with montelukast 10 mg once daily with low-dose inhaled beclometasone (200 µg twice daily) and placebo in patients with mild-moderate asthma who awakened 5–6 times per week with asthma symptoms, at baseline (Malmstrom *et al.*, 1999). Both montelukast and beclometasone significantly improved mean morning FEV_1 compared to placebo (placebo 0.7% (−2.3 to 3.7); montelukast 7.4% (4.6 to 10.1); beclometasone 13.1% (10.1 to 16.2); $p < 0.001$ for active treatments versus placebo), but beclometasone was statistically superior to montelukast ($p < 0.01$). Similarly, both montelukast and beclometasone significantly reduced nocturnal awakenings, but beclometasone was superior to montelukast ($p < 0.01$). In a separate study, Busse *et al.* (2001) compared the efficacy of montelukast 10 mg once daily to low-dose inhaled fluticasone (88 µg twice daily) in 533 subjects with mild-moderate persistent asthma who, on average, awakened once per night with asthma symptoms. After 24 weeks of treatment with either montelukast or inhaled fluticasone, they found that low-dose inhaled fluticasone resulted in significantly greater improvements in morning FEV_1 (fluticasone 22.9%; montelukast 14.5%; $p < 0.001$) and fewer nocturnal awakenings due to asthma (fluticasone 62% decrease; montelukast 47.5% decrease; $p = 0.023$), in addition to significantly greater improvements in evening PEFR, overall asthma symptom scores and use of rescue β_2-adrenoreceptor agonists. Adverse event and asthma exacerbation profiles were similar for the two groups. Considered together, the findings of these studies suggest that, although montelukast does reduce nocturnal asthma symptoms and improve morning lung function, the effects are not as great as those seen with inhaled corticosteroids dosed at conventional times during the day.

7.3.6 Sodium cromoglicate and nedocromil

Sodium cromoglicate and nedocromil are relatively weak inhaled anti-inflammatory agents (compared to inhaled corticosteroids). The low incidence of side effects has made these agents first-line controllers for use in asthmatic children (National Asthma Education and Prevention Program, 1997) although more recent evidence suggests they are less effective than inhaled corticosteroids (Szefler et al., 2000).

By preventing pulmonary mast cell degranulation, inhaled sodium cromoglicate and nedocromil inhibit release of histamine and other bronchoconstrictive and inflammatory mediators into airway tissue. Plasma histamine has been shown to vary in a circadian fashion in asthmatics, but not in healthy controls, with levels peaking at 04.00, the nadir in pulmonary function (Barnes et al., 1980).

Several studies (Hetzel et al., 1985; Morgan et al., 1986; Williams and Stableforth, 1986) have shown that sodium cromoglicate and nedocromil have a modest effect, at best, on nocturnal asthma. The effects of these drugs, particularly nedocromil, are best appreciated after 4–12 weeks of therapy, suggesting that the response is due to actions other than mast cell membrane stabilisation.

7.3.7 Acetylcholine muscarinic-receptor (ACh mR) antagonists

Vagolytic medications such as intravenous atropine are more effective bronchodilators during the night than during the daytime. Morrisson et al. (1988) administered intravenous atropine to asthmatics at 16.00 and 04.00. A significant increase in PEFR was observed at both 16.00 (PEFR 400–440 L/min) and 04.00 (260–390 L/min), but there was a greater percentage increase at 04.00.

Inhaled ACh mR antagonists may improve overnight lung function in some nocturnal asthmatics. Coe and Barnes (1986) studied the effect of the inhaled quaternary ACh mR antagonist oxitropium bromide, 0.2 mg at bedtime. They found a group of subjects who had good improvement in overnight lung function, but they could not otherwise differentiate this group from the nonresponders.

In general, ACh mR antagonists do not have a great impact on the overnight decrement in lung function seen in nocturnal asthma.

Moreover, currently available inhaled ACh mR antagonists, such as ipratropium bromide, are too short-acting (4–6 h) to cover the entire night. If given at bedtime, higher doses than used during the day must be given to lengthen the effective duration of action of the drug. Alternatively, if a patient awakens during the night, inhalation of ipratropium bromide at that time can be beneficial. The efficacy of newer, longer-lasting ACh mR antagonists, such as tiotropium, which can be dosed once every 24 h (Barnes, 2000), in the treatment of nocturnal asthma remains to be evaluated.

7.4 Summary

In conclusion, nocturnal worsening of asthma is a serious clinical problem that often goes unrecognised by patients and health care professionals alike. Circadian changes in lung function in asthmatics result from a complex interaction of pathophysiological processes, including worsening of distal lung function and inflammation at night. A chronotherapeutic approach to asthma takes these circadian changes into account in the selection and dosage timing of medications. In the future, optimal treatment of nocturnal asthma, and of asthma in general, will target temporal changes in asthma severity as well as the location of lung inflammation.

References

Ballard R D, Saathoff M C, Patel D K, *et al.* (1989). Effect of sleep on nocturnal bronchoconstriction and ventilatory patterns in asthmatics. *J Appl Physiol* 67: 243–249.

Ballard R D, Tan W C, Kelly P L, *et al.* (1990). Effect of sleep and sleep deprivation on ventilatory response to bronchoconstriction. *J Appl Physiol* 69: 490–497.

Barnes P J (1998). Current issues for establishing inhaled corticosteroids as the antiinflammatory agents of choice in asthma. *J Allergy Clin Immunol* 101: S427–S433.

Barnes P J (2000). The pharmacological properties of tiotropium. *Chest* 117: 63S–66S.

Barnes P, FitzGerald G, Brown M, *et al.* (1980). Nocturnal asthma and changes in circulating epinephrine, histamine, and cortisol. *N Engl J Med* 303: 263–267.

Beam W R, Ballard R D, Martin R J (1992a). Spectrum of corticosteroid sensitivity in nocturnal asthma. *Am Rev Respir Dis* 145: 1082–1086.

Beam W R, Weiner D E, Martin R J (1992b). Timing of prednisone and alterations of airways inflammation in nocturnal asthma. *Am Rev Respir Dis* 146: 1524–1530.

Bellia V, Bonanno A, Cibella F, *et al.* (1996). Urinary leukotriene E4 in the assessment of nocturnal asthma. *J Allergy Clin Immunol* 97: 735–741.

Bender B G, Annett R D (1999). Neuropsychological outcomes of nocturnal asthma. *Chronobiol Int* 16: 695–710.

Bogin R M, Ballard R D (1992). Treatment of nocturnal asthma with pulsed-release albuterol. *Chest* 102: 362–366.

Britton M G, Earnshaw J S, Palmer J B (1992). A twelve month comparison of salmeterol with salbutamol in asthmatic patients. European Study Group. *Eur Respir J* 5: 1062–1067.

Brugman S M, Larsen G L, Henson P M, *et al*. (1993). Increased lower airways responsiveness associated with sinusitis in a rabbit model. *Am Rev Respir Dis* 147: 314–320.

Busse W W, Lemanske R F Jr (2001). Asthma. *N Engl J Med* 344: 350–362.

Busse W, Raphael G D, Galant S, *et al*. (2001). Low-dose fluticasone propionate compared with montelukast for first- line treatment of persistent asthma: a randomized clinical trial. *J Allergy Clin Immunol* 107: 461–468.

Butchers P R, Vardey C J, Johnson M (1991). Salmeterol: a potent and long-acting inhibitor of inflammatory mediator release from human lung. *Br J Pharmacol* 104: 672–676.

Ceresa F, Angeli A, Boccuzzi G, *et al*. (1969). Once-a-day neurally stimulated and basal ACTH secretion phases in man and their response to corticoid inhibition. *J Clin Endocrinol Metab* 29: 1074–1082.

Chan C S, Woolcock A J, Sullivan C E (1988). Nocturnal asthma: role of snoring and obstructive sleep apnoea. *Am Rev Respir Dis* 137: 1502–1504.

Chen W Y, Chai H (1982). Airway cooling and nocturnal asthma. *Chest* 81: 675–680.

Clark T J, Hetzel M R (1977). Diurnal variation of asthma. *Br J Dis Chest* 71: 87–92.

Coe C I, Barnes P J (1986). Reduction of nocturnal asthma by an inhaled anticholinergic drug. *Chest* 90: 485–488.

Crompton G K, Ayres J G, Basran G, *et al*. (1999). Comparison of oral bambuterol and inhaled salmeterol in patients with symptomatic asthma and using inhaled corticosteroids. *Am J Respir Crit Care Med* 159: 824–828.

Cuttitta G, Cibella F, Visconti A, *et al*. (2000). Spontaneous gastroesophageal reflux and airway patency during the night in adult asthmatics. *Am J Respir Crit Care Med* 161: 177–181.

D'Alonzo G E, Smolensky M H, Feldman S, *et al*. (1990). Twenty-four hour lung function in adult patients with asthma. Chronoptimized theophylline therapy once-daily dosing in the evening versus conventional twice-daily dosing. *Am Rev Respir Dis* 142: 84–90.

Davies R J, Green M, Schofield N M (1976). Recurrent nocturnal asthma after exposure to grain dust. *Am Rev Respir Dis* 114: 1011–1019.

Dethlefsen U, Repgas R (1985). Ein Neues therapieprinzip bei nachtilchen asthma. *Klin Med* 80: 44–47.

De Vries K, Goei J T, Boody-Noord H, Orie N G M (1962). Changes during 24 hours in the lung function and histamine reactivity of the bronchial tree in asthmatics and bronchitic subjects *Int Arch Allergy* 20: 93–101.

Diette G B, Markson L, Skinner E A, *et al*. (2000). Nocturnal asthma in children affects school attendance, school performance, and parents' work attendance. *Arch Pediatr Adolesc Med* 154: 923–928.

DiRaimondo V C, Forsham P H (1956). Some clinical implications of spontaneous diurnal variation in adrenal cortical secretory activity. *Am J Med* 21: 321–323.

Fairfax A J, McNabb W R, Davies H J, et al. (1980). Slow-release oral salbutamol and aminophylline in nocturnal asthma: relation of overnight changes in lung function and plasma drug levels. *Thorax* 35: 526–530.

Faurschou P, Engel A M, Haanaes O C (1994). Salmeterol in two different doses in the treatment of nocturnal bronchial asthma poorly controlled by other therapies. *Allergy* 49: 827–832.

Field S K, Sutherland L R (1998). Does medical antireflux therapy improve asthma in asthmatics with gastroesophageal reflux? A critical review of the literature. *Chest* 114: 275–283.

Field S K, Gelfand G A, McFadden S D (1999). The effects of antireflux surgery on asthmatics with gastroesophageal reflux. *Chest* 116: 766–774.

Fitzpatrick M F, Mackay T, Driver H, et al. (1990). Salmeterol in nocturnal asthma: a double blind, placebo controlled trial of a long acting inhaled beta 2 agonist. *Br Med J* 301: 1365–1368.

Fitzpatrick M F, Engleman H, Whyte K F, et al. (1991). Morbidity in nocturnal asthma: sleep quality and daytime cognitive performance. *Thorax* 46: 569–573.

Fitzpatrick M F, Engleman H M, Boellert F, et al. (1992). Effect of therapeutic theophylline levels on the sleep quality and daytime cognitive performance of normal subjects. *Am Rev Respir Dis* 145: 1355–1358.

Gardiner P V, Ward C, Booth H, et al. (1994). Effect of eight weeks of treatment with salmeterol on bronchoalveolar lavage inflammatory indices in asthmatics. *Am J Respir Crit Care Med* 150: 1006–1011.

Georges G, Bartelson B B, Martin R J, et al. (1999). Circadian variation in exhaled nitric oxide in nocturnal asthma. *J Asthma* 36: 467–473.

Gervais P, Reinberg A, Gervais C, et al. (1977). Twenty-four-hour rhythm in the bronchial hyperreactivity to house dust in asthmatics. *J Allergy Clin Immunol* 59: 207–213.

Grant S D, Forsham P H, DiRaimondo V C (1965). Suppression of 17-hydroxycorticosteroids in plasma and urine by single and divided doses of triamcinolone. *N Engl J Med* 273: 1115–1118.

Greening A P, Ind P W, Northfield M, et al. (1994). Added salmeterol versus higher-dose corticosteroid in asthma patients with symptoms on existing inhaled corticosteroid. Allen & Hanburys Limited UK Study Group. *Lancet* 344: 219–224.

Hetzel M R, Clark T J (1980). Comparison of normal and asthmatic circadian rhythms in peak expiratory flow rate. *Thorax* 35: 732–738.

Hetzel M R, Clarke J H, Gillam S J, et al. (1985). Is sodium cromoglycate effective in nocturnal asthma?, *Thorax* 40: 793–794.

Horwitz R J, McGill K A, Busse W W (1998). The role of leukotriene modifiers in the treatment of asthma. *Am J Respir Crit Care Med* 157: 1363–1371.

Irvin C G, Pak J, Martin R J (2000). Airway–parenchyma uncoupling in nocturnal asthma. *Am J Respir Crit Care Med* 161: 50–56.

Israel E, Cohn J, Dube L, et al. (1996). Effect of treatment with zileuton, a 5-lipoxygenase inhibitor, in patients with asthma. A randomized controlled trial. Zileuton Clinical Trial Group. *J Am Med Assoc* 275: 931–936.

Kallenbach J M, Webster T, Dowdeswell R, *et al.* (1985). Reflex heart rate control in asthma. Evidence of parasympathetic overactivity. *Chest* 87: 644–648.
Kaplan J, Fredrickson P A, Renaux S A, *et al.* (1993). Theophylline effect on sleep in normal subjects. *Chest* 103: 193–195.
Kidney J, Dominguez M, Taylor P M, *et al.* (1995). Immunomodulation by theophylline in asthma. Demonstration by withdrawal of therapy. *Am J Respir Crit Care Med* 151: 1907–1914.
Kiljander T O, Salomaa E R, Hietanen E K, *et al.* (1999). Gastroesophageal reflux in asthmatics: A double-blind, placebo-controlled crossover study with omeprazole. *Chest* 116: 1257–1264.
Kraft M, Djukanovic R, Wilson S, *et al.* (1996a). Alveolar tissue inflammation in asthma. *Am J Respir Crit Care Med* 154: 1505–1510.
Kraft M, Pak J, Borish L, *et al.* (1996b). Theophylline's effect on neutrophil function and the late asthmatic response. *J Allergy Clin Immunol* 98: 251–257.
Kraft M, Torvik J A, Trudeau J B, *et al.* (1996c). Theophylline: potential antiinflammatory effects in nocturnal asthma. *J Allergy Clin Immunol* 97: 1242–1246.
Kraft M, Wenzel S E, Bettinger C M, *et al.* (1997). The effect of salmeterol on nocturnal symptoms, airway function, and inflammation in asthma. *Chest* 111: 1249–1254.
Kraft M, Pak J, Martin R J (1998a). Serum cortisol in asthma: marker of nocturnal worsening of symptoms and lung function? *Chronobiol Int* 15: 85–92.
Kraft M, Striz I, Georges G, *et al.* (1998b). Expression of epithelial markers in nocturnal asthma. *J Allergy Clin Immunol* 102: 376–381.
Kraft M, Martin R J, Wilson S, *et al.* (1999a). Lymphocyte and eosinophil influx into alveolar tissue in nocturnal asthma. *Am J Respir Crit Care Med* 159: 228–234.
Kraft M, Vianna E, Martin R J, *et al.* (1999b). Nocturnal asthma is associated with reduced glucocorticoid receptor binding affinity and decreased steroid responsiveness at night. *J Allergy Clin Immunol* 103: 66–71.
Kraft M, Hamid Q, Chrousos G P, *et al.* (2001a). Decreased steroid responsiveness at night in nocturnal asthma. Is the macrophage responsible? *Am J Respir Crit Care Med* 163: 1219–1225.
Kraft M, Pak J, Martin R J, *et al.* (2001b). Distal lung dysfunction at night in nocturnal asthma. *Am J Respir Crit Care Med* 163: 1551–1556.
Laitinen L A, Laitinen A, Haahtela T (1993). Airway mucosal inflammation even in patients with newly diagnosed asthma. *Am Rev Respir Dis* 147: 697–704.
Laviolette M, Malmstrom K, Lu S, *et al.* (1999). Montelukast added to inhaled beclomethasone in treatment of asthma. Montelukast/Beclomethasone Additivity Group. *Am J Respir Crit Care Med* 160: 1862–1868.
Liu M C, Dube L M, Lancaster J (1996). Acute and chronic effects of a 5-lipoxygenase inhibitor in asthma: a 6-month randomized multicenter trial. Zileuton Study Group. *J Allergy Clin Immunol* 98: 859–871.
Lockey R F, DuBuske L M, Friedman B, *et al.* (1999). Nocturnal asthma: effect of salmeterol on quality of life and clinical outcomes. *Chest* 115: 666–673.
Lundback B, Rawlinson D W, Palmer J B (1993). Twelve month comparison of salmeterol and salbutamol as dry powder formulations in asthmatic patients. European Study Group. *Thorax* 48: 148–153.

Malmstrom K, Rodriguez-Gomez G, Guerra J, et al. (1999). Oral montelukast, inhaled beclomethasone, and placebo for chronic asthma. A randomized, controlled trial. Montelukast/Beclomethasone Study Group. *Ann Intern Med* 130: 487–495.

Mannino D M, Homa D M, Pertowski C A, et al. (1998). Surveillance for asthma – United States, 1960–1995. *MMWR CDC Surveill Summ* 47: 1–27.

Martin R J (1993). *Nocturnal Asthma: Mechanisms and Treatment*. Mount Kisco. NY: Futura.

Martin R J, Pak J (1991). Nasal CPAP in nonapneic nocturnal asthma. *Chest* 100: 1024–1027.

Martin R J, Cicutto L C, Ballard R D, et al. (1989). Circadian variations in theophylline concentrations and the treatment of nocturnal asthma. *Am Rev Respir Dis* 139: 475–478.

Martin R J, Cicutto L C, Ballard R D (1990). Factors related to the nocturnal worsening of asthma. *Am Rev Respir Dis* 141: 33–38.

Martin R J, Cicutto L C, Smith H R, et al. (1991). Airways inflammation in nocturnal asthma. *Am Rev Respir Dis* 143: 351–357.

Martin R J, Pak J, Irvin C G (1993). Effect of lung volume maintenance during sleep in nocturnal asthma. *J Appl Physiol* 75: 1467–1470.

Martin R J, Kraft M, Beaucher W N, et al. (1999). Comparative study of extended release albuterol sulphate and long-acting inhaled salmeterol xinafoate in the treatment of nocturnal asthma. *Ann Allergy Asthma Immunol* 83: 121–126.

Milledge J S, Morris J (1979). A comparison of slow-release salbutamol with slow-release aminophylline in nocturnal asthma. *J Int Med Res* 7, 106–110.

Mohiuddin A A, Martin R J (1990). Circadian basis of the late asthmatic response. *Am Rev Respir Dis* 142: 1153–1157.

Moore-Gillon J (1988). Volmax (salbuterol CR 8 mg) in the management of nocturnal asthma: a placebo-controlled study. *Eur J Respir Dis Suppl* 1: 306S

Morgan A D, Connaughton J J, Catterall J R, et al. (1986). Sodium cromoglycate in nocturnal asthma. *Thorax* 41: 39–41.

Morrison J F, Pearson S B, Dean H G (1988). Parasympathetic nervous system in nocturnal asthma. *Br Med J (Clin Res Ed)*, 296: 1427–1429.

National Asthma Education and Prevention Program (1997). *Expert Panel Report 2: Guidelines for the Diagnosis and Management of Asthma*. Bethesda, MD: National Institutes of Health National Heart Lung and Blood Institute.

National Center for Health Statistics (1991). *Utilization of Short-stay Hospitals: United States, 1991 Annual Summary. Vital and Health Statistics Series 13*, No. 72. DHHS Publication No. (PHS)83–1733. Washington, DC: US Government Printing Office, 36.

National Center for Health Statistics (1993). *National Hospital Discharge Survey: Annual Summary, 1991*. DHHS Publication No. (PHS)93–1775. Washington DC: US Government Printing Office.

Nichols T, Nugent C, Tyler F (1965). Diurnal variation in suppression of adrenal function by glucocorticoids. *J Clin Endocrinol* 25: 343–349.

Noonan M J, Chervinsky P, Brandon M, et al. (1998). Montelukast, a potent leukotriene receptor antagonist, causes dose-related improvements in chronic asthma. Montelukast Asthma Study Group. *Eur Respir J* 11: 1232–1239.

Noonan M, Karpel J P, Bensch G W, et al. (2001). Comparison of once-daily to twice-daily treatment with mometasone furoate dry powder inhaler. *Ann Allergy Asthma Immunol* 86: 36–43.

O'Byrne P (1998). Asthma pathogenesis and allergen-induced late responses. *J Allergy Clin Immunol* 102: S85–S89.

Pearlman D S, Chervinsky P, LaForce C, *et al.* (1992). A comparison of salmeterol with albuterol in the treatment of mild-to-moderate asthma. *N Engl J Med* 327: 1420–1425.

Pedersen B, Dahl R, Larsen B B, *et al.* (1993). The effect of salmeterol on the early- and late-phase reaction to bronchial allergen and postchallenge variation in bronchial reactivity, blood eosinophils, serum eosinophil cationic protein, and serum eosinophil protein X. *Allergy* 48: 377–382.

Pincus D J, Szefler S J, Ackerson L M, *et al.* (1995). Chronotherapy of asthma with inhaled steroids: the effect of dosage timing on drug efficacy. *J Allergy Clin Immunol* 95: 1172–1178.

Pincus D J, Humeston T R, Martin R J (1997). Further studies on the chronotherapy of asthma with inhaled steroids: the effect of dosage timing on drug efficacy. *J Allergy Clin Immunol* 100: 771–774.

Postma D S, Koeter G H, Meurs H, *et al.* (1984). Slow release terbutaline in nocturnal bronchial obstruction: relation of terbutaline dosage and blood levels with circadian changes in peak flow values. *Annu Rev Chronopharmacol* 1: 101–104.

Proud D, Reynolds C J, Lichtenstein L M, *et al.* (1998). Intranasal salmeterol inhibits allergen-induced vascular permeability but not mast cell activation or cellular infiltration. *Clin Exp Allergy* 28: 868–875.

Reicin A, White R, Weinstein S F, *et al.* (2000). Montelukast, a leukotriene receptor antagonist, in combination with loratadine, a histamine receptor antagonist, in the treatment of chronic asthma. *Arch Intern Med* 160: 2481–2488.

Reinberg A, Halberg F, Falliers C J (1974). Circadian timing of methylprednisolone effects in asthmatic boys. *Chronobiologia* 1: 333–347.

Reinberg A, Guillet P, Gervais P, *et al.* (1977). One month chronocorticotherapy (Dutimelan 8 15 mite). Control of the asthmatic condition without adrenal suppression and circadian rhythm alteration. *Chronobiologia* 4: 295–312.

Reinberg A, Gervais P, Chaussade M, *et al.* (1983). Circadian changes in effectiveness of corticosteroids in eight patients with allergic asthma. *J Allergy Clin Immunol* 71: 425–433.

Reiss T F, Chervinsky P, Dockhorn R J, *et al.* (1998). Montelukast, a once-daily leukotriene receptor antagonist, in the treatment of chronic asthma: a multicenter, randomized, double-blind trial. Montelukast Clinical Research Study Group. *Arch Intern Med* 158: 1213–1220.

Richter J E (2000). Gastroesophageal reflux disease and asthma: the two are directly related. *Am J Med* 108(Suppl 4a): 153S–158S.

Rivington R N, Calcutt L, Child S, *et al.* (1985). Comparison of morning versus evening dosing with a new once-daily oral theophylline formulation. *Am J Med* 79: 67–72.

Robertson C F, Rubinfeld A R, Bowes G (1990). Deaths from asthma in Victoria: a 12-month survey. *Med J Aust* 152: 511–517.

Sears M R, Taylor D R, Print C G, *et al.* (1990). Regular inhaled beta-agonist treatment in bronchial asthma. *Lancet* 336: 1391–1396.

Selby C, Engleman H M, Fitzpatrick M F, *et al.* (1997). Inhaled salmeterol or oral theophylline in nocturnal asthma? *Am J Respir Crit Care Med* 155: 104–108.

Silkoff P E, Ballard R D, Pak J, *et al.* (2001). Circadian variability in blood melatonin in nocturnal asthma. *Am J Crit Care Med* 163: A585.

Siracusa A, Curradi F, Abbritti G (1978). Recurrent nocturnal asthma due to tolylene di-isocyanate: a case report. *Clin Allergy* 8, 195–201.

Sontag S J (2000). Why do the published data fail to clarify the relationship between gastroesophageal reflux and asthma? *Am J Med* 108 (Suppl 4a): 159S–169S.

Soutar C A, Costello J, Ijaduola O, *et al.* (1975). Nocturnal and morning asthma. Relationship to plasma corticosteroids and response to cortisol infusion. *Thorax* 30: 436–440.

Spector S L, Smith L J, Glass M (1994). Effects of 6 weeks of therapy with oral doses of ICI 204,219: a leukotriene D4 receptor antagonist, in subjects with bronchial asthma. ACCOLATE Asthma Trialists Group. *Am J Respir Crit Care Med* 150: 618–623.

Storms W W, Nathan R A, Bodman S F, *et al.* (1992). The effect of repeat action albuterol sulphate (Proventil Repetabs) in nocturnal symptoms of asthma. *J Asthma* 29: 209–216.

Sullivan P, Bekir S, Jaffar Z, *et al.* (1994). Anti-inflammatory effects of low-dose oral theophylline in atopic asthma. *Lancet* 343: 1006–1008.

Sutherland E R, Martin R J, Rex M D, *et al.* (2001). Melatonin stimulates increased production of IL-1 and IL-6 by peripheral blood mononuclear cells at 4 am versus 4 pm in asthmatic subjects. *Am J Respir Crit Care Med* 163: A867.

Szefler S J, Ando R, Cicutto L C, *et al.* (1991). Plasma histamine, epinephrine, cortisol, and leukocyte beta-adrenergic receptors in nocturnal asthma. *Clin Pharmacol Ther* 49: 59–68.

Szefler S J, Weiss S, Tonascia J, *et al.* (2000). Long-term effects of budesonide or nedocromil in children with asthma. *N Engl J Med* 343: 1054–1063.

Tan W C, Martin R J, Pandey R, *et al.* (1990). Effects of spontaneous and simulated gastroesophageal reflux on sleeping asthmatics. *Am Rev Respir Dis* 141: 1394–1399.

Taylor A J, Davies R J, Hendrick D J, *et al.* (1979). Recurrent nocturnal asthmatic reactions to bronchial provocation tests. *Clin Allergy* 9, 213–219.

ten Hacken N H, van der Vaart H, van der Mark T W, *et al.* (1998). Exhaled nitric oxide is higher both at day and night in subjects with nocturnal asthma. *Am J Respir Crit Care Med* 158: 902–907.

Turki J, Pak J, Green S A, *et al.* (1995). Genetic polymorphisms of the beta 2-adrenergic receptor in nocturnal and nonnocturnal asthma. Evidence that Gly16 correlates with the nocturnal phenotype. *J Clin Invest* 95: 1635–1641.

Turner-Warwick M (1988). Epidemiology of nocturnal asthma *Am J Med* 85: 6–8.

Weersink E J, Douma R R, Postma D S, *et al.* (1997a). Fluticasone propionate, salmeterol xinafoate, and their combination in the treatment of nocturnal asthma. *Am J Respir Crit Care Med* 155: 1241–1246.

Weersink E J, van Zomeren E H, Koeter G H, *et al.* (1997b). Treatment of nocturnal airway obstruction improves daytime cognitive performance in asthmatics. *Am J Respir Crit Care Med* 156: 1144–1150.

Weiss K B, Sullivan S D (2001). The health economics of asthma and rhinitis. I. Assessing the economic impact. *J Allergy Clin Immunol* 107: 3–8.

Wenzel S E, Trudeau J B, Kaminsky D A, *et al.* (1995). Effect of 5-lipoxygenase inhibition on bronchoconstriction and airway inflammation in nocturnal asthma. *Am J Respir Crit Care Med* 152: 897–905.

Wiegand L, Mende C N, Zaidel G, *et al.* (1999). Salmeterol vs theophylline: sleep and efficacy outcomes in patients with nocturnal asthma. *Chest* 115: 1525–1532.

Williams A J, Stableforth D (1986). The addition of nedocromil sodium to maintenance therapy in the management of patients with bronchial asthma. *Eur J Respir Dis Suppl* 147: 340–343.

Zwillich C W, Neagley S R, Cicutto L, *et al.* (1989). Nocturnal asthma therapy. Inhaled bitolterol versus sustained-release theophylline. *Am Rev Respir Dis* 139: 470–474.

8

Rhythms in therapeutics of cardiovascular disease

Björn Lemmer

8.1 Introduction

8.1.1 Chronobiology of the cardiovascular system

Though day–night variations in both heart rate (HR) and blood pressure (BP) were first described in the 17th century (see Lemmer, 1996a,b, 2000), the recent development of easy-to-use devices to monitor BP and HR continuously in man (ABPM, ambulatory blood pressure monitoring) has demonstrated that BP in normotensive and in hypertensive patients is clearly dependent on the time of day (see Lemmer, 1996a). Moreover, different forms of hypertension may exhibit different circadian patterns. In primary hypertension due, for example, to renal disease, gestation, diabetes mellitus or Cushing's disease, the rhythm in BP is abolished or even reversed in about 70% of cases, with highest values at night (Cugini *et al.*, 1989; Middeke and Schrader, 1994; for review see Lemmer, 1996b; Lemmer and Portaluppi, 1997). This is of particular interest since it correlates with increased end-organ damage in cardiac, cerebral, vascular and renal tissues.

While the rhythms in HR and BP are the best-known periodic functions of the cardiovascular system, other parameters have also been shown to exhibit circadian variations: stroke volume, cardiac output, blood flow, peripheral resistance; parameters of ECG recordings; plasma concentrations of pressor hormones such as noradrenaline (norepinephrine), renin, angiotensin, aldosterone; atrial natriuretic hormone and plasma cAMP concentration; blood viscosity, aggregability and fibrinolytic activity; and others. Figure 8.1 gives a simplified scheme of the physiological parameters involved in the regulation of BP, including the main target tissues on which antihypertensive drugs act.

In addition, as has already been described in Chapter 4, other bodily functions involved in absorption, distribution, metabolism and renal elimination of drugs are not constant throughout the 24 h day, giving rise to daily variations in a drug's pharmacokinetics.

8.1.2 Rhythms in cardiovascular events

Pathophysiological events within the cardiovascular system also do not occur at random (Master, 1960; for review see Willich and Muller, 1996; Lemmer and Portaluppi, 1997). Thus, the onset of nonfatal or fatal myocardial infarction predominates around 06.00–12.00. A similar circadian time pattern has been shown for sudden cardiac death, stroke, ventricular arrhythmias and arterial embolism. Symptoms in coronary heart disease patients such as myocardial ischaemia, angina attacks or silent ischaemia are also significantly more frequent during the day than at night, whereas the onset of angina attacks in variant angina peaks at night around 04.00.

During the early morning hours, not only do cardiovascular events predominate but there is also a rapid rise in BP, a rapid increase in sympathetic tone and in the concentrations of pressor hormones, and the highest values in peripheral resistance (Lemmer, 1996b). Thus, it appears that the early morning hours are the hours of highest cardiovascular risk.

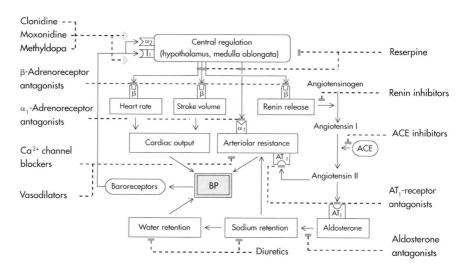

Figure 8.1 Simplified scheme of the mechanisms involved in the regulation of blood pressure (BP) and the main target mechanisms by which drugs lower blood pressure. Modified from Lemmer (1996b).

8.2 Chronopharmacology of hypertension

Having in mind the organisation in time of living systems, including humans, it is easy to conceive that not only must the *right* amount of the *right* substance be at the *right* place, but this must also occur at the *right* time. This is the more important when an organism or individual has to act or react in favourable biotic or environmental conditions that are themselves highly periodic. Thus, it is easy to understand that exogenous compounds, including drugs, may challenge the individual differently depending on the *time* of administration.

Here we will focus on data pertinent to the field of BP regulation and its pharmacological manipulation (Lemmer, 1989, 1991, 1999). Drug treatment of hypertension includes various types of drugs such as diuretics, β- and α-adrenoreceptor blocking drugs, calcium channel blockers, angiotensin-converting enzyme inhibitors, angiotensin AT_1-receptor blockers, and others that differ in their sites of action as well as in their physicochemical characteristics, pharmacokinetic properties and pharmaceutical formulation and, thus, dosing interval. Despite the great number of published studies evaluating the efficacy of antihypertensive drugs, the time of day of drug administration has rarely been a specific point of investigation. In this chapter emphasis will be put on crossover studies with antihypertensive drugs (administered in the morning versus the evening) to highlight this point more clearly.

For adequate comparison of the results obtained with drugs that lower high BP, it is important to note that ABPM is now regarded as the method of choice for evaluating BP profiles.

8.2.1 β-Adrenoreceptor antagonists

β-Adrenoreceptor antagonistic drugs can be divided into four main groups: nonselective (e.g. propranolol, oxprenolol); $β_1$-selective (e.g. bisoprolol, metoprolol, atenolol); compounds with intrinsic sympathomimetic/agonist activity (ISA; e.g. pindolol, carteolol); and β-adrenoreceptor antagonists with additional activities, e.g. α-adrenoreceptor blockade (carvedilol).

Unfortunately, no crossover (morning versus evening) study has been published with β-adrenoreceptor antagonists in hypertensive patients. It is difficult to draw definite conclusions on the importance of the circadian time of drug dosing for antihypertensive drug efficacy from studies performed without time-specified drug dosing. A resumé of 20 'conventionally' performed studies showed that β-adrenoreceptor antagonists – whether $β_1$-selective, nonselective, or with ISA – may not affect, may reduce or may even abolish the rhythmic pattern in BP. In general, however, there is a tendency for β-adrenoreceptor antagonists predominately to reduce daytime BP levels and not to greatly affect night-time values, being less effective or totally ineffective in reducing the early morning rise in BP (Stanton and O'Brien, 1994). Consistently with this, decreases in HR caused by β-adrenoreceptor antagonists are more pronounced during daytime hours. In healthy subjects, a crossover study with propranolol similarly showed a more pronounced decrease in HR and BP during daytime hours than at night (Langner and Lemmer, 1988). Interestingly, the agent with partial agonist activity, pindolol, even increased HR at night (Quyyumi et al., 1984).

As has already been described in Chapter 4, the pharmacokinetics of β-adrenoreceptor antagonists can also vary over 24 h. Table 8.1 summarises the circadian variation in pharmacokinetics of drugs affecting the cardiovascular system.

In conclusion, available clinical data indicate that β-adrenoreceptor-mediated regulation of BP dominates during daytime hours and is of minor importance during the night and the early morning hours. This correlates well with the circadian rhythm in sympathetic tone as indicated by the rhythm in plasma noradrenaline and cAMP.

8.2.2 Calcium channel blockers

Calcium channel blockers are not a homogenous group of drugs. Vasodilatation by calcium channel blockers occurs at lower concentrations than the cardiodepressant effects. However, the difference between vasodilating and cardiodepressant effects is greater with the 1,4-dihydropyridines (e.g. nifedipine, nitrendipine, isradipine, amlodipine) than with compounds such as verapamil and diltiazem. Moreover, these drugs differ in their kinetics, with amlodipine having a particularly long half-life.

The effects of calcium channel blockers were analysed mainly by visual inspection of the BP profiles. In primary hypertensives, three times a day dosing of nonretarded verapamil did not greatly change the BP profile, and was less effective at night (Gould et al., 1982). A single morning dose of a sustained-release verapamil showed a good 24 h BP control (Caruana et al., 1987), whereas a sustained-release formulation of diltiazem was less effective at night (Lemmer et al., 1994). Dihydropyridine derivatives of varying pharmacokinetics seem to reduce BP to varying degrees during day and night; pharmaceutical formulation and dosing interval may also have affected the results.

Only nine studies using a crossover design have been published

Table 8.1 Pharmacokinetic parameters of cardiovascular active drugs determined in crossover studies

Drug[a]	Dose[b] (mg) and duration	C_{max} (ng/mL) Morning	C_{max} (ng/mL) Evening	t_{max} (h) Morning	t_{max} (h) Evening	Reference
Digoxin	0.5, sd	3.6*	1.8	1.2	3.2	Bruguerolle et al. (1988)
Enalaprilat						
after Enalapril	10, sd	33.8	41.9	4.4	4.5	Witte et al. (1993)
after Enalapril	10, daily × 3 weeks	46.7	53.5	3.5*	5.6	
IS-5-MN i.r.	60, sd	1605.0	1588.0	0.9*	2.1	Scheidel and Lemmer (1991), Lemmer et al. (1989)
IS-5-MN s.r.	60, sd	509.0	530.0	5.2	4.9	Lemmer et al. (1991c)
Molsidomine	8, sd	27.0	23.5	1.7	1.9	Nold and Lemmer (1998)
Nifedipine i.r.	10, sd	82.0*	45.7	0.4*	0.6	Lemmer et al. (1991a,b)
Nifedipine s.r.	2 × 20, daily × 1 week	48.5	50.1	2.3	2.8	Lemmer et al. (1991a,b)
Atenolol	50, sd	440.0	391.8	3.2	4.0	Shiga et al. (1993)
Oxprenolol[c]	80, sd	507.0	375.0	1.0	1.1	Koopmans et al. (1993)
Propranolol[d]	80, sd	38.6*	26.2	2.5	3.0	Langner and Lemmer (1988)
Propranolol (±)	80, sd	68	60	2.3	2.7	Semenowicz-Siuda et al. (1984)
Verapamil s.r.	360, daily × 2 weeks	389.0	386.0	7.2*	10.6	Jespersen et al. (1989)
Verapamil	80, sd	59.4*	25.6	1.3	2.0	Hla et al. (1992)

At least two dosing times (around 06.00–08.00 and 18.00–20.00) were studied, in some studies up to six circadian times were included. Only the parameters C_{max} = peak drug concentration, and t_{max} = time to C_{max}, are given.

[a] IS-5-MN, isosorbide mononitrate; i.r., immediate-release preparation; s.r., sustained-release preparation.
[b] sd = single dose.
[c] Significant difference in half-life.
[d] (±)-Propranolol was given, kinetic data for (−)-propranolol.
*p at least <0.05, morning versus evening.

(Table 8.2). In essential hypertensives, amlodipine, isradipine and nifedipine delayed-release, and in normotensives immediate-release nifedipine, did not affect the 24 h BP profile differently after once-morning or once-evening dosing (Mengden *et al.*, 1992; Fogari *et al.*, 1993; Greminger *et al.*, 1994; Lemmer *et al.*, 1991a), whereas with nitrendipine and lacidipine the profile remained unaffected or only slightly changed after evening dosing (Umeda *et al.*, 1994; van Montfrans *et al.*, 1998) (Table 8.2). In primary hypertensive patients, twice-daily nifedipine lowered the BP throughout a 24 h period (Lemmer *et al.*, 1991a). Most interestingly, the greatly disturbed BP profile in those with secondary hypertension due to renal failure was normalised only after evening dosing of isradipine (Portaluppi *et al.*, 1995) (Table 8.2).

A time-of-day effect has also been described for the kinetics of various calcium channel blockers (Table 8.2). The bioavailability of an immediate-release formulation of nifedipine was found to be reduced by about 40% after evening compared to morning dosing, with C_{max} being higher and t_{max} being shorter after morning dosing (Lemmer *et al.*, 1991a, b). No such circadian time-dependent kinetics were observed with a sustained-release formulation of nifedipine (Lemmer *et al.*, 1991a). Also, regular (Hla *et al.*, 1992) as well as sustained-release verapamil (Jespersen *et al.*, 1989) displayed higher C_{max} and/or shorter t_{max} values after morning dosing. Similar chronokinetics have been reported after oral dosing of other cardiovascular active drugs such as enalapril, propranolol and others (for review see Lemmer and Bruguerolle, 1994). Conversely, intravenously infused nifedipine did not display daily variations in its pharmacokinetics (Hla *et al.*, 1992), indicating that gastrointestinal mechanisms must be involved in the chronokinetics of the drug. This assumption is supported by the observation that gastric emptying time for solids is faster (Goo *et al.*, 1987) and hepatic blood flow (representative for gastrointestinal tract perfusion) is higher during the early morning than in the afternoon (Lemmer and Nold, 1991). Both mechanisms could nicely explain higher C_{max} and shorter t_{max} values after oral ingestion of nonretarded drugs in the morning (see Lemmer and Bruguerolle, 1994).

8.2.3 Angiotensin-converting enzyme (ACE) inhibitors

Captopril is a representative short-acting ACE inhibitor; longer-acting ones are enalapril, benazepril, quinapril and ramipril. These drugs are hydrolysed to active metabolites. ACE inhibitors are not only effective antihypertensive drugs but can also increase life expectancy in congestive heart failure.

Table 8.2 Effects of calcium channel blockers on the 24 h pattern in blood pressure[a]

Drug	Dose (mg/day)	Duration and time of dosing	Patients (n) and diagnosis[b]	Effect on 24 h blood pressure			Reference
				Day	Night	24 h profile	
Amlodipine	5	4 weeks a.m. p.m.	20, EH	++ ++	++ ++	Preserved Preserved	Mengden et al. (1992)
Amlodipine	5	3 weeks 08.00 20.00	12, EH	+ +	+ +	Preserved Preserved	Nold et al. (1998)
Nifedipine delayed release	30	1 or 2 weeks 10.00 22.00	10, EH	++ ++	++ ++	Preserved Preserved	Greminger et al. (1994)
Nitrendipine	20	4 weeks 07.00 19.00	41, EH	+ +	+ +	Preserved Preserved	Meilhac et al. (1992)
Nitrendipine	10	3 days 06.00 18.00	6, EH	++ +	++ ++	Preserved Changed	Umeda et al. (1994)
Isradipine	5	4 weeks 07.00 19.00	18, EH	++ ++	++ ++	Preserved Preserved	Fogari et al. (1993)
Isradipine	5	4 weeks 08.00 20.00	16, RH[c]	++ +	++ +++	Not normalised Normalised	Portaluppi et al. (1995)
Nifedipine immediate-release	10	Single dose 08.00 19.00	12, NT	++ +	++ +	Preserved Preserved	Lemmer et al. (1991a)
Lacidipine	4	6 weeks 07.00–09.00 22.00–24.00	33, EH	++ (+)	(+) +(+)	Preserved Preserved	van Montfrans et al. (1998)

[a]Table includes only data from crossover studies.
[b]EH, essential (primary) hypertensives; RH, renal (secondary) hypertensives; NT, normotensives.
[c]Abolished 24 h blood pressure profile (nondipping) under control conditions.

Several studies with ACE inhibitors, dosed once in the morning or twice daily, showed that these drugs did not greatly modify the 24 h BP pattern (Stanton and O'Brien, 1994). However, intra-arterial studies with enalapril or ramipril have shown that, while they cause sustained daytime reduction in BP, these drugs had only marginal effects on night-time BP (Gould *et al.*, 1982). Thus, the findings obtained with ACE inhibitors in conventional clinical studies are inconsistent.

Five crossover studies (morning versus evening dosing) with ACE inhibitors in essential hypertensive patients have been published (Table 8.3). They demonstrate that evening dosing in contrast to morning dosing of benazepril (Palatini *et al.*, 1993) and enalapril (Witte *et al.*, 1993) resulted in a more pronounced night-time drop and that the 24 h BP profile was distorted by evening enalapril (Witte *et al.*, 1993). Evening dosing of quinapril (Palatini *et al.*, 1992) also resulted in a more pronounced effect than morning dosing; the BP pattern, however, was not greatly modified. A fixed combination of captopril and the diuretic hydrochlorothiazide attenuated the BP profile after morning dosing, with a slightly more pronounced reduction in night-time BP after evening dosing (Middeke *et al.*, 1991). In the light of the reduced cardiac reserve of at-risk patients with hypertension (Perloff *et al.*, 1983; Pringle *et al.*, 1992), too pronounced a night-time drop in BP after evening dosing might be a potential risk factor for the occurrence of ischaemic events.

8.2.4 Other antihypertensives

Antihypertensives of other classes have rarely been studied in relation to possible circadian variation. Once-daily (morning) dosing of the diuretics xipamide (Raftery *et al.*, 1981) and indapamide (Ocon and Mora, 1990) reduced BP in essential hypertensives without changing the 24 h BP pattern. On twice-daily dosing, the α-adrenoreceptor antagonists indoramin (Gould *et al.*, 1981) and prazosin (Weber *et al.*, 1987) also did not change the BP profile. A single night-time dose of the α-adrenoreceptor antagonist doxazosin reduced both systolic and diastolic BP throughout day and night, but the greatest reduction occurred in the morning (Pickering *et al.*, 1994). Since α-adrenoreceptor blockade more effectively reduced the peripheral resistance during the early morning hours than at others times of the day (Panza *et al.*, 1991), these findings point to the importance of α-adrenoreceptor-mediated regulation of BP during this time of day. In addition, the peak treatment effect after night-time dosing of doxazosin was later than predicted from the drug's pharmacokinetics, an observation that nicely supports similar findings on a circadian

Table 8.3 Effects of ACE inhibitors on the 24 h pattern in blood pressure[a]

Drug	Dose (mg/day)	Duration and time of dosing	Patients (n) and diagnosis	Effect on 24 h blood pressure			Reference
				Day	Night	24 h profile	
Benazepril	10	Single dose 09.00 21.00	10, EH	+++ +	++ ++	Preserved Changed	Palatini et al. (1993)
Enalapril	10	Single dose 07.00 19.00 3 weeks 07.00 19.00	10, EH	++ ++ ++ +	+ +++ + ++	Preserved Changed Preserved Changed	Witte et al. (1993)
Quinapril	20	4 weeks 08.00 22.00	18, EH	++ ++	+ ++	Preserved Preserved	Palatini et al. (1992)
Ramipril	2.5	4 weeks 08.00 20.00	33, EH	+ (+)	(+) +	Preserved Preserved	Myburgh et al. (1995)
Perindopril	2	4 weeks 09.00 21.00	18, EH	++ +	+ ++	Preserved Changed	Morgan et al. (1997)

[a]Table includes only data from crossover studies.
EH, essential (primary) hypertensives.

phase dependency in the dose–response relationship of nifedipine (Lemmer *et al.*, 1991a), enalapril (Witte *et al.*, 1993) and propranolol (Langner and Lemmer, 1988).

8.2.5 Duration of the antihypertensive effect

Restriction of BP monitoring by ABPM to 24 h may carry potential pitfalls when evaluating the duration of the antihypertensive effect (Lemmer 1995, 1996a). After chronic morning dosing, atenolol produced inadequate 24 h control of BP, but the pattern of reduction persisted one day after the drug was redrawn, as demonstrated by ABPM for a 48 h span (Gould *et al.*, 1991). Very similar findings were reported after single-morning and single-evening dosing of enalapril (Witte *et al.*, 1993). These observations clearly support the notion that regulatory mechanisms of the 24 h BP profile may predominate at certain times of day and that, to avoid false conclusions, ABPM should not be restricted to 24 h. These observations also are in favour of a circadian time-dependent dose–response relationship, as mentioned above.

8.3 Chronopharmacology of coronary heart disease

8.3.1 Oral nitrates

As already described, the onset of angina pectoris and the frequency of angina attacks exhibit a circadian rhythm, a finding that has been confirmed by many authors (see Willich and Muller, 1996). It is therefore not surprising that anti-anginal drugs may also exert circadian-phase-dependent effects in patients with variant (Prinzmetal) or stable angina pectoris. The first evidence for this was provided by the classic paper of Yasue *et al.* (1979); in this study 13 patients with Prinzmetal's variant angina performed treadmill exercise tests in the early morning (05.00–08.00) and again in the afternoon (13.00–16.00). Exercise-induced angina attacks as well as ECG abnormalities occurred in all patients in the morning, but in only two patients in the afternoon. Interestingly, administration of glyceryl trinitrate in the morning prevented the angina attacks and had a greater dilating effect on the coronary vessel than when given in the afternoon, indicating temporal variations in vasomotor tone. Essentially the same results were obtained with the calcium channel blocker diltiazem, whereas propranolol aggravated the symptoms when given in the early morning. This study clearly demonstrated for the first time that exercise-induced coronary spasm was dependent on the time of day at which the exercise challenge was performed. Supporting this hypothesis and the time-dependency of medication, the authors of the glyceryl trinitrate study demonstrated an increase in patency of the great coronary arteries induced by this drug when it was taken in the morning but not when it was taken in the afternoon. Similar findings in patients with stable angina pectoris with and without treatment with propranolol are mentioned below. Further support for a daily variation in arterial hyperreactivity comes from the study on patients with variant angina pectoris in whom the threshold of angina attacks provoked by ergonovine (an α-adrenoreceptor-stimulating secale alkaloid) was lower in the morning than in the afternoon (Waters et al., 1984).

The chronopharmacology of the oral nitrates isosorbide dinitrate (ISDN) and isosorbide mononitrate (IS-5-MN) has been investigated only in preliminary studies after acute dosing in healthy subjects (Lemmer et al., 1991b; Scheidel and Lemmer, 1991; see Table 8.1). ISDN-induced decreases in BP and increases in HR (standing position) were found after night-time dosing, even though peak concentrations of ISDN occurred after morning dosing.

8.3.2 β-Adrenoreceptor antagonists

β-Adrenoreceptor antagonists have been shown to reduce the risk of recurrent myocardial infarction (Norwegian Multicenter Study Group, 1981). Several studies addressed the question whether these drugs could influence the circadian pattern in myocardial infarction. In the MILIS study (Muller et al., 1985), the morning peak in the onset of myocardial infarction was abolished by β-adrenoreceptor blockade. This finding was supported by several other studies (for review see Lemmer and Portaluppi, 1997). The number of sudden cardiac deaths between 05.00 and 11.00 was reduced by propranolol, whereas the occurrence of deaths at other circadian times was almost equal in untreated and treated patients (Peters et al., 1989). Interestingly, non-Q-wave infarction did not show circadian variation, either in untreated patients or in those receiving β-adrenoreceptor antagonists (Kleiman et al., 1990). From analysis of the high- and low-frequency components it was shown that HR variability after acute myocardial infarction was significantly affected by atenolol or metoprolol, reducing sympathetic activation and increasing vagal tone, with the effects being more pronounced during daytime hours (Sandrone et al., 1994).

Most studies concerning the rhythmicity of ischaemic episodes and of the incidence of angina attacks demonstrated an abolition or attenuation of the morning peak by propranolol or the more β_1-selective antagonists metoprolol and atenolol (for example Mulcahy *et al.*, 1988; Imperi *et al.*, 1987). In other studies with atenolol, atenolol/propranolol or bevantolol, no influence on the circadian distribution of angina attacks or of ischaemic episodes was observed, although their total number was reduced at both times of day (for example Quyyumi *et al.*, 1984, 1987; Benhorin *et al.*, 1993; Gillis *et al.*, 1992). Pindolol, a β-adrenoreceptor antagonist with intrinsic sympathomimetic activity, even increased HR at night and did not reduce the incidence of angina attacks (Quyyumi *et al.*, 1984). Despite the preservation of circadian rhythmicity in silent ischaemia, Benhorin *et al.* (1993) demonstrated that the time-dependent variation in ischaemic threshold (i.e. HR at the onset of ischaemia), was completely abolished by treatment with β-adrenoreceptor antagonists.

Whether β-adrenoreceptor antagonists influence circadian rhythms in angina could also depend on the subgroup of patients studied. A significant circadian variation, with a morning increase, was shown in the frequency of ventricular premature complexes; this was abolished by β-adrenoreceptor antagonists (Gillis *et al.*, 1992). However, a baseline circadian rhythm could be demonstrated only in patients with left ventricular ejection fraction (LVEF) greater than 0.30, whereas in those with more pronounced ventricular dysfunction (LVEF < 0.30) no rhythmicity was observed. Analysis of all patients without subdivisions would have obscured the circadian time-dependent effects of drug treatment of ventricular arrhythmias. In myocardial infarction, ventricular arrhythmias were also suppressed more effectively during early morning and daytime hours than during the night (Lichtstein *et al.*, 1983). A detailed analysis of data from the ASIS study (Andrews *et al.*, 1993) showed the importance of subclassification of ischaemic events with regard to concomitant changes in HR. In that study, propranolol markedly reduced those ischaemic events that occurred while or directly after HR increased, whereas the proportion of episodes not related to HR changes was even more pronounced under β-adrenoreceptor antagonist treatment. Since HR-related ischaemia was shown to have a different circadian pattern from non-HR-related episodes, anti-ischaemic treatment (Andrews *et al.*, 1993) affecting only one type of ischaemia can definitely be expected to show a circadian time-dependency.

Unfortunately, prospective chronopharmacological studies comparing the effects of β-adrenoreceptor antagonists on ischaemic events after dosing at different times of the day have not been performed so far.

At least in myocardial infarction and in stable angina pectoris, the attenuation of the increased sympathetic tone during the early morning hours and during the day seems to be an important therapeutic aspect in the use of the β-adrenoreceptor antagonists (see Lemmer and Witte, 1995; Lemmer and Portaluppi, 1997).

In conclusion, it appears that treatment of coronary heart disease by β-adrenoreceptor antagonists effectively reduces ischaemic events at any time of day. β-Adrenoreceptor antagonists seem to be of special therapeutic value in the early morning hours, which are the hours of high risk (see Lemmer and Witte, 1995).

8.3.3 Calcium channel blockers

Controversial results have been reported concerning the effects of calcium channel blockers on the circadian pattern in ischaemic episodes or in myocardial infarction (for review see Lemmer and Witte, 1995). Mulcahy *et al.* (1988) reported that nifedipine, in contrast to atenolol, did not alter the circadian profile in the episodes and the duration of ischaemia. However, reductions by nifedipine in the morning incidence of ischaemic episodes have been observed in several other studies (Lemmer and Portaluppi, 1997; Lemmer and Witte, 1995).

In conclusion, calcium channel blockers, mainly short-acting and nonretarded preparations, seem to be less effective than β-adrenoreceptor antagonists in reducing ischaemic events during the night and early morning hours. However, the role of formulation and/or subclasses of the calcium channel blockers remains to be elucidated. Finally, the early morning peak in myocardial infarction seems less influenced, if at all, by these drugs.

8.4 Summary

There is sound evidence to show that the cardiovascular system is well organised over time. Mechanisms of regulation and pathophysiological events are not constant over 24 h. Moreover, certain diseases may even alter the normal physiological circadian pattern. This observation has implications for drug treatment in terms of drug formulation and dosing intervals. Pitfalls may arise from neglecting circadian phase-dependency in pharmacokinetics and in the concentration–effect relationship. Moreover, different classes of drugs may be affected by circadian influences to a greater or lesser extent. Without doubt, 'time-of-day' has to be included in our diagnostic and therapeutic strategies.

References

Andrews T C, Fenton T, Toyosaki N, *et al*. (1993). Subsets of ambulatory myocardial ischemia based on heart rate activity: circadian distribution and response to anti-ischemic medication. *Circulation* 88: 92–100.

Benhorin J, Banai S, Moriel M, *et al*. (1993). Circadian variations in ischemic threshold and their relation to the occurrence of ischemic episodes. *Circulation* 87: 808–814.

Bruguerolle B, Bouvenot G, Bartolin R, Manolis J. (1988). Chronopharmacocinétique de la digoxine chez le sujet de plus de soixante-dix ans. *Therapie* 43: 251–253.

Caruana M, Heber M, Bridgen G, Raftery E B (1987). Assessment of 'once daily' verapamil for the treatment of hypertension using ambulatory, intra-arterial pressure recording. *Eur J Clin Pharmacol* 32: 549–553.

Cugini P, Kawasaki T, DiPalma L, *et al*. (1989). Preventive distinction of patients with primary or secondary hypertension by discriminant analysis of chronobiologic parameters on 24-h blood pressure patterns. *Jpn Circ J* 53: 1363–1370.

Fogari R, Malocco E, Tettamanti F, *et al*. (1993). Evening vs morning isradipine sustained release in essential hypertension: a double-blind study with 24 h ambulatory monitoring. *Br J Clin Pharmacol* 35: 51–54.

Gillis A M, Peters R W, Mitchell L B, *et al*. (1992). Effects of left ventricular dysfunction on the circadian variation of ventricular premature complexes in healed myocardial infarction. *Am J Cardiol* 69: 1009–1014.

Goo R H, Moore J G, Greenberg E, Alazraki N P. (1987). Circadian variation in gastric emptying of meals in humans. *Gastroenterology* 93: 515–518.

Gould B A, Raftery E B. (1991).Twenty-four-hour blood pressure control: an intra-arterial review. *Chronobiol Int* 8: 495–505.

Gould B A, Mann S, Davies A, *et al*. (1981). Indoramin: 24-hour profile of intra-arterial ambulatory blood pressure, a double-blind placebo controlled crossover study. *Br J Clin Pharmacol* 12(Suppl): 67s–73s.

Gould B A, Mann S, Kieso H, *et al*. (1982). The 24-hour ambulatory blood pressure profile with verapamil. *Circulation* 65: 22–27.

Greminger P, Suter P M, Holm D, *et al*. (1994). Morning versus evening administration of nifedipine gastrointestinal therapeutic system in the management of essential hypertension. *Clin Invest* 72: 864–869.

Hla K K, Latham A N, Henry J A. (1992). Influence of time of administration on verapamil pharmacokinetics. *Clin Pharmacol Ther* 51: 366–370.

Imperi G A, Lambert C R, Coy K, *et al*. (1987). Effects of titrated beta blockade (metoprolol) on silent myocardial ischemia in ambulatory patients with coronary artery disease. *Am J Cardiol* 60: 519–524.

Jespersen C M, Frederiksen M, Hansen J F, *et al*. (1989). Circadian variation in the pharmacokinetics of verapamil. *Eur J Clin Pharmacol* 37: 613–615.

Kleiman N S, Schechtman K B, Young P M, *et al*. (1990). Lack of diurnal variation in the onset of non-Q wave infarction. *Circulation* 81: 548–555.

Koopmans R, Oosterhuis B, Karemaker J M, *et al*. (1993). The effect of oxprenolol dosage time on its pharmacokinetics and haemodynamic effects during exercise in man. *Eur J Clin Pharmacol* 44: 171–176.

Langner B, Lemmer B (1988). Circadian changes in the pharmacokinetics and cardiovascular effects of oral propranolol in healthy subjects. *EurJ Clin Pharmacol* 33: 619–624.

Lemmer B (1989). Temporal aspects in the effects of cardiovascular active drugs in man. In: Lemmer B, ed. *Chronopharmacology – Cellular and Biochemical Interactions*. New York: Marcel Dekker, 525–542.

Lemmer B (1991). The cardiovascular system and daily variation in response to antihypertensive and antianginal drugs: recent advances. *Pharmacol Ther* 1991; 51: 269–274.

Lemmer B (1995). Timing of cardiovascular medications – pitfalls and challenges. *Br J Cardiol* 2: 303–309.

Lemmer B (1996a). Differential effects of antihypertensive drugs on circadian rhythm in blood pressure from the chronobiological point of view. *Blood Press Monit* 1: 161–169.

Lemmer B (1996b). Circadian rhythm in blood pressure: signal transduction, regulatory mechanisms and cardiovascular medication. In: Lemmer B, ed. *From the Biological Clock to Chronopharmacology*. Stuttgart: Medpharm, 91–117.

Lemmer B (1999). Chronopharmacokinetics – implications for drug treatment. *J Pharm Pharmacol* 51: 887–890.

Lemmer B (2000). Cardiovascular chronobiology and chronopharmacology: importance of timing of dosing. In: White W B, ed. *Cardiovascular Chronobiology and Variability in Clinical Research and Clinical Practice*. Totowa, NJ: Humana Press, 255–272.

Lemmer B, Bruguerolle B (1994). Chronopharmacokinetics – are they clinically relevant? *Clin Pharmacokinet* 26: 419–427.

Lemmer B, Nold G (1991). Circadian changes in estimated hepatic blood flow in healthy subjects. *Br J Clin Pharmacol* 32: 627–629.

Lemmer B, Portaluppi F (1997). Chronopharmacology of cardiovascular diseases. In: Redfern P, Lemmer B, eds. *Handbook of Experimental Pharmacology*, vol. 125, *Physiology and Pharmacology of Biological Rhythms*. New York: Springer, 251–297.

Lemmer B, Witte K (1995). Chronopharmacological aspects of coronary heart disease. In: Willich S N, Muller J E, eds. *Triggering of Acute Coronary Heart Disease – Implications for Prevention*. Kluwer: Dordrecht, 311–320.

Lemmer B, Scheidel B, Stenzhorn G, et al. (1989). Clinical chronopharmacology of oral nitrates. *Z Kardiol* 87(Suppl 2): 61–63.

Lemmer B, Nold G, Behne S, Kaiser R (1991a). Chronopharmacokinetics and cardiovascular effects of nifedipine. *Chronobiol Int* 8: 485–494.

Lemmer B, Scheidel B, Behne S (1991b). Chronopharmacokinetics and chronopharmacodynamics of cardiovascular active drugs: propranolol, organic nitrates, nifedipine. *Ann N Y Acad Sci* 618: 166–181.

Lemmer B, Scheidel B, Blume H, Becker HJ (1991c). Clinical chronopharmacology of oral sustained-release isosorbide-5-mononitrate in healthy subjects. *Eur J Clin Pharmacol* 40: 71–75.

Lemmer B, Sasse U, Witte K, Hopf R (1994). Pharmacokinetics and cardiovascular effects of a new sustained-release formulation of diltiazem. *Naunyn Schmiedebergs Arch Pharmacol* 349: R141.

Lichtstein E, Morganroth J, Harrist R, et al. (1983). Effect of propranolol on ventricular arrhythmia. The beta-blocker heart attack trial experience. *Circulation* 67: 5–10.

Master A M (1960). The role of effort and occupation (including physicians) in coronary occlusion. *J Am Med Assoc* 174: 942–948.

Meilhac B, Mallion J M, Carre A, *et al*. (1992). Etude de l'influence de l'horaire de la prise sur l'effet antihypertenseur et la tolerance de la nitrendipine chez des patients hypertendus essentiels legers a moderes. *Therapie* 47: 205–210.

Mengden T, Binswanger B, Gruene S (1992). Dynamics of drug compliance and 24-hour blood pressure control of once daily morning vs evening amlodipine. *J Hypertens* 10(Suppl 4): S136.

Middeke M, Schrader J (1994). Nocturnal blood pressure in normotensive subjects and those with white coat, primary, and secondary hypertension. *Br Med J* 308: 630–632.

Middeke M, Kluglich M, Holzgreve H (1991). Chronopharmacology of captopril plus hydrochlorothiazide in hypertension: morning versus evening dosing. *Chronobiol Int* 8: 506–510.

Morgan T, Anderson A, Jones E (1997). The effect on 24 h blood pressure control of an angiotensin converting enzyme inhibitor (perindopril) administered in the morning or at night. *J Hypertens* 15: 205–211.

Mulcahy D, Keegan J, Cunningham D, *et al*. (1988). Circadian variation of total ischaemic burden and its alteration with anti-anginal agents. *Lancet* ii: 755–759.

Muller J E, Stone P H, Turin Z G, *et al*. (1985). The MILIS study group: circadian variation in the frequency of onset of acute myocardial infarction. *N Engl J Med* 313: 1315–1322.

Myburgh D P, Verho M, Botes J H, *et al*. (1995). 24-Hour pressure control with ramipril: comparison of once-daily morning and evening administration. *Curr Ther Res* 56: 1298–1306.

Nold G, Lemmer B (1998). Pharmacokinetics of sustained-release molsidomine after morning versus evening application in healthy subjects. *Naunyn Schmiedebergs Arch Pharmacol* 357: R173.

Nold G, Strobel G, Lemmer B (1998). Morning versus evening amlodipine treatment: effect on circadian blood pressure profile in essential hypertensive patients. *Blood Press Monit* 3: 17–25.

Norwegian Multicenter Study Group (1981). Timolol-induced reduction in mortality and reinfarction in patients surviving acute myocardial infarction. *N Engl J Med* 304: 801–807.

Ocon J, Mora J (1990). Twenty-four-hour blood pressure monitoring and effects of indapamide. *Am J Cardiol* 65: 58H–61H.

Palatini P, Racioppa A, Raule G, *et al*. (1992). Effect of timing of administration on the plasma ACE inhibitory activity and the antihypertensive effect of quinapril. *Clin Pharmacol Ther* 52: 378–383.

Palatini P, Mos L, Motolese M, *et al*. (1993). Effect of evening versus morning benazepril on 24-hour blood pressure: a comparative study with continuous intraarterial monitoring. *Int J Clin Pharmacol Ther Toxicol* 31: 295–300.

Panza J A, Epstein S E, Quyyumi A A (1991). Circadian variation in vascular tone and its relation to alpha-sympathetic vasoconstrictor activity. *N Engl J Med* 325: 986–990.

Perloff D, Sokolow M, Cowan R (1983). The prognostic value of ambulatory blood pressures. *J Am Med Assoc* 249: 2792–2798.

Peters R W, Muller J E, Goldstein S, *et al*. (1989). Propranolol and the morning increase in the frequency of sudden cardiac death (8HAT study). *Am J Cardiol* 63: 1518–1520.

Pickering T G, Levenstein M, Walmsley P, *et al.* (1994). Nighttime dosing of doxazosin has peak effect on morning ambulatory blood pressure. Results of the HAL T study. *Am J Hypertens* 7: 844–847.

Portaluppi F, Vergnani L, Manfredini R, *et al.* (1995). Time-dependent effect of isradipine on the nocturnal hypertension of chronic renal failure. *Am J Hypertens* 8: 719–726.

Pringle S D, Dunn F G, Tweddel A C, *et al.* (1992). Symptomatic and silent myocardial ischaemia in hypertensive patients with left ventricular hypertrophy. *Br Heart J* 67: 377–382.

Quyyumi A A, Wright C, Mockus L, *et al.* (1984). Effect of partial agonist activity in β-blockers in severe angina pectoris: a double blind comparison of pindolol and atenolol. *Br Med J* 289: 951–953.

Quyyumi A A, Crake T, Wright C M, *et al.* (1987). Medical treatment of patients with severe exertional and rest angina: double blind comparison of β-blocker, calcium antagonist, and nitrate. *Br Heart J* 57: 505–511.

Raftery E S, Melville D I, Gould S A, *et al.* (1981). A study of the antihypertensive action of xipamide using ambulatory intra-arterial monitoring. *Br J Clin Pharmacol* 12: 381–385.

Sandrone G, Mortara A, Torzillo D, *et al.* (1994). Effects of beta blockers (atenolol or metoprolol) on heart rate variability after acute myocardial infarction. *Am J Cardiol* 74: 340–345.

Scheidel B, Lemmer B (1991). Chronopharmacology of oral nitrates in healthy subjects. *Chronobiol Int* 8: 409–419.

Semenowicz-Siuda K, Markiewicz A, Korczynska-Wardecka J (1984). Circadian bioavailability and some effects of propranolol in healthy subjects and liver cirrhosis. *Int J Clin Pharmacol Ther Toxicol* 22: 653–658.

Shiga T, Fujimura A, Tateishi T, *et al.* (1993). Differences of chronopharmaco-kinetic profiles between propranolol and atenolol in hypertensive subjects. *J Clin Pharmacol* 33: 756–761.

Stanton A, O'Brien E (1994). Auswirkungen der Therapie auf das zirkadiane Blutdruckprofil. *Kardio* 3: 1–8.

Umeda T, Naomi S, Iwaoka T, *et al.* (1994). Timing for administration of an antihypertensive drug in the treatment of essential hypertension. *Hypertension* 23(Suppl 1): 1211–1214.

van Montfrans G A, Schelling A, Buurke E J, *et al.* (1998). Dosing time of lacidipine and the circadian blood pressure curve: the MOTIME study. *J Hypertens* 16(Suppl 9): 815–819.

Waters D D, Miller D D, Bouchard A, *et al.* (1984). Circadian variation in variant angina. *Am J Cardiol* 54: 61–64.

Weber M A, Tonkon M J, Klein R C (1987). Effect of antihypertensive therapy on the circadian blood pressure pattern. *Am J Med* 82(Suppl 1A): 50–52.

Willich S N, Muller J E, eds. (1996). *Triggering of Acute Coronary Syndromes*. Dordrecht: Kluwer Academic.

Witte K, Weisser K, Neubeck M, *et al.* (1993). Cardiovascular effects, pharmacokinetics, and converting enzyme inhibition of enalapril after morning versus evening administration. *Clin Pharmacol Ther* 54: 177–186.

Yasue H, Omote S, Takizawa A, *et al.* (1979). Circadian variation of exercise capacity in patients with Prinzmetal's variant angina: role of exercise-induced coronary arterial spasm. *Circulation* 59: 938–948.

9

Rhythms, pain and pain management

Gaston Labrecque and Marie-Claude Vanier

9.1 Introduction

Pain is one of the most common symptoms for which patients seek advice and help from health professionals. Pain is a complex phenomenon characterised as an unpleasant sensation that often disturbs the normal patterns of patients' activity, sleep and thoughts. It is also a subjective phenomenon influenced by factors such as anxiety, fatigue, suggestion or emotion as well as prior experience. Thus, the patient is the only person who can describe the intensity of his or her pain and determine the extent of pain relief produced by analgesics. Therefore, clinicians must rely on the patient's evaluation of pain intensity in deciding which drug to use, how much and when.

Evaluation of pain intensity is achieved mainly by quantifying patients' behaviour or by using tools that can quantify the subjective reports of pain by patients. The tools most commonly used are daily activity diaries, sleep patterns, food intake, demand for or intake of medication, and amount of time spent standing, sitting or reclining. The observation of patients' behavioural changes provides the objective and practical method for inferring that they are experiencing pain (Chapman and Jones, 1944). This technique is especially important in infants and in adults who have difficulty in communicating (McGrath, 1990; Reading, 1989). Although behavioural procedures are important, subjective techniques such as rating scales (e.g. Visual Analog Scale (VAS); Jensen et al., 1986) or pain questionnaires (e.g. McGill Pain Questionnaire (MPQ); Melzac, 1975, 1983), are still the most widely used tools in clinical settings. The techniques used to determine pain intensity will not be discussed here as they have been reviewed elsewhere (Labreque et al., 1997).

The objectives of this chapter are to present the main findings in the chronobiology of pain and the chronopharmacology of analgesics. This chapter focuses mainly on biological rhythms that have a direct bearing on therapeutics and outlines guidelines that can be used in the clinical situation to individualise the analgesic regimen in order to maximise the beneficial effects of analgesic drugs.

9.2 Chronobiology of pain

9.2.1 Rhythms in endogenous opioid peptide levels

In the last 25 years, data obtained in laboratory animals indicate the existence of 24 h variations in the plasma and brain concentrations of β-endorphin or enkephalins: peak values were obtained late during the resting period or at the beginning of the activity period (for review, see Labrecque et al., 1997). For instance, Naber et al. (1981) used a radioimmunological assay to determine opioid activity. They sampled human plasma and monkey cerebrospinal fluid at 2 h intervals over a 24 h period and found an episodic secretion of an opioid substance with a morning peak (10.00) and an evening dip (22.00). There was a 40% difference between the peak and the trough values in human plasma. These data are in rather good agreement with those of Petraglia et al. (1983), who reported circadian variation of β-endorphin concentrations in six healthy volunteers with peak and trough levels occurring at 08.00 and 20.00, respectively. Similar data were obtained by Rolandi et al. (1992), who found a circadian rhythm of plasma β-endorphin levels in six adult healthy volunteers (28–37 years): the highest and lowest levels were found at 06.00 and 00.00, respectively. A morning peak and an afternoon trough were also found in the plasma β-endorphin levels of neonates with mean gestational ages of 31.7 and 32.5 weeks, respectively (Hindmarsh et al., 1989; Sankaran et al., 1989).

It is also interesting to note that diurnal variations were reported in the plasma levels of β-endorphin in 62 healthy pregnant women and in 11 healthy nonpregnant women. Räisänen (1988) found a 2-fold rise in plasma β-endorphin levels in the last semester of the pregnancy. In five women near term and again in the early puerperium, a circadian pattern of β-endorphin was found in weeks 38–41 weeks of pregnancy: the highest and lowest values were found at 08.00 and 00.00, respectively. This time-dependent variation was not found on the 4th day after delivery.

Finally, it should be noted that Von Knorring et al. (1982) identified a circannual variation in the concentrations of endorphins in the cerebrospinal fluid (CSF) of 90 patients with chronic pain syndrome of psychogenic and organic aetiology: the highest concentrations were found in January and February and the lowest CSF endorphin levels occurred in July and August. To our knowledge, no other data are available on the circannual variation in the levels of opioid peptides.

In summary, human studies show clearly that the levels of endogenous opioid peptides are higher at the beginning of the day and lower in the evening. Thus, we should expect that pain level will fluctuate around the clock and that the requirements for analgesia will also vary as a function of the time of day.

9.2.2 Rhythms in experimental pain

The scientific and standardised study of temporal variations in experimental pain began in the early part of the 20th century. Grabfield and Martin (Grabfield and Martin, 1912; Martin *et al.*, 1914) were the first to show that electrical stimulation produced the highest pain level when administered at 10.30, and lowest pain was found during the resting period of the volunteers. Many other investigators attempted to determine time-dependent variations in experimental pain. Table 9.1 illustrates that contradictory results were obtained in these studies: the time of peak or trough in pain intensity differed from one study to another, while in others no significant variation with time was found. These divergent findings may be due to methodological differences since the pain stimuli used varied between studies. It is also interesting to note that pain was measured in different parts of the body and many parameters were used to determine either the threshold or the intensity of pain. Furthermore, the investigators did not differentiate between sharp or dull pain, or between epicritic and protopathic pain, nor did they take into account the affective reaction to nociceptive stimuli. In addition, the sleep–activity pattern of the volunteers was not indicated in most studies, although the reader would assume that the subjects were diurnally active and were restrained from cigarette smoking and coffee drinking. Further research is needed in the area of experimental pain.

9.2.3 Pain rhythms in clinical situations

Circadian variation in pain is observed every day in hospital settings, but clinicians are not generally aware of the significance of these data. Table 9.2 summarises the main data on biological rhythms of pain in different clinical situations.

A circadian pattern of pain was found in 543 patients with toothache caused by dental caries (Pöllmann, 1981a). Figure 9.1a illustrates that pain due to toothache increases during the late night hours

Table 9.1 Studies on circadian variations in experimental pain in human volunteers[a]

Painful stimuli	Sites	Parameters studied	Data obtained	Reference
Electrical stimulation		Irritability	Peak: 10.30 Trough: 23.30–01.00 and 04.00–05.00	Grabfield and Martin (1912)
		Irritability	No circadian rhythm detected	Martin and Grabfield (1914)
Cold stimulation	Hand, foot	Pain threshold	No circadian rhythm detected	Macht et al. (1916)
Heat	Forehead	Pain threshold	No circadian rhythm detected	Hardy et al. (1940)
Heat	Forehead	Pain threshold	No circadian rhythm detected	Schumacher et al. (1940)
Heat	Forehead	Pain and pain tolerance thresholds	No circadian rhythm detected	Chapman and Jones (1944)
Radiant heat		Pain intensity	Peak: 06.30 Trough: 18.30 No circadian rhythm detected (women)	Proccacci et al. (1973, 1974)
Electrical shocks		Pain levels	Peak pain: morning Trough: evening	Davis et al. (1978)
Electrical current	Forearm	Pain and pain tolerance thresholds	No circadian rhythm detected	Stacher et al. (1982)
Electrical and cold stimuli	Tooth	Pain threshold	Peak: 15.00–18.00 Trough: 00.00	Pöllmann (1984)
Electrical current	Forearm	Detection and pain thresholds Pain tolerance threshold	No circadian rhythm detected	Morawetz et al. (1984)
Cold stimulation	Hand, foot	Detection and pain thresholds	No circadian rhythm detected	Strian et al. (1989)
Inflatable cuff for 20 s	Head	Pain intensity	Peak headache: 02.00 Trough: 14.00	Göbel and Cordes (1990)
Nociceptive flexion reflex	Sural nerve	Muscular twitch and pain intensity	Peak pain: 05.00 Trough pain: 09.00–13.00:	Bourdallé-Badie et al. (1990)
Carbon dioxide	Left nostril	Pain intensity	Peak: 02.00	Hummel et al. (1992, 1994); Kobal et al. (1992)

[a]In most studies, the number of subjects was about 10–15, except in one (Schumacher et al., 1940) where 150 volunteers participated in the trial.

Table 9.2 Biological rhythms of pain in patients

Causes of pain	No. of patients	Time of peak	Time of trough	Reference
Anginal pain	7789	06.00–12.00	00.00–06.00	Cannon et al. (1997)
Unstable angina	2586	08.00–10.00	02.00–04.00	Cannon et al. (1997)
Myocardial infarction	1229	05.00–09.00		Master (1960)
	703	05.00–10.00		Muller et al. (1985)
Backache	60	Morning		Pownall and Pickvance (1985)
	19	20.00[a]	08.00	Pednault and Parent (1993)
Biliary colic	50	23.00–03.00	09.00–13.00	Rigas et al. (1990)
Cancer	130	18.00	04.00–10.00	Sittl et al. (1990)
Intractable pain	41	22.00	08.00	Folkard et al. (1976)
Migraine	15	10.00	00.00	Solomon (1992)
	117	08.00–12.00	00.00	Waters and O'Connors (1971)
	114	04.00–08.00	12.00	Ostfeld (1963)
Osteoarthritis	20	22.00[b]	02.00–06.00	Bellamy et al. (1990)
	57	14.00–22.00[b]		Lévi et al. (1985)
	4	19.00–23.00	07.00	Job-Deslandre et al. (1983)
Rheumatoid arthritis	19	06.00–08.00	18.00	Kowanko et al. (1982)
Toothache	543	08.00	15.00	Pöllmann (1981)

[a]A secondary peak was detected at 20.00 in this study.
[b]Inter-individual variations were found in these studies.

and peaks at 07.00; the trough was found at 15.00. An interesting correlation exists between the chronobiology of electrically induced or cold-induced pain (Pöllmann, 1984) and that of toothache: the threshold of electrical stimulation was lowest between midnight and 06.00, while the acrophase of patients' toothache increased during the latter part of the night to reach a peak at 08.00.

In 41 patients suffering from intractable pain, Glynn et al. (1975) and Folkard et al. (1976) determined the pattern of patient's pain using

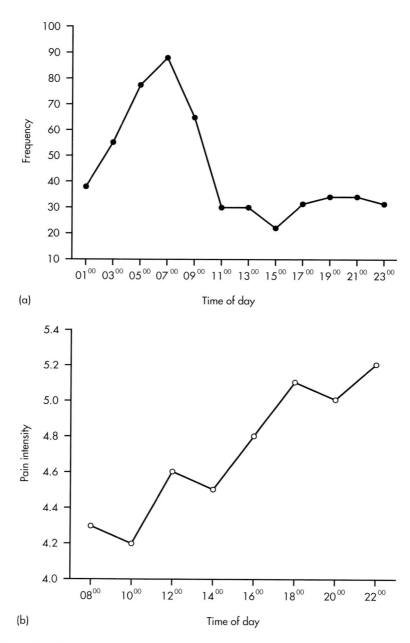

Figure 9.1 Temporal variations in pain in patients. (a) The circadian variation of dental pain in 543 patients with caries. (b) Pain intensity rating/hour of day; the increase in intractable pain observed in 41 hospitalised patients. Redrawn from Pöllmann (1981a); Folkard et al. (1976).

a VAS every 2 h for 7 consecutive days during the waking span. Figure 9.1b illustrates that the intensity of intractable pain increased throughout the day and the highest pain intensity was found at the last evaluation of the day, i.e. at 22.00.

Rigas *et al.* (1990) studied the circadian rhythm of biliary colic in 50 patients who underwent emergency or elective cholecystectomy for symptomatic biliary tract stones. All patients were interviewed during their first postoperative week about the following aspects of their biliary pain: time of onset and pain features such as location, duration, number and time of prior pain episodes, and relationship to meals and body position. Patients who could not recall the exact time of pain with confidence were asked to describe their painful episode as occurring during the 'daytime' (06.00–20.00), 'night-time' (20.00–06.00) or 'anytime'. The pain pattern was compared to that of a control group consisting of 27 patients with renal colic. The data indicated that 76% of the 50 patients with biliary pain were able to report a specific clock time or a 2 h time span during which the painful episodes occurred. Biliary pain displayed a significant circadian rhythm, with a peak between 23.00 and 03.00. In contrast, no periodicity was found in patients with renal pain or in those with nonbiliary abdominal pain secondary to miscellaneous causes.

A circadian pattern was also found in the time of onset of migraine (Solomon, 1992). A total of 214 migraine attacks were reported by 15 patients during the 20-week study period. There was a marked increase in attacks between 08.00 and noon, with a peak at 10.00 and a trough around midnight. These data are in agreement with the early work of Waters and O'Connors (1971) and with the data of Ostfeld (1963), who reported that migraine attacks were most common at the end of the night (04.00) or early morning (0800–12.00), but were rather rare during the night.

The chronopathology of backache has been studied by two groups of investigators. The first study was carried out by Pownall and Pickvance (1985) in 60 patients with either persistent but recent (less than a year) or chronic (three or more years) low back pain. Patients whose back pain was due to malignancy, infection or abnormal metabolism were excluded from the study. After at least one week free from use of nonsteroidal anti-inflammatory drugs (NSAIDs), patients evaluated the circadian variation of their backache. The data showed that 33 patients indicated that their pain was worst in the morning and 18 that it was worse in the afternoon, compared with only 6 experiencing pain in the evening or night period. Completely different data were obtained by Pednault and Parent (1993) in a study on 19 outpatients with chronic low back pain caused by discal degeneration, fusion pseudoarthrose or discoidectomy. The patients were synchronised on a diurnal activity pattern (sleep 00.00–07.00) and they used a VAS every 4–5 h during the activity span for 7 consecutive days to self-evaluate pain intensity. Although some inter-individual differences were noted, data analysis clearly indicated a circadian variation of chronic low back pain with highest and lowest pain occurring at 20.00 and 08.00, respectively. It is not possible to explain the discrepancy between these two studies, but it may be that differences in the causes of pain studied by the two groups of investigators could account for the differences reported in the pain pattern. These data indicate again why authors must give complete information on the potential causes of pain in the patients included in their studies.

The chronobiology of pain reported by arthritic patients is of interest because the occurrence of pain varies according to the cause of arthritis. In patients with rheumatoid arthritis, the acrophase of pain was found mostly at the beginning of the day (Kowanko *et al.*, 1981, 1982), whereas it occurred at the end of the day in patients with osteoarthritis (OA) (Bellamy *et al.*, 1990; Lévi *et al.*, 1985; Job-Deslandre *et al.*, 1983). However, it must be pointed out that important inter-individual differences can be found in the time of highest pain intensity. Table 9.3 indicates that while most OA patients reported pain at the end of the day, others reported that peak pain occurred early in the morning and no rhythm of pain could be detected in a few patients (Bellamy *et al.*, 1990; Lévi *et al.*, 1985). These inter-individual differences should be taken into account when prescribing analgesic and anti-arthritic medication.

Although pain is regularly reported after surgery and by patients with cancer, the chronobiology of pain has rarely been studied directly

Table 9.3 Inter-individual differences in time of peak pain perception in patients with osteoarthritis

Time of day	Lévi et al. *(1985)*	Bellamy et al. *(1990)*
08.00–14.00	3	0
14.00–20.00	19	8
20.00–00.00	ND	5
00.00–08.00	8	3
2 peaks/24 h	23	0
No circadian profile	4	4
No. of patients in the studies	57	20

ND, not determined in the study.

in these patients. To our knowledge, the only study identifying the temporal variation of pain in cancer patients was done by Sittl *et al.* (1990), who used a VAS at eight times during the 24 h period. Their data showed that peak pain occurred at 18.00. Most other studies were carried out in patients receiving opioid analgesics and were mainly concerned with the relief produced by these drugs. These data will be reviewed later.

The data on the chronobiology of clinical pain can be summarised as follows:

- Pain intensity is not constant over the 24 h period.
- Pain patterns differ according to the cause of pain.
- Important inter-individual differences are found in the pattern of pain. This was particularly the case in patients with backache or arthritis.
- Pain rating should be carried out at different times of day in each patient.
- The chronobiology of pain needs to be taken into account when determining the optimal time for analgesic administration.

Very few clinicians consider the chronobiology of pain when preparing pain treatment strategies for their patients. This may explain in part why pain is often present even though analgesics are administered continuously after surgery.

9.3 Chronopharmacology of analgesics

9.3.1 Local anaesthetics

Local anaesthetics prevent or relieve pain by interrupting nerve conduction. They bind within the pore of the sodium ion channel in axons and block ion movement through the pores. The effect of these agents rapidly reverses upon their diffusion from the site of action and the anaesthetic action is usually restricted to the site of application (Catterall and Mackie, 1996).

Reinberg and Reinberg (1977) were the first to demonstrate a significant circadian rhythm in the action of lignocaine (lidocaine) or betoxycaine applied to human teeth and skin. The longest duration of anaesthesia was found at 15.00, with a 100% difference in the peak and trough values. Pöllmann (1981b) found similar data with mepivacaine. Lemmer and Wiemers (1989) used an electronic pulptester in 28 healthy volunteers and in 55 patients who were treated for dental filling as a result of caries. Figure 9.2 illustrates that a significant circadian rhythm in the duration of local anaesthetic effect was obtained in 36 patients receiving articaine (32 mg) plus adrenaline (epinephrine) (0.0048 mg) (Ultracain D-S) at 08.00, 11.00, 14.00 or 17.00: peak and trough were found after drug application at 14.00 and 17.00, respectively.

These data can be explained in part by time-dependent changes in the pharmacokinetics of local anaesthetics. For instance, Bruguerolle and colleagues (Bruguerolle and Isnardon, 1985; Bruguerolle et al., 1991) reported that highest plasma levels occurred at 16.00, both in 24 men receiving lignocaine injections and in 29 children receiving a cream containing 25 mg/L lignocaine and 25 mg/L prilocaine (Emla). They also reported that there was a 3-fold difference in the bupivacaine plasma clearance at different times of the day. Figure 9.3 illustrates that the bupivacaine clearance was lowest in the afternoon and much higher at the beginning of the day (Bruguerolle and Prat, 1988). Finally, it is surprising to note that the chronobiology of the transcutaneous passage of local anaesthetics has not been investigated to any great extent. Bruguerolle

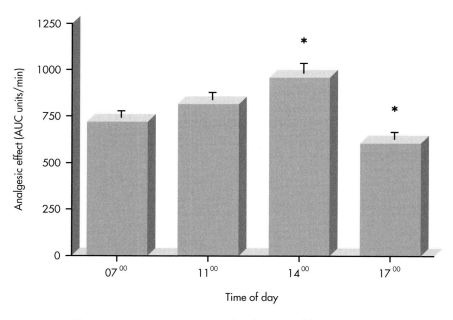

Figure 9.2 Time-dependent variations in the duration of local anaesthetic analgesia (AUC units/min). Articaine–adrenaline solution was injected para-apically in one of the upper teeth of 36 patients with dental caries. The asterisk indicates a statistically significant difference ($p < 0.05$) between the 14.00 and 17.00 data. Redrawn from Lemmer and Wiemers (1989).

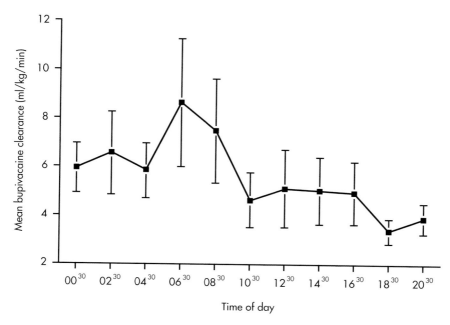

Figure 9.3 Circadian variations in bupivacaine plasma clearance after a peridural constant-rate infusion in 13 men. Mean bupivacaine clearance ± SE (mL/kg/min). Redrawn from Bruguerolle *et al.* (1988).

et al. (1988) used the penetration of bupivacaine, etidocaine and mepivacaine into red blood cells as a model to study the passage of lignocaine through biological membranes. A circadian variation of the passage of the three drugs was demonstrated, with a maximum occurring in the middle of the activity period. Further research is needed in this area.

9.3.2 Anti-inflammatory agents

Morning pain is a very characteristic symptom of rheumatoid arthritis (RA) and NSAIDs are known to produce significant side effects on the gastrointestinal (GI) tract. Thus, it is not surprising that studies have been done on the time-dependent variation in the effectiveness and toxicity of NSAIDs. The first and also the largest multicentre chronotherapeutic study on NSAIDs was carried out by French rheumatologists in 517 patients suffering from osteoarthritis (OA) affecting the hip ($n = 240$), the knee ($n = 240$) or both joints ($n = 37$) (Lévi *et al.*, 1985; Bouchancourt and Le Louarn, 1982; Simon *et al.*, 1982). The objective of these studies was to answer the simple question often asked by patients: When should the once-daily medication be taken? Each patient was his or her own control and took a 75 mg sustained-release indometacin capsule daily at 08.00 for one week, at 12.00 for another week and at 20.00 for the last week. Each patient was asked to report side effects and to perform a VAS self-rating of pain intensity every 2 h from 07.00 to 23.00 for 1–2 days during the washout period on days 4–6 of each test week. In these studies, the effectiveness of indometacin was found by both physicians and patients to be very good or excellent. When the data of all patients were grouped per hour of indometacin ingestion, no significant differences were found in the effectiveness of indometacin as a function of the time of indometacin administration. Morning ingestion time was considered best by 28% of the patients, whereas 22% and 35% of the patients preferred the noon and evening administration, respectively. However, marked inter-individual differences were observed in the time of peak symptoms of arthritis (see Table 9.2) and in when patients preferred to take their NSAID. The data were used to determine the optimal time of NSAID administration: the evening administration was most effective in subjects with a predominantly nocturnal or early morning pain, while administration in the morning or at noon was most effective in subjects with peak pain occurring in the afternoon or evening. The analgesic effect of the drug was increased by about 60% over previous values when the NSAID was taken at the time preferred by the patients (Lévi et al., 1985).

In 17 patients suffering from RA, Kowanko et al. (1981) and Swannel (1983) showed that a 100 mg dose of flurbiprofen administered twice a day was more effective than a 50 mg dose ingested four times daily. However, subjective measurements of pain and stiffness indicated that part of the twice-daily flurbiprofen dose must be administered at night to control morning stiffness and pain. Rejholec et al. (1984) confirmed these data in a larger double-blind multicentre study.

Reinberg et al. (1991) also completed a smaller study on the chronopharmacology of tenoxicam in patients with spondylarthritis ($n = 11$), OA ($n = 7$) and RA ($n = 8$). This NSAID has interesting pharmacokinetic properties as its half-life is about 3 days and it is highly bound (99.3%) to plasma proteins (Heintz, 1989). In the double-blind, crossover and randomised study, patients were given tenoxicam 20 mg daily by mouth for 2 weeks at 08.00, 12.00 or 20.00 and pain and stiffness were self-monitored throughout the day. The results from this preliminary study suggested that optimal effectiveness occurred with the noon administration of tenoxicam. Further research is needed on NSAIDs with similar pharmacokinetic profiles.

Patients using non-selective NSAIDs reported GI or neurosensory side effects that were so troublesome that they often discontinued their medication. Thus, it was not surprising to find that the frequency of side effects was the most striking difference observed in the effects of indometacin administered at different times of the day. The frequency of side effects produced by indometacin was studied in a double-blind crossover trial with 66 patients lasting 3 weeks (Lévi et al., 1985). The data indicated that 33% of the patients reported undesirable effects when indometacin was ingested at 08.00 in comparison with 7% of the patients reporting these undesirable effects when the drug was taken at 20.00 (Simon et al., 1982). During the three weeks of the study, the side effects were consistently higher after the 08.00 administration of the drug than at any other time of day. About 75% of the undesirable effects were CNS-related (vertigo, headache, anxiety), while the others were of GI origin (nausea, gastric pain, diarrhoea). There was no difference in the frequency of side effects in males and females. Figure 9.4 illustrates that similar data were obtained in another double-blind trial with 118 outpatients with OA of the hip or the knee receiving 200 mg slow-release tablets of ketoprofen. In patients taking ketoprofen in the morning, total and GI side effects were twice those observed after dosing at night (Boissier et al., 1990). These data are in agreement with the findings of

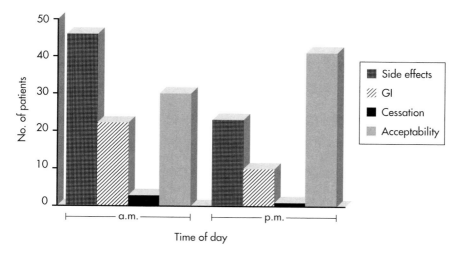

Figure 9.4 Temporal changes in side effects produced by sustained-release ketoprofen in osteoarthritic patients. A 200 mg dose of ketoprofen was given at 08.00 (n = 59) or 20.00 (n = 59) and the tolerance to this NSAID was evaluated as total side effects reports, number of patients with GI problems, cessation of drug use and overall assessment by patients. Redrawn from Boissier et al. (1990).

Moore and Goo (1987), who observed that the number of gastric mucosal lesions produced by oral administration of aspirin at 10.00 was twice that obtained after 22.00 dosing. In a recent crossover study, 16 healthy male volunteers ingested single oral doses of 75 mg and 1000 mg aspirin at either 08.00 or 20.00. Two hours after dosing, gastric lesions were objectively scored by direct and video-endoscopic examinations. No circadian phase dependency could be confirmed for either dose of aspirin (Nold et al., 1995). It is of interest to note that no patient complained of GI pain, although lesions were present after the 1000 mg aspirin dose. In most studies, the data suggest that NSAIDs were better tolerated, and fewer patients stopped the use of NSAIDs, when the drugs were administered in the evening rather than in the morning. However, large inter-individual differences were observed.

Although the studies described above focused on the well-established NSAIDs in arthritic patients, we would expect that they could be applied equally to the newer NSAIDs such as celecoxib (Celebrex) and rofecoxib (Vioxx), although the frequency of side effects appears to be smaller with the COX-2 agents.

9.3.3 Opioid analgesics

Very few investigators have studied the temporal variation in the effects of morphine and other opioid analgesics in patients with postoperative pain. Table 9.4 summarises the data currently available. In the first chronopharmacological study with patient-controlled analgesia (PCA), Graves et al. (1983) found a significant circadian variation in intravenous morphine dosing requirements in patients undergoing elective abdominal surgery or after gastric bypass surgery. These investigators reported that maximal demands for morphine occurred at 09.00, whereas the lowest frequency of requests for the drug was observed at 03.00. In fact, the morning doses were 15% larger than those administered during the night. Dunn et al. (1993) also studied the morphine requirements of 28 diurnally active patients undergoing abdominal surgery. In this PCA study, two peaks of morphine demand were found, with acrophases occurring between 08.00 and 12.00 and between 16.00 and 20.00. Completely different data were obtained in a study of 32 patients undergoing thoracic or abdominal surgery (Labrecque et al., 1988). The objective of this 3-day double-blind study was to compare the analgesia produced by a constant infusion of morphine or by subcutaneous injections of the drug every 4 h. Pain was assessed with a VAS one hour after each subcutaneous injection for 3 days. A circadian

Table 9.4 Time-dependent variations in the effect of analgesics in postoperative and cancer patients

Causes of pain	No. of patients	Modes of administration; drugs used; parameters studied	Peak	Trough	Reference
Gastric bypass, elective surgery	14	i.v. PCA pump Morphine demands	09.00	03.00	Graves et al. (1983)
Abdominal surgery	28	i.v. PCA pump Morphine demands	08.00–20.00	16.00–20.00	Dunn et al. (1993)
Thoracic or abdominal surgery	17	i.v. perfusion	18.30–19.20	–	Labrecque et al. (1988)
	15	s.c. perfusion Pain levels, VAS			
Postsurgical gynaecological cancer	19	i.v. PCA Morphine demands HM demands	04.00–08.00	12.00–16.00	Auvil-Novak et al. (1990)
Postsurgical gynaecological cancer	45	i.v.a. PCA pump Morphine demands	04.00–08.00 08.00–12.00	00.00–04.00 00.00–04.00	Auvil-Novak et al. (1990)
Cancer patients	20	i.v. Morphine every 4 h Pain intensity	08.00–20.00	00.00–08.00	Wilder-Smith et al. (1992)
Cancer patients	42	VAS score, % pain intensity			Wilder-Smith et al. (1992)
		SR Morphine	22.00	10.00	Wilder-Smith and Wilder-Smith (1992)
		Bruprenorphine twice daily Pain intensity, VAS	22.00–02.00	06.00	
Cancer patients	8	HM CP + PCA pump	22.00	14.00	Vanier et al. (1992)
		HM basal rate + PCA	22.00	14.00–02.00	
		PCA bolus demands; pain intensity, VAS	18.00–22.00	06.00	
Cancer patients	61	Extra doses of Morphine	10.00–22.00	At night	Bruera et al. (1992)
Cancer patients	16	i.v. or s.c. narcotic demands	06.00–18.00	18.00–06.00	Citron et al. (1992)

PCA, patient controlled analgesia; VAS, visual analogue scale; SR, slow-release; HM, hydromorphone; CP, continuous perfusion.

variation was detected on day 1 of morphine administration, no matter how the opiate was given. The acrophase of pain was found between 18.30 and 19.15. It is not yet possible to explain the discrepancies between these studies.

In cancer patients, the first PCA study was done by Auvil-Novak *et al.* (1988) in 19 postsurgical gynaecological cancer patients receiving either morphine sulphate or hydromorphone. Pump records were collected to ascertain the number of PCA attempts at self-medication within 4 h periods, the total dosage and the number of demands undelivered owing to the lockout of the pump. The number of attempts was significantly greater between 04.00 and 08.00 for both morphine and hydromorphone and was significantly lower between 12.00 and 16.00 for morphine and between 00.00 and 04.00 for hydromorphone. The morning demands were about 60% larger than those administered at other times of day. There was also a temporal variability in drug demands that exceeded what was allowed by the pump: peak demands in this situation occurred between 00.00 and 04.00 for morphine and between 04.00 and 08.00 for hydromorphone. The lowest demands were found between 12.00 and 16.00 for morphine and between 16.00 and 20.00 for hydromorphone. The other PCA study of Auvil-Novak *et al.* (1990) was done in 45 postsurgical gynaecological patients receiving morphine. These data showed that the highest doses of morphine were administered between 08.00 and 12.00, whereas the lowest doses were administered between midnight and 04.00.

Wilder-Smith *et al.* (1992) used a verbal rating scale in 20 diurnally active cancer patients and recorded the effect of 16 mg of morphine administered at 4 h intervals. The pain intensity scores were highest between 08.00 and 20.00 and were significantly smaller during the night. These investigators also studied the diurnal pattern of pain in 42 cancer patients receiving either slow-release morphine or buprenorphine (Wilder-Smith and Wilder-Smith, 1992). The data indicate that there is a distinct circadian rhythm of pain intensity during treatment with long-acting opioids: maximal effectiveness was found in the evening (22.00) or at night (02.00). while the smallest change of pain intensity occurred early in the day (10.00).

Finally, Vanier *et al.* (1992) measured pain intensity in eight cancer patients receiving hydromorphone by a basal-rate injection associated with PCA demands or continuous infusion of the opioid. Pain intensity was measured with a VAS at 4 h intervals: the data indicated that pain intensity was twice as high at 22.00 (1.33 ± 0.8 arbitrary pain units) as at 14.00 (0.64 ± 0.7 arbitrary pain units). When the PCA bolus demands were summed with the rescue doses of analgesics, the mean number of requests for the medication was largest between 18.00 and 22.00 and smallest between 02.00 and 06.00.

Bruera *et al.* (1992) reviewed the circadian distribution in the extra doses of opioids received by 61 patients admitted to a palliative care unit. All patients had pain due to advanced cancer and were bed-ridden. The patients received regular parenteral narcotics every 4 h, but they could request intermittent subcutaneous injections of narcotics by calling the nurse. The number of extra doses of opioids was determined at 4 h intervals over a 24 h period. A total of 1322 extra doses were administered during 610 patient days. Figure 9.5 illustrates that the largest

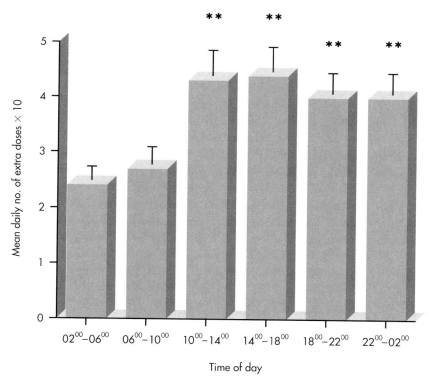

Figure 9.5 Distribution of extra doses of opioid agonist in hospitalised cancer patients during a 24 h period. Patients received parenteral narcotics every 4 h, but they could request extra doses at any time simply by calling the nurse. Data are presented as mean extra dose × 10 ± SE. The asterisks indicate a statistically significant difference ($p < 0.01$) between the data obtained during the day and the evening and those found at night (02.00–06.00) and at the beginning of the morning (06.00–10.00). Redrawn from Bruera *et al.* (1992).

number of extra doses was administered between 10.00 and 22.00 and this number reduced to about 60% of its original value during the night-time or in the early morning (02.00–10.00). Forty-five of 61 patients (76%) received most of their extra dose of narcotics between 10.00 and 22.00. Citron et al. (1992) also reported that the demands for analgesic were 48% higher during the day than at night.

The large differences found in the intensity of pain and in the amount of opioids administered throughout the day cannot be explained at present. Pain is a complex phenomenon and investigators must always take into account that many factors can influence the intensity of pain and its evaluation by the patients. More research is needed before a consensus can be reached on the time of drug administration that should be recommended to obtain optimal pain relief.

9.3.4 Placebo

It is well accepted that pain can easily be altered by many factors, including previous experiences or anxiety, that may vary throughout the 24 h period. Thus, it is interesting to determine whether biological rhythms can be found in the effect of placebo. To our knowledge, Pöllmann (1987) is the only investigator who has evaluated the importance of psychological factors on the evaluation of pain intensity and pain relief. Pöllmann reported a circadian variation in the analgesic potency of placebo. In healthy individuals, a sugar-coated placebo tablet increased pain threshold in healthy teeth by 25–30% when ingested during the day (i.e. between 09.00 and 21.00). When administered at night, the placebo did not produce any analgesic effect. Although ethical problems make this type of experiment difficult, it is clear that further work is needed in this area.

9.4 Guidelines for chronotherapeutic use

The time-dependent changes in pain and in the analgesic effect of drugs are relevant for the daily practice of clinicians or clinical pharmacists. The data summarised in this chapter could be used in clinical situations to optimise and to individualise the drug treatment of patients with pain and to reduce the frequency of side effects induced by drug administration. To get maximal analgesic effect, clinicians should be able to:

- *Determine when pain level is highest and lowest.* Using a VAS or a quick questionnaire, the practitioner is able to determine quickly whether pain is constant over the 24 h period, when the pain is

most intense and how many peaks of pain are obtained within 24 h. The chronobiological approach gives information on the temporal pattern of disease-induced pain and on the inter-individual variations in pain intensity.
- *Administer analgesics to produce highest blood levels when pain is highest.* In practice, drugs are commonly administered at regular intervals (such as 4 h after the last dose), but this approach does not take into account that pain intensity and/or the duration of the analgesic effect can vary throughout the day. The temporal pattern of pain will be useful in optimising and individualising drug treatment as well as deciding the frequency of drug administration over 24 h.
- *Refrain from administering drugs at time of highest toxicity or when the frequency of side effects is highest.* Chronopharmacological studies with established NSAIDs indicate that the side effects are more pronounced early in the morning than in the evening. All things being equal, one would avoid or reduce administration at times of day when side effects are most prominent. However, this may not be so simple in practice, because in many patients this coincides with the greatest need for analgesia.

9.5 Conclusions

The main studies dealing with time-dependent variations in pain and in pain management have been reviewed here. Pain is a very complex phenomenon influenced by anxiety, fatigue, suggestion or emotion, as well as prior experiences. The practitioner must rely on the patient's capacity to evaluate pain intensity and the pain relief produced by analgesics. Thus, it is not surprising to find time-dependent variations in pain and in analgesia, but practitioners do not always take these temporal variations into account when evaluating pain and prescribing analgesics. It is also interesting to note that a physiological basis for the temporal changes in pain has been identified, namely circadian variation in the plasma or brain levels of enkephalins or β-endorphin. Furthermore, human data indicate that the intensity of pain varies throughout the day, and that the times for the highest and lowest pain intensity are not the same for each disease. Inter-individual variations were also observed in pain pattern and it is possible that inadequate pain management associated with drug overdosing or underdosing, which occur frequently in clinical practice, can be explained in part by these differences in the chronobiology of diseases.

The data obtained so far suggest that practitioners must take into account the findings on biological rhythms in pain and in pain management. Clinicians must accept the concept that doses of drugs must be modulated throughout the day to obtain maximal analgesia. They must also realise that inter-individual variations in the temporal pattern of pain are important. Further research is needed in the field, mainly in patients with cancer, arthritis or chronic pain. The final goal of this work is to find the optimal pharmacological regimen to reduce pain with minimal side effects. Knowledge of biological rhythms in pain and in pain management is certainly a valuable element in reaching this goal.

References

Auvil-Novak S E, Novak R D, Smolensky M H, *et al.* (1988). Twenty-four hour variation in self-administration of morphine sulfate and hydromorphone by post-surgical gynecologic cancer patients. *Annu Rev Chronopharmacol* 5: 343–346.

Auvil-Novak S E, Novak R D, Smolensky M H, *et al.* (1990). Temporal variation in the self-administration of morphine sulfate via patient-controlled analgesia in post-operative gynecologic cancer patients. *Annu Rev Chronopharmacol* 7: 253–256.

Bellamy N, Sothern R B, Campbell J (1990). Rhythmic variations in pain perception in osteoarthritis of the knee. *J Rheumatol* 17: 364–372.

Boissier C, Decousus H, Perpoint B, *et al.* (1990). Timing optimizes sustained-release ketoprofen treatment of osteoarthritis. *Annu Rev Chronopharmacol* 7: 289–292.

Bouchancourt P, Le Louarn C (1982). Étude chronothérapeutique de l'indométacine à effet prolongé dans l'arthrose des membres inférieurs. *Tribune Médicale* (Suppl): 32–35.

Bourdallé-Badie C, Andre M, Pourquier P, *et al.* (1990). Circadian rhythm of pain in man: study by measure of nociceptive flexion reflex. *Annu Rev Chronopharmacol* 7: 249–252.

Bruera E, Macmilland K, Kuehn N, Miller M J (1992). Circadian distribution of extra doses of narcotic analgesics in patients with cancer pain: a preliminary report. *Pain* 49: 311–314.

Bruguerolle B, Isnardon R (1985). Daily variations in plasma levels of lidocaine during local anaesthesia in dental practice. *Ther Drug Monit* 7: 369–370.

Bruguerolle B, Prat M (1988). Temporal variations of membrane permeability to local anaesthetic agents: bupivacaine and mepivacaine documented by their erythrocytic passage in mice. *Annu Rev Chronopharmacol* 5: 227–230.

Bruguerolle B, Dupont M, Lebre P, Legre G (1988). Bupivacaine chronokinetics in man after a peridural constant rate infusion. *Annu Rev Chronopharmacol* 5: 223–226.

Bruguerolle B, Giauffre E, Prat M (1991). Temporal variations in transcutaneous passage of drugs: the example of lidocaine in children and rats. *Chronobiol Int* 8: 277–282.

Cannon C P, McCabe C H, Stone P H, *et al.* (1997). Circadian variation in the onset of unstable angina and non-Q-wave acute myocardial infarction (The TIMI III registry and TIMI IIIB). *Am J Cardiol* 79: 253–258.
Catterall W, Mackie K (1996). Local anaesthetics. In: Hardman J G, Limbird L E, Molinoff P B, *et al.*, eds. *Goodman and Gilman's. The Pharmacological Basis of Therapeutics*, 9th Edn. New York: McGraw-Hill, 331–347.
Citron M L, Kaira J M, Seltzer VL, *et al.* (1992). Patient-controlled analgesia for cancer pain: a long-term study of inpatient and outpatient use. *Cancer Invest* 10: 335–341.
Chapman P, Jones C M (1944) Variations in cutaneous and visceral pain sensitivity in normal subjects. *J Clin Invest* 23: 81–91.
Davis G C, Buchsbaum M S, Bunnew W E (1978). Naloxone decreases diurnal variation in pain sensitivity and somatosensory evoked potentials. *Life Sci* 23: 1449–1460.
Dunn M, Blouin D, Labrecque G (1993). Time-dependent variations in morphine-requirements in patients with post-operative pain. *Proc 21st Conf Int Soc Chronobiology*, Quebec, abstract III-10.
Folkard S, Glynn C J, Lloyd J W (1976). Diurnal variation and individual differences in the perception of intractable pain. *J Psychosom Res* 20: 289–301.
Glynn C J, Lloyd J W, Folkard S (1975). The diurnal variation in perception of pain. *Proc R Soc Med* 69: 369–372.
Göbel H, Cordes P (1990). Circadian variation of pain sensitivity in pericranial musculature. *Headache* 30: 418–422.
Grabfield G P, Martin E G (1912). Variations in the sensory threshold for faradic stimulation in normal human subjects. I. The diurnal rhythm. *Am J Physiol* 31: 300–308.
Graves D A, Batenhorst R L, Bennett J G, *et al.* (1983). Morphine requirements using patient-controlled analgesia: influence of diurnal variation and morbid obesity. *Clin Pharm* 2: 49–53.
Hardy D J, Wolff H G, Goodell H (1940). Studies on pain. A new method for measuring pain: observations on spatial summation of pain. *J Clin Invest* 19: 649–657.
Heintz R C (1989). Pharmacocinétique du ténoxicam. In: Gaucher A, Pourel J, Netter P, Kessler M, eds. *Actualités en Physiopathologie et Pharmacologie Articulaire*. Paris: Masson.
Hindmarsh K W, Tan L, Sankaran K, Laxdal V D (1989). Diurnal rhythms of cortisol, ACTH and β-endorphin levels in neonates and adults. *West J Med* 151: 153–156.
Hummel T, Hepper M, Kaiser R, *et al.* (1992). Investigation of circadian effects on the plasma levels of dihydrocodeine and tramadol after peroral administration. *Proc 5th Int Conf Biological Rhythms and Medications*, Amelia Island, abstract XIII-4.
Hummel T, Kraetsch H G, Lötsch J, *et al.* (1994). Analgesic effects of dihydrocodeine and tramadol when administered either in the morning or evening. *Chronobiol Int* 12: 62–72.
Jensen M P, Karoly P, Braver S (1986) The measurement of clinical pain intensity: a comparison of six methods. *Pain* 27: 117–126.
Job-Deslandre C, Reinberg A, Delbarre F (1983). Chronoeffectiveness of indomethacin

in four patients suffering from an evolutive osteoarthritis of the hip or knee. *Chronobiologia* 10: 245–254.

Kobal G, Hummel T, Kraetsch H G, Lötsch J (1992). Circadian analgesic effects of dihydrocodeine and tramadol. *Proc 5th Int Conf Biological Rhythms and Medications*, Amelia Island, abstract XIII–5.

Kowanko I C, Pownall R, Knapp M S, et al. (1981). Circadian variations in the signs and symptoms of rheumatoid arthritis and in the therapeutic effectiveness of flurbiprofen at different times of the day. *Br J Clin Pharmacol* 11: 477–484.

Kowanko I C, Knapp M S, Pownall R, Swannel A J (1982). Domiciliary self-measurement in rheumatoid arthritis and the demonstration of circadian rhythmically. *Ann Rheum Dis* 41: 453–455.

Labrecque G, Lepage-Savary D, Poulin E (1988). Time-dependent variations in morphine-induced analgesia. *Annu Rev Chronopharmacol* 5: 135–138.

Labrecque G, Kazazi M, Vanier M C (1997). Biological rhythms in pain and analgesia. In: Redfern P H, Lemmer B, eds. *Physiology and Pharmacology of Biological Rhythms*. Berlin: Springer, 619–649.

Lemmer B, Wiemers R (1989). Circadian changes in stimulus threshold and in the effect of a local anaesthetic drug in human teeth: studies with an electronic pulptester. *Chronobiol Int* 6: 157–162.

Lévi F, Le Louarn C, Reinberg A (1985). Timing optimized sustained indomethacin treatment of osteoarthritis. *Clin Pharmacol Ther* 37: 77–84.

Macht D J, Herman N B, Levy C S (1916). A quantitative study of the analgesia produced by opium alkaloids, individually and in combination with each other in normal man. *J Pharmacol Exp Ther* 8: 1–37.

Martin E G, Bigelow G H, Grabfield G B (1914). Variations in the sensory threshold for faradic stimulation in normal human subjects. II. The nocturnal rhythm. *Am J Physiol* 33: 415–422.

Master A M (1960). The role of effort and occupation (including physicians) in coronary occlusion. *J Am Med Assoc* 174: 942–948.

Melzac R (1975). The McGill questionnaire: major properties and scoring methods. *Pain* 1: 277–299.

Melzack R (1983). The McGill Pain Questionnaire. In: Melzack R, ed. *Pain Measurement and Assessment*, New York: Raven Press, 111–117.

McGrath P A (1990). *Pain in Children: Nature, Assessment, and Treatment*. New York: Guilford Press.

Morawetz R F, Parth P, Pöppel E (1984). Influence of the pain measurement technique on the diurnal variation of pain perception. In: Bromm B, ed. *Pain Measurement in Man*, Amsterdam: Elsevier, 409–415.

Moore J G, Goo R (1987). Day and night aspirin-induced gastric mucosal damage and protection by ranitidine in man. *Chronobiol Int* 4: 111–116.

Muller J E, Stone P H, Turin Z G, et al. (1985). The Milis Study Group: circadian variation in the frequency of onset of acute myocardial infarction. *N Engl J Med* 313: 1315–1322.

Naber D, Cohen R M, Pickar D, et al. (1981) Episodic secretion of opioid activity in human plasma and monkey CSF. *Life Sci* 28: 931–935.

Nold G, Drossard K, Lehman K, Lemmer B (1995). Gastric mucosal lesions after morning versus evening application of 75 mg or 1000 mg acetylsalicylic acid (ASA). *Naunyn Schmiedebergs Arch Pharmacol* 351: R17.

Ostfeld A M (1963). The natural history and epidemiology of migraine and muscle contraction headache. *Neurology* 13: 11–15.
Pednault L, Parent M (1993). Circadian rhythm of chronic low back pain caused by discal degeneration, fusion pseudoarthrose or discoidectomy. *Proc 21st Conf Int Soc Chronobiology*, Quebec, abstract III-9.
Petraglia F, Facchinetti F, Parrini D, *et al.* (1983). Simultaneous circadian variation of plasma ACTH, β-lipoprotein, β-endorphin and cortisol. *Horm Res* 17: 147–152.
Pöllmann L (1981a). Étude de la chronobiologie des dents. *Rev Stomatol Chir Maxillofac* 82: 201–203.
Pöllmann L (1981b). Circadian changes in the duration of local anaesthesia. *J Interdiscip Cycle Res* 12: 187–191.
Pöllmann L (1984). Duality of pain demonstrated by the circadian variation in tooth sensitivity. In: Haus E, Kabat H, eds. *Chronobiology 1982–1983*. Basle: Karger, 225–228.
Pöllmann L (1987). Circadian variation of potency of placebo as analgesic. *Funct Neurol* 22: 99–103.
Pownall R, Pickvance NJ (1985). Does treatment matter? A double blind crossover study of ibuprofen 2400 mg per day in different dosage schedules in treatment of chronic low back pain. *Br J Clin Pract* 39: 267–275.
Proccaci P, Moretti R, Zoppi M, *et al.* (1973). Rythmes circadiens et circatrigindiens du seuil de la douleur cutanée chez l'homme. *Bull Groupe Et Rythmes Biol* 5: 65–75.
Proccaci P, Dellacorte M, Zoppi M, Maresca M (1974). Rhythmic changes of the cutaneous pain threshold in man. A general review. *Chronobiologia* 1: 77–87.
Räisänen I (1988). Plasma levels and diurnal variation of β-endorphin, β-lipotropin and corticotropin during pregnancy and early puerperium. *Eur J Obstet Gynecol Reprod Biol* 27: 13–20.
Reading A E (1989). Testing pain mechanisms in persons in pain. In: Wall P D, Melzack R, eds. *Textbook of Pain*, 2nd edn. Edinburgh: Churchill Livingstone, 269.
Reinberg A, Reinberg M A (1977). Circadian changes of the duration of action of local anaesthetic agents. *Naunyn Schiedebergs Arch Pharmacol* 297: 149–152.
Reinberg A, Manfredi R, Khan M F, *et al.* (1991). Chronothérapie de ténoxicam. *Thérapie* 46: 101–108.
Rejholec V, Vitulova V, Vachtenheim J (1984). Preliminary observations from a double-blind crossover study to evaluate the efficacy of flurbiprofen given at different times of day in the treatment of rheumatoid arthritis. *Annu Rev Chronopharmacol* 1: 357–360.
Rigas B, Torosis J, Mcdougall C J, *et al.* (1990). The circadian rhythm of biliary colic. *J Clin Gastroenterol* 12: 409–414.
Rolandi E, Gandolfo C, Franceschini R, *et al.* (1992). Twenty-four-hour beta-endorphin secretory pattern in Alzheimer's disease. *Neuropsychobiology* 25: 188–192.
Sankaran K, Hindmarsh K W, Tan L (1989). Diurnal rhythm of β-endorphin in neonates. *Dev Pharmacol Ther* 12: 1–6.
Schumacher G A, Goodell H, Hardy J F, Wolff H G (1940). Uniformity of pain threshold in man. *Science* 92: 110–112.

Simon L, Herisson P, Le Louarn C, Lévi F (1982). Etude hospitalière de chronothérapeutique avec 75 mg d'indométacine à effet prolongé en pathologie rhumatismale dégénérative. *Tribune Médicale (Suppl.)*: 43–47.

Sittl R, Kamp HD, Knoll R (1990). Zirkadiane Rhythmik des Schmerzempfindens bei Tumorpatienten. *Nervenheilkunde* 9: 22–24.

Solomon G D (1992). Circadian rhythms and migraine. *Cleve Clin J Med* 59: 326–329.

Stacher G, Bauer P, Schneider C, et al. (1982). Effects of combination of oral naproxen sodium and codeine in experimentally induced pain. *Eur J Clin Pharmacol* 21: 485–490.

Strian F, Lautenbacher S, Galfe G, Hölzl R (1989). Diurnal variations in pain perception and thermal sensitivity. *Pain* 36: 125–131.

Swannel A J (1983). Biological rhythms and their effect in the assessment of disease activity in rheumatoid arthritis. *Br J Clin Pract* 38(Suppl 33): 16–19.

Vanier M C, Labrecque G, Lepage-Savary D (1992). Temporal changes in the hydromorphone analgesia in cancer patients. *Proc 5th Int Conf Biological Rhythms and Medications*, Amelia Island, abstract XIII-8.

Von Knorring L, Almay B G L, Johansson F, et al. (1982). Circannual variation in concentrations of endorphins in cerebrospinal fluid. *Pain* 12: 265–272.

Waters W E, O'Connors P J (1971). Epidemiology of headache and migraine in women. *J Neurol Neurosurg Psychiatry* 34: 148–153.

Wilder-Smith C H, Schimke J, Bettiga A (1992). Circadian pain responses with tramadol (T), a short-acting opioid and alpha-adrenergic agonist, and morphine (M) in cancer pain. *Proc 5th Int Conf Biological Rhythms and Medications*, Amelia Island, abstract X111-6.

Wilder-Smith C H, Wilder-Smith O H (1992). Diurnal patterns of pain in cancer patients during treatment with long-acting opioid analgesics. *Proc. 5th Int Conf Biological Rhythms and Medications*, Amelia Island, abstract XIII-7.

10

Rhythms of cancer chemotherapy

Christian Focan

10.1 Introduction

In this chapter we address the importance of circadian rhythms to the biology of cancer and to cancer therapy. There have been a number of recent reviews on this subject (Hrushesky and Bjarnason, 1993; Focan, 1995a; Wood and Hrushesky, 1996; Lévi, 1997, 1999, 2000, 2001, 2002). Accordingly, we will focus essentially on the most recent developments both at experimental and clinical level. We will particularly develop and review the following.

1. The growing importance of genetic and molecular biology to cancer biology
2. Some consideration of the importance of disturbance of circadian structure for carcinogenesis
3. Recent data on chronopharmacology and chronotoxicology of anti-cancer drugs
4. Recent examples of chronoefficacy
5. The influence of both the cancer and its treatment on the circadian rhythms of the host
6. The probable influence of circadian structure of the host on quality of life, fatigue, psychological issues and prognosis

Finally, various human cancers (colorectal, lung, and head and neck) will be discussed as possible models for the implementation of chronotherapy of tumours in clinical practice.

10.2 Genetic and molecular biology

The molecular biology of the circadian clock has been reviewed in detail in Chapter 3. What is less well understood is how the central oscillator controls rhythms in peripheral cells and, more pertinent to the current discussion, how these peripheral rhythms interact with mechanisms controlling cell cycling and cell death or survival. Bjarnason recently described circadian variation of expression of four clock genes in human oral mucosa and skin (Bjarnason et al., 2001) (Figure 10.1). Sequencing programmes and the development of new technologies based on deoxyribonucleic acid (DNA) microarrays have opened new possibilities of identifying genes regulated by the circadian clock (Grundschober et al., 2001; Delaunay et al., 2002). For example, studies performed by Delaunay et al. (2002) on *Arabidopsis thaliana*, *Drosophila* or cultured fibroblasts identified 1–6% cycling genes, sometimes clustering on a given chromosome, under clock control. Circadian cycling genes have been observed in cells of many peripheral tissues including liver, muscles and heart (Ikeda, 2000; Storch et al., 2002).

However, work from the group of Balsalobre, using the protein shock method, has clearly shown a hierarchy between the central clock located in the suprachiasmatic nuclei (SCN) and peripheral clocks, the latter susceptible to resetting in case of loss of central control (Balsalobre et al., 1998).

Also recently identified was the circadian expression of cell cycle genes and of the various cyclins controlling each phase of the cell cycle (Bjarnason et al., 1999, 2001; Bjarnason and Jordan, 2000). In particular, in the oral mucosa of normal human subjects, Bjarnason and colleagues (Bjarnason et al., 1999, 2001; Bjarnason and Jordan, 2000) have established circadian variations of these cyclins by simple immunohistochemical methods. They were able to relate those variations to those of the circadian genes (Figure 10.1). They also found a temporal link between the activity of S-phase-committed cyclins and that of thymidylate synthase, an enzyme implicated in DNA synthesis (Bjarnason et al., 2001; Bjarnason and Jordan, 2000). Finally, they observed a circadian variation in apoptosis, as evidenced by the circadian expression of P53 (peak at noon) and Bcl-2 proteins (peak at 01.00). Interestingly, the same observations, but with a 12 h delay, have been reported for BAX protein (peak at 16 HALO (hours after light onset)) and Bcl-2 protein in bone marrow of mouse (peaks at 3 to 9 HALO). In contrast, Bcl-2 expression in mammary adenocarcinoma MA13/C and P53 expression both in bone marrow or tumour had no circadian rhythmicity (Liu et al., 2001a; Granda et al., 2002a). Relationships with transcription of clock genes have been investigated by Liu et al. (2001a,b).

Recent studies in animals have focused on the circadian expression of clock genes both in tumours (e.g. Glasgow osteosarcoma) and in healthy tissues of animals bearing tumours (Delaunay et al., 2002).

Genetic and molecular biology

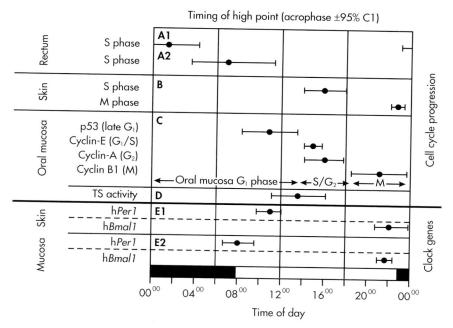

Figure 10.1 The timing of cell cycle phases in human rectum, skin, and oral mucosa: correlation with the expression of clock genes in skin and oral mucosa. Panel A: Acrophases for S phase in human rectal mucosa based on two human studies by Marra et al. (1994) (A1) and Buchi et al. (1991) (A2) are depicted along a 24 h time scale. The 95% confidence limits for each variable are shown. Panel B: Acrophases for S phase and M phase in human skin based on pooled data from 12 studies of M phase and 14 studies of S phase are depicted with reference to the 24 h time scale. The 95% confidence limits for each variable are shown. Panel C: Acrophases for the best-fitting cosine for P53, cyclin-E, cyclin-A and cyclin-B1 are depicted along the 24 h time scale, based on a study of human oral mucosa. The 95% confidence limits for each variable are shown. The cell cycle phase, for which each protein is a marker, is shown in parentheses on the vertical axis. Panel D: Acrophase for the peak activity of TS (thymidylate synthetase) (S phase) in oral mucosa is depicted along the 24 h time scale. The 95% confidence limits for each variable are shown. Panel E1: Acrophases for the peaks of expression for hPer1 and hBmal1 in skin are depicted along the 24 h time scale. The 95% confidence limits for each variable are shown. Panel E2: Acrophases for the peaks of expression for hPer1 and hBmal1 in oral mucosa are depicted along the 24 h time scale. The 95% confidence limits for each variable are shown. Reproduced with permission from Bjarnason and Jordan (2002), courtesy of Marcel Dekker, Inc.

10.3 Circadian coordination, carcinogenesis and tumour outcome

As outlined in previous chapters, circadian coordination is physiologically controlled by a central clock located in the SCN of the hypothalamus. The consequence of destruction of the SCN upon circadian organisation (as gauged, for example, by the rest–activity and temperature cycles) and on tumour growth were investigated by Filipski *et al.* (2002a) in Glasgow osteosarcoma-bearing mice. Stereotaxic SCN destruction clearly suppressed the rest–activity and temperature cycles, while tumour growth was faster and survival of animals was shortened (Filipski *et al.*, 2002a). While the growth of the same osteosarcoma was not influenced by keeping the animals in continuous darkness or light, an experimental jet-lag produced by an 8 h advance of the light–dark (L:D 12:12 cycle) markedly altered the circadian structure and also facilitated tumour growth (Filipski *et al.*, 2002b). In those animals, the circadian rhythmicity of corticosterone and lymphocytes was altered, with a tendency to an inversion of rhythms (Filipski *et al.*, 2002b).

Li and Xu (1997) have observed that repeated light–dark shifting in Ehrlich carcinoma-bearing or sarcoma 180-bearing rodents could disrupt the pattern of circadian immune rhythms (with lowering of blood leukocytes and lymphocytes) together with a reduction of survival durations and acceleration of tumour growth (Li and Xu, 1997).

Other studies were performed on rat hepatic carcinogenesis induced by nitrosamines (Van den Heiligenberg *et al.*, 1999). Under constant light, the risk of developing hepatic tumour nodules (number and size) was potentiated; similarly, the circadian rhythm of urinary melatonin excretion was completely abolished (Van den Heiligenberg *et al.*, 1999). However, other authors working on other models failed to confirm this relationship. As an example, no promotion of cancer could be observed in dimethylbenzanthracene- (DMBA-)induced breast carcinomas of rats exposed to constant light (Anderson *et al.*, 2000). Travlos *et al.* (2001) failed to enhance methylnitrosourea-induced breast cancers through short intermittent light (with five 1 min exposures to incandescent light every 2 h after the start of each dark phase) despite an inhibition of melatonin production.

It has been suggested that in humans exposure to light at night might increase the risk of breast cancer by suppressing the normal nocturnal production of melatonin by the pineal gland, which in turn could increase the release of oestrogen by the ovaries (Stevens and Rea, 2001). This relationship was illustrated by the work of Davis *et al.* (2001), who were able to correlate the risk of breast cancer with either significant light exposure at night (brightest bedrooms) or 'graveyard' shift work. Similarly, Schernhammer *et al.* (2001) established a relationship between the risk of breast cancer and the duration of rotating night-shift work in a cohort of over 78 000 nurses: the higher the number of years, the higher the breast cancer risk (p-trend 0.02). The same increased risk had already been described in postmenopausal radiotelegraph operators and in Nordic flight attendants (Ballard *et al.*, 2000; Rafnsson *et al.*, 2000). Conversely, impaired vision with, it is assumed, reduced suppression of melatonin secretion could reduce the risk of developing a breast cancer (Stevens and Rea, 2001).

Various mechanisms implicating the melatonin pathway (and light-induced inhibition of nocturnal melatonin secretion) have been postulated. These include altered circadian structure favouring the emergence of tumour clones (effectively, people who had poor adaptation to shift work had the lowest melatonin serum titres); reduced expression of P53 (with reduced apoptosis); reduced free-scavenger activity of melatonin (and reduced prevention of oxidative damage of DNA); and reduced immunomodulating effect (Davis *et al.*, 2001; Schernhammer *et al.*, 2001).

Other studies have shown that hormonal circadian rhythms were altered in women at high risk of breast cancer compared to women at low risk (Halberg *et al.*, 1981; Ticher *et al.*, 1996). These observations suggest that the endocrine time structure between individual women can be used as an indicator of breast cancer risk.

The importance of conserving a strong circadian structure during tumour development has also been emphasised. Sephton *et al.* (2000) investigated the diurnal cortisol slope after titration in saliva of 104 patients with metastatic breast cancer. The cortisol slope predicted subsequent survival up to 7 years; an almost doubled survival rate was observed at 4 years in patients with conserved diurnal gradient (Figure 10.2). Earlier mortality occurred among patients with relatively flat rhythms, indicating a lack of normal diurnal structure; also, flattened profiles were linked with low counts and suppressed activity of NK cells (Sephton *et al.*, 2000). Previously, Bartsch *et al.* (1997) had also reported that nocturnal urinary 6-sulphatoxymelatonin excretion is negatively correlated with tumour-size in primary breast cancer. Similar observations were reported for patients suffering from urogenital tract cancer (Taverna *et al.*, 1997).

In the same way, Mormont *et al.* (2000) studied a group of 200 patients with metastatic colorectal cancer, in whom the individual

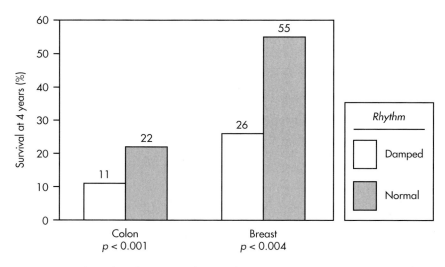

Figure 10.2 Circadian factors predictive of tumour outcome (per cent survival at 4 years): rest–activity cycle (metastatic colon cancer) (Mormont et al., 2000); cortisol diurnal rhythm (metastatic breast cancer) (Sephton et al., 2000).

circadian organisation was evaluated by actometric recording of the rest–activity rhythms. Here also, the survival at 4 years was doubled (Figure 10.2) in patients with a strong circadian rhythmicity; most interestingly, in multivariate analysis, a conserved circadian structure became a prognostic factor totally independent of other more usual prognostic determinants, such as performance status, liver involvement, tumour burden or number of tumour sites. In non-Hodgkin lymphomas, Micke et al. (1997) have suggested that patients with undisturbed circadian rhythms of interleukin-2 receptors, serum thymidine kinase and β-2-microglobulin had a better chance of achieving a complete response than patients with disturbed rhythms.

10.4 Biological rhythms in host and/or tumours

In this section we review pertinent recent data on the circadian biology of hosts or tumours that may influence scheduling of chronotherapy.

10.4.1 Animal data

The circadian clock in the SCN is thought to drive daily rhythms of behaviour by secreting factors that act locally within the hypothalamus.

Kramer *et al.* (2001) identified transforming growth factor α (TGF-α) as a likely inhibitor of mouse locomotion. TGF-α activity, as measured by its messenger RNA (mRNA), is expressed rhythmically in the SCN with a peak at 6 HALO; when infused into the third ventricle, it reversibly inhibits locomotor activity and disrupts circadian sleep–wake cycles. These actions are interestingly mediated by epidermal growth factor (EGF) receptors on neurons in the hypothalamic supraventricular zone. Mice with a hypomorphic EGF receptor mutation exhibited excessive daytime locomotor activity and failed to suppress activity when exposed to light; this implicates EGF receptor signalling in the daily control of motor activity and identifies a neural circuit in the hypothalamus that likely mediates the regulation of behaviour both by the SCN and by the retina. This regulation of behaviour by light and by the SCN could be considered as comprising two different inputs to a single hypothalamic circuit (Kramer *et al.*, 2001).

Takane *et al.* (2000, 2001) investigated whether the diurnal rhythm of cell cycling is associated with that of interferon α receptor (IFNAR) (or interferon β) expression in implanted tumour cells. The expression of IFNAR mRNA significantly increased when the proportion of tumour cells in DNA synthesis (S) phase increased *in vitro*. A diurnal rhythm was observed for cell cycle distribution in implanted tumour cells. The specific binding of interferon α (IFN-α) to receptor and IFNAR mRNA expression both increased when the proportion of tumour cells in S phase increased *in vivo*. The time-dependent expression of IFNAR was supported by that of transcription factor level induced by IFN-β. These results suggest that the rhythm of IFNAR expression is closely related to that of cell cycle distribution in implanted tumour cells (Takane *et al.*, 2001).

Several recent studies have focused on the circadian variation of pro- and anti-apoptotic protein expression. Changes throughout the day–night cycle in Bcl-2 and BAX protein levels in mouse bone marrow and their relation with the transcription of circadian clock genes have been investigated (Liu *et al.*, 2001a,b). A circadian rhythm in the mRNAs of m*Per2* and m*Bmal1* genes was shown, whereas m*Clock* and m*Tim* gene expression remained constant. Bcl-2 and BAX proteins varied 2.8-fold and 5-fold, respectively, throughout 24 h. The peak in Bcl-2 expression occurred near the middle of the rest period (at 4–7 HALO), when pro-apoptotic anticancer agents, such as docetaxel or fluorouracil are best tolerated, while BAX peaked 12 h later (at 16 HALO), in a synchronous fashion with m*per2* and m*Bmal1*. Considering the well-established role of BAX and Bcl-2 in apoptosis, these results suggest that apoptotic pathways are regulated by the circadian clock in mouse bone marrow (Liu et al., 2001b).

Columbo et al. (2000) reported day–night differences of spontaneous apoptosis in two murine tumour models. The times of maximal and minimal apoptosis and mitosis were inversely correlated (peak of apoptosis matched with the trough of mitosis).

Comparative assessments of rhythms in cell cycle distribution in bone marrow and in MA13/C mammary tumours of C3H/HEN mice confirmed the circadian rhythmicity in bone marrow and revealed the persistence of circadian variations of cell cycle phases in 21-day-old tumours (Granda et al., 2001a). However, the cell cycle-related rhythms were regulated differently in bone marrow and in tumour. The maxima of the cell cycle phase distribution rhythms occurred earlier in the tumour, i.e. by 7 h for G_1 phase, by 12 h for S phase and by 2 h for G_2/M phase (Granda et al., 2001a). Could these observations account for the frequent coincidence discussed below between the time of best tolerability and that of best efficacy of anticancer drugs?

Blank et al. (1995) also investigated circadian variation in mitotic activity of bone marrow and tumour in sarcoma-bearing rats. Whereas the circadian rhythm in mitotic activity is similar in bone marrow of healthy and tumour-bearing animals, with a peak occurring shortly after the onset of the dark (activity) span, the circadian rhythm in mitotic activity of the tumour had a much smaller amplitude and a different acrophase occurring late in the dark span. The authors suggest that this difference in acrophase between bone-marrow and tumour may have implications for scheduling of oncotherapy.

Natural killer (NK) cell cytotoxicity was reported to manifest a circadian rhythm peaking during wakefulness in both human blood and rat spleen. F344 rats bearing MADB106 (a NK-sensitive mammary adenocarcinoma that metastasises only to the lungs) were used to investigate fluctuations in NK cell numbers or in cytotoxicity (Shakhar et al., 2001). The results indicated that, during the dark phase, splenic NK cell cytotoxicity increased mostly as a result of a higher percentage of NK cells in the spleen. In contrast, blood NK cell cytotoxicity significantly decreased and this decline was independent of circulating NK cell numbers, which remained constant. Lung tumour clearance increased in the dark span, but no corresponding changes in the number of metastases were observed 3 weeks later. It was concluded that circadian changes in rat NK cell cytotoxicity are organ-specific, involve changes in both cell distribution and activity, and may affect short-term *in vivo* indices of NK tumoricidal activity (Shakhar et al., 2001).

10.4.2 Human data

The importance of recent work from the team of Bjarnason has already been highlighted (Bjarnason *et al.*, 1999, 2001; Bjarnason and Jordan, 2000) (Figure 10.1). Beyond the confirmation of a circadian rhythmicity of cell division in oral mucosa and in skin (as gauged by immunohistochemical determination of cyclins activity implicated in the progression of cell-cycling), they were also able to provide the first demonstration of the circadian expression of five clock genes. A link between the activity of those peripheral clock genes and the succession of cell cycle phases was also demonstrated (Bjarnason *et al.*, 2001; Bjarnason and Jordan, 2000). Moreover, Bjarnason *et al.* showed the circadian expression in relation to S phase of the thymidylate synthase activity as well as in the circadian control of apoptosis (P53 and Bcl-2 expression) (Figure 10.1).

Circadian variations of epithelial cell proliferation in human rectal crypts were re-assessed by repeated (every 4 h) rectal biopsies in 23 subjects (Marra *et al.*, 1994). The labelling indices after incorporation of tritiated thymidine (H3tdR) into the DNA revealed important temporal variations with peaks during the night and a trough in the afternoon. These variations were confined to the area of the crypt normally associated with replication. These observations confirmed previous work of Buchi *et al.* (1991). Combining their observations of circadian cell division in rectal mucosa and in bone marrow Buchi and Smaaland also showed a circannual variation (see Sothern *et al.*, 1995).

The circadian rhythm of DNA synthesis of a poorly differentiated nasopharyngeal human cell line (CNE2) was investigated by flow cytophotometry in nude mice (Xian *et al.*, 2002). The average proportion of tumour cells in G_1, S or G_2-M phase varied according to circadian time, with a maximum occurring at 9 HALO for G_1 phase, 2 HALO for S phase and 21 HALO for G_2-M phase cells; a bimodal temporal pattern was also observed for S phase cells (Xian *et al.*, 2002).

Cytotoxic agents such as chloroethylnitrosoureas mostly alkylate DNA at the guanine O-6 position. This highly mutagenic lesion can be repaired by O^6-alkylguanine-DNA alkyltransferase (AGT). AGT is known to display a circadian rhythm in mouse liver, coincident with that of nitrosurea tolerability (Lévi 2001, 2002). Using 12 healthy human volunteers whose circadian synchronisation was verified by rest–activity cycle assessed with wrist actigraphy and by plasma cortisol and melatonin rhythms, a study was done to identify a circadian rhythm in circulating mononuclear cells (MNC) (Marchenay *et al.*, 2001). Effectively, a circadian rhythm was statistically validated with an acrophase at 00.30.

These results suggest the desirability of further investigation of AGT rhythmicity both in circulating MNC and in tumour target tissues as a prerequisite for clinical testing of chronotherapy with alkylating agents.

The diurnal rhythmicity of serum erythropoietin (EPO) was analysed in 20 healthy subjects and in 20 patients with myeloma without (10 cases) or with (10 cases) impaired renal function (Pasqualetti et al., 1996). The physiological circadian rhythm in serum EPO concentrations with a maximum in the afternoon was confirmed in normal subjects and in myeloma patients without renal insufficiency; this rhythm was damped and/or abolished in the myeloma group with renal impairment (Pasqualetti et al., 1996). In renal cancers secreting EPO the diurnal variation of EPO was also abolished (Buemi et al., 1997).

Zubelewicz et al. examined the dynamics of circadian fluctuations of TNF-α and TNF-α P55 and P75 soluble receptors in serum of healthy controls and of advanced gastrointestinal cancer patients (Zubelewicz et al., 1995, 2001). In both groups, while the typical cortisol peak was found at 08.00, both the cytokine and its soluble receptors displayed statistically significant circadian fluctuations (peak for TNF-α at 00.36; peak for P55 at 22.00 when P75 reached its minimum).

A circadian rhythm of dihydropyrimidine dehydrogenase (DPD) enzyme implicated in the catabolism of fluorinated pyrimidines has been unequivocally observed in bone marrow and liver of laboratory rodents (Diasio and Harris, 1989; Focan, 1995a; Lévi, 2001). However, controversies persisted in humans (Harris et al., 1990). Barrat et al. (2002) failed to find a difference in the DPD activity in oral mucosa in a two-time-point study in 20 patients with metastatic colorectal cancer; but they confirmed a significant increase of DPD activity at night in the lymphocytes of the same subjects (Barrat et al., 2002).

Peripheral cytokine activity is under neuroendocrine control, basically under the influence of the pineal gland and its circadian secretion of melatonin (Weinberg, 1996). Circadian rhythmicity of various interleukins (IL-2, IL-6, IL-10, IL-12) or the growth factor granulocyte-macrophage colony-stimulating factor (GM-CSF) remains disputed (Akbulut et al., 1999). Similarly, Chevalier et al. (2002) failed to demonstrate a circadian rhythm of IL-15 and/or IL-9 in plasma of synchronised healthy subjects (synchronisation gauged by cortisol or melatonin rhythms). However, Haus et al. (2001) found, in metastatic breast cancer, circadian variations in plasma cytokines (basic fibroblast growth factor, epidermal growth factor, insulin-like growth factor (ILGF1) or the receptor protein of ILGF1) with peaks between 11.00 and 14.20; a 12.00 component was also seen.

10.5 Chronotoxicity and chronopharmacology in animals

Extensive reviews have detailed the universal phenomenon of chronotolerance to almost all anticancer drugs tested (Focan, 1995; Lévi, 1997; Wood and Hrushesky, 1996) (Figure 10.3). Circadian variation in their pharmacology has also been described.

Here we consider recent developments in these domains, with 'old' anticancer drugs (mitozantrone (mitoxantrone), platinum derivatives) and newer compounds such as docetaxel, vinorelbine, irinotecan (CPT-11), Tumour necrosis factor α (TNF-α) and anti-cyclooxygenase-2 (COX-2) drugs (Figure 10.3). Also, some results in humans with possible impact on clinical practice will be discussed.

10.5.1 Chronopharmacology of single agents

10.5.1.1 Fluoropyrimidine derivatives

The pharmacokinetics of fluorouracil (5-FU) was assessed in mice after subcutaneous implantation of pellets that release 5-FU over a 3-week-period, mimicking a continuous intravenous infusion (Codacci-Pisanelli *et al.*, 1995). Interestingly, beyond individual variation in plasma titres from 0.1 to 1.1 μmol/L, 5-FU levels displayed a circadian rhythm with a peak at 6 HALO.

Continuous hepatic artery infusions using floxuridine (FUDR) was tested in rats bearing hepatic metastases from a colon adenocarcinoma. 'Day-cycled' infusion with 60% of FUDR-dose delivered between 15.00 and 21.00 was shown to substantially reduce lethality: at the same dose (10 mg/kg/day for 14 days) or at higher dose (up to 15 mg/kg/day), mortality rate was reduced from 80% to 0% when compared to continuous infusion at a constant rate (Kemeny *et al.*, 1994).

10.5.1.2 Mitozantrone (mitoxantrone)

Mitozantrone (mitoxantrone) is an anthracenedione (an anthracycline-related class of compounds) used in clinical practice. Different groups of male B6D2 F_1 mice synchronised to a 12:12 L:D cycle, received a single intravenous injection of mitozantrone at 4 times or 6 times equally spaced over 24 h (Lévi *et al.*, 1994). A dose–response relationship characterised body weight loss and survival rate. The dose–toxicity relationship further depended closely upon dosing time. Thus, a dose of 16 mg/kg killed 100% of the mice injected at 3 HALO, and none at 11

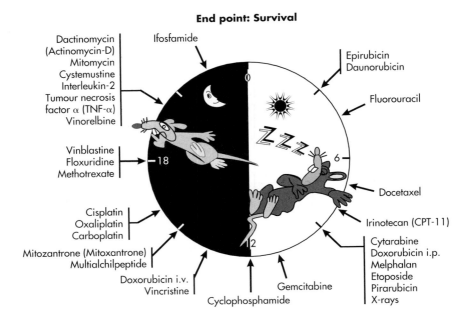

Figure 10.3 Circadian chronotolerance for anticancer drugs in laboratory mice or rats. Animals are synchronised with an alternation of 12 h of light, when they mainly rest, and 12 h of darkness, when they are mostly active. The time of best tolerability is indicated by an arrow. Modified with permission from Lévi (2002), courtesy of Marcel Dekker, Inc.

or at 15 HALO ($p < 0.001$). Least haematological toxicity and fastest recovery of a normal circulating leukocyte count corresponded to mitozantrone injection near the middle of the dark (activity) span, at 16 HALO. Similar findings characterised colonic and splenic lesions. Moreover, mitozantrone was both distributed and eliminated faster after injection at 16 HALO. If such data apply to cancer patients, as was the case for other drugs investigated with this methodology, an afternoon infusion should enable high-dose mitozantrone to be better tolerated (Lévi *et al.*, 1999, 2000, 2001).

10.5.1.3 Platinum derivatives

Carboplatin Circadian chronotolerance and chronopharmacology of platinum derivatives have been reported (Boughattas *et al.*, 1989, 1990). Lu and Yin (1994) revisited the circadian chronotoxicity of carboplatin in mice. Four groups received a single intraperitoneal injection of 192 mg/kg (estimated LD_{50}) at four different times of day. A difference in mortality according to the time of administration was found (28% for mice receiving the drug at 09.00 versus 71% for those treated at 21.00). The susceptibility to the drug was also found to be circadian stage-dependent, with a better tolerance in the early sleep phase.

Cisplatin A circadian rhythm in intracellular reduced glutathione (GSH) was statistically validated in liver, jejunum, colon and bone marrow but not in the kidney of male B6D2 F_1 mice (Boughattas *et al.*, 1996). In the kidney, cysteine – a final product of GSH catabolism – displayed a variation with a 12 h period (Li *et al.*, 1997). Tissue GSH concentration increased in the dark (activity) span and decreased in the light (rest) period of mice. The minimum and maximum of tissue GSH content corresponded respectively to the maximum and minimum of cisplatin toxicity.

The role of GSH rhythms with regard to cisplatin (*cis*-dichlorodiammineplatinum(II)) toxicity was further investigated, using a specific inhibitor of GSH biosynthesis, buthionine sulfoximine (BSO) (Boughattas *et al.*, 1996). Its effects were assessed on both tissue GSH levels and cisplatin toxicity at three times of day. BSO resulted in a 10-fold decrease of the 24 h-mean GSH in kidney and a moderate GSH decrease in liver (with a period changed from 24 h to a composite (24 + 12 h) period) and jejunum. Pretreatment with BSO largely enhanced cisplatin toxicity (as assessed by survival, leukopenia and histological damage in kidney and bone marrow), which varied according to a circadian rhythm but with a modified period: survival followed a 12 h rhythm instead of a 24 h rhythm. Although BSO partly and/or totally abolished the tissue GSH rhythms, it did not modify those in cisplatin toxicity. It was concluded that GSH, cycling concentrations of which result from a balance between synthesis and catabolism, has an important influence on cisplatin toxicity but not on the circadian mechanism of platinum chronotoxicity (Boughattas *et al.*, 1996).

The effects of BSO on the rhythm of cisplatin toxicity in mice bearing a transplantable pancreatic adenocarcinoma were also investigated (Li *et al.*, 1998). Survival rate was 97% in mice treated with cisplatin alone and 47% in those previously receiving BSO. As in healthy mice, BSO pretreatment further shifted the survival or body weight change periods from 24 h to (10 + 24 h) (Li *et al.*, 1998).

10.5.1.4 Docetaxel

Docetaxel chronotolerance was measured by Tampellini *et al.* (1998) in male B6D2 F_1 mice. Taking as end points survival and body weight change, docetaxel in various schedules was administered at different times of day. Dosing in the second half of the dark period resulted in significantly increased toxicity compared to that observed following administration during the light period. The survival rate ranged from 56% in the mice treated at 23 HALO to 94% and 88% in those injected at 3 or 11 HALO, respectively. Also, granulocytopenia at nadir was 48% ± 14% at 7 HALO, versus 84% ± 3% at 19 HALO; severe jejunal mucosa necrosis occurred in 5/18 mice treated at 23 HALO as opposed to 2/18 treated at 7, 11 or 19 HALO. Finally, the time of least docetaxel toxicity corresponded to the circadian nadir in S or G_2-M phase and to the circadian maximum of Bcl-2 immunofluorescence in bone marrow. According to these results, docetaxel is better tolerated in the light (rest) period of rodents, i.e. when fewest bone marrow or intestinal cells are actively dividing (Tampellini et al., 1998).

The chronotolerance of docetaxel was revisited by Granda et al. (2001b) in C3H/HEN mice. Again, a dose-dependent mortality was observed; however, irrespective of dose level, mean body weight loss was 3-fold larger in mice treated near the middle of the dark period (19 HALO) compared with those treated near the middle of the light period (7 HALO).

10.5.1.5 Vinorelbine

The relevance of chronopharmacology for improving tolerability of the antimitotic drug vinorelbine was investigated in female B6D2 F_1 mice housed under 12:12 L:D (Filipski et al., 1999). The drug at a single dose of 20–30 mg/kg was better tolerated around 19 HALO and worse tolerated around 7 HALO. At 19 HALO, bone marrow necrosis and leukopenia were nearly halved when compared to those observed in animals treated at 7 HALO. There were no differences in results from control or tumour-bearing (P388 leukaemia) mice.

10.5.1.6 Cystemustine

Cystemustine is a new nitrosourea with high antitumour activity and a short plasma half-life in mice. The influence of circadian dosing time upon its toxicity was first investigated in L:D synchronised male B6D2 F_1 mice (Martineau-Pivoteau et al., 1996). Late survival rate varied from 4% in mice receiving a single dose of cystemustine (conventional lethal dose 50%) at 7 HALO up to 88% in mice treated at 15 or at 19 HALO. Target organ toxicities (bone marrow, circulating blood cells, spleen, colon and duodenum) were studied following a single slightly lower dose of cystemustine. Leukopenia was the major haematological effect encountered. Leukocyte count nadir occurred 7 days after injection and was lowest following cystemustine at 7 HALO as compared to 13 or 19 HALO. Recovery was faster after cystemustine at 19 HALO as compared to other dosing times. Bone marrow necrotic lesions were more pronounced one day after cystemustine at 7 HALO than after cystemustine at 19 HALO. Thus, a large-amplitude circadian rhythm characterised the toxicity of this nitrosourea in mice. The lowest cystemustine toxicity was found near the middle of the active span of the rest–activity circadian cycle of mice.

10.5.1.7 Irinotecan (CPT-11)

Two studies have evaluated the chronotoxicity of a new compound in the treatment of colorectal cancer, irinotecan hydrochloride (CPT-11).

Filipski *et al.* (1997) injected CPT-11 (60–75 mg/kg) weekly at six times of day into male B6D2 F_1 mice. Immediate mortality occurred in 21% of the mice treated at 7 HALO and 48% of those injected at 19 or 23 HALO ($p = 0.02$), with less than 10% late mortality. Body weight loss was maximal 24 h after the tenth dose and followed a similar time course to jejunal epithelial necrosis: it was 10% at 11 HALO and 18% at 23 HALO ($p < 0.001$). Histological bone marrow necrosis, femoral depletion in nucleated bone marrow cells, leukopenia and granulopenia nadirs of short duration were significantly more pronounced at 19, 23 or 3 HALO. The number of granulocyte-macrophage colony-forming units (CFU-GM) gradually increased up to 2.3-fold over controls 24 h after the tenth dose of CPT-11; this indicates that the drug exerts bone marrow toxicity downstream of CFU-GM progenitors. Both acute and delayed toxicities were significantly reduced by CPT-11 dosing at 7 to 11 HALO, i.e. in the second half of the rest span in mice, possibly as a result of rhythms in enzymatic activity.

Using ICR male mice, Ohdo *et al.* (1997) investigated the mechanisms underlying the circadian rhythm of CPT-11 toxicity from the viewpoint of the host response and the pharmacokinetics of the drug. The loss of body weight after an intraperitoneal injection of CPT-11 (100 mg/kg) was greater in the late dark and the early light period and less marked late in the light period and the early dark. The CPT-11-induced leukopenia was greater in the late dark and less marked in the late light. The lower toxicity of CPT-11 was observed when DNA synthesis and type 1 DNA topoisomerase activity in bone marrow cells decreased, and the higher toxicity was observed when these activities began to increase.

There were circadian stage-dependent changes in the concentrations of CPT-11 and its major metabolite (SN-38; 7-ethyl-10-hydroxycamptothecin) in plasma. The highest concentrations of CPT-11 and SN-38 in plasma were observed when the level of CPT-11-induced toxicity increased.

Both studies are consistent in identifying the end of the rest period as the time at which CPT-11 is least toxic.

10.5.1.8 Tumour necrosis factor α (TNF-α)

The narrow therapeutic index of TNF-α both in animals and in humans has limited its anticancer utility. Hrushesky *et al.* (1994) examined the effect of time of day of TNF-α administration on lethal toxicity in BALB/c female mice. The probability of dying varied up to 9-fold across all times of day. The frequency of induced lethality with shorter times to death was greatest when the drug was administered just before waking; the survival probability was highest when TNF-α was given in the second half of the activity period, roughly corresponding, for humans, to late afternoon and evening.

10.5.1.9 Celecoxib

Celecoxib inhibits the enzyme cyclooxygenase-2 (cox-2). The drug is cytotoxic in a variety of solid tumour cell lines. Chronotoxicity of celecoxib was examined by Blumenthal *et al.* (2001) after intraperitoneal administration to nude mice bearing subcutaneous breast xenografts. The best tolerance (survival 100%; highest maximum tolerated dose, MTD) was found at 7 HALO and the worst (only 10% survival with an MTD 2.5 times lower) occurred at 17 HALO.

10.5.2 Chronopharmacology of combination treatment

10.5.2.1 Fluorouracil (5-FU) and cisplatin

Circadian variations of organ-specific and lethal toxicities were investigated following intravenous injection of 5-FU and/or cisplatin in female C3H mice (Shakil *et al.*, 1993). Circadian variation in toxicity was similar for the drugs given singly or together: lowest toxicity at 9 HALO; less severe leukopenia at 15 HALO; maximal reduction in spleen size after treatment at 3 HALO (Shakil *et al.*, 1993). In part these results are not consistent with previous evaluations of circadian toxicity of 5-FU and platinum derivatives in which best tolerance was observed if the drugs were administered 12 h apart (Boughattas *et al.*, 1989, 1990; Lévi, 1997).

10.5.2.2 Docetaxel and doxorubicin

Chronotolerance of a combination of docetaxel and doxorubicin was tested by Granda *et al.* (2001b). Singly or in combination, docetaxel and doxorubicin were significantly better tolerated near the middle of the rest period (7 HALO), irrespective of dose (Granda *et al.*, 2001b).

10.5.2.3 Irinotecan (CPT-11) and oxaliplatin

The haematological and intestinal toxicities of oxaliplatin were significantly reduced by the administration of the drug near the middle of the dark period in mice (Boughattas *et al.*, 1989; Li *et al.*, 1998). Similarly, tolerability of CPT-11 as gauged by body weight loss and intestinal or hematological toxicities was least following drug delivery in the second half of the rest span in mice.

The relevance of circadian rhythms in irinotecan and oxaliplatin tolerability was confirmed when both drugs were administered to animals. They were best tolerated at the same period of day whether given singly or in combination (Granda, 2002b) (Figure 10.3).

10.6 Chronotoxicity and chronopharmacology in humans

10.6.1 Fluoropyrimidine derivatives

Tolerance in the pharmacokinetics of Fluorouracil (5-FU) given in bolus intravenous injection was studied by Novakowska Dulawa (1990). Dosing 5-FU at 01.00 resulted in the longest half-life, the largest distribution volumes and AUCs and the lowest clearance, as well as in the least depression in leukocyte counts. This reduced plasma clearance of 5-FU at night resulted in an accumulation of the drug between 00.00 and 06.00 during continuous infusion at a constant rate. This finding was observed in eight studies involving various types of cancer patients (bladder, GI tract, lung or head and neck) (Figure 10.4) with various treatment schedules (5-FU given at high doses for 1–5 days), as single agent or combined with other drugs (Lévi, 2002). In two other studies, using protracted low doses, the plasma peak occurred around 12.00, possibly as a result of the nonlinear pharmacokinetic disposition of the drug (Lévi, 2002).

Interpatient variability in 24 h mean 5-FU levels made it necessary to express each concentration relative to the 24 h mean of the same patient, in order to demonstrate synchronised circadian rhythmicity at the group level (Bressolle *et al*., 1999; Takimoto *et al*., 1999; Kwiatowski, 2001; Milano, 2002). Furthermore, differences in individual circadian time structure, including the prominence of ultradian rhythms, may influence the results (Takimoto *et al*., 1999; Milano, 2002).

Chronomodulated delivery of 5-FU (± folinic acid (leucovorin; LV), sparfosic acid or oxaliplatin) with peaks flow rates at 04.00 was associated with a significant reduction of interpatient variability in the plasma levels of 5-FU as compared to constant-rate infusion or to chronomodulated administration with peak flow rates at 13.00 or 19.00 (Figure 10.5). These three latter schedules displayed increased toxicity as compared to the first (Lévi, 2002).

Chronomodulated 5-FU with folinic acid (leucovorin; LV) was

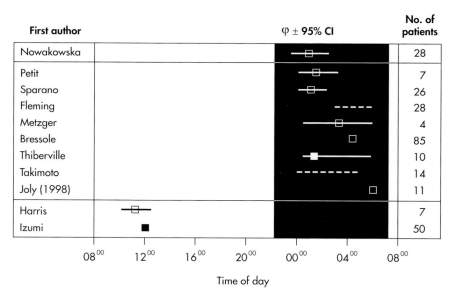

Figure 10.4 Circadian changes in plasma 5-FU disposition in cancer patients related to day–night cycle in 11 separate studies. For each study, the time of maximum as estimated by the acrophase (open square) is shown with its 95% confidence limits, whenever they are reported. Otherwise, the time of reported average peak (filled square) or the range of individual peak times (dotted line) is shown. The time of lowest 5-FU clearance following bolus administration at different times is shown on the top line. The times of maximum concentration of 5-FU are shown in the middle panel for 2- to 5-day constant rate infusions and in the lower panel for 14-day constant rate infusions. Courtesy of F. Lévi.

given near MTD to 22 patients with metastatic colorectal cancer. Plasma levels of 5-FU were measured semi-continuously for 24 h during the first and fourth days of infusion. The patients with altered 5-FU circadian disposition (i.e. reduction of nocturnal clearance) displayed increased toxicity (Kwiatkowsky, 2001). The need for specific chronotherapy schedules for patients with disruption of circadian organisation may be considered as such patients are also those with the poorer outcome (partly related to high tumour burden and poor performance status) (Focan, 1995a; Mormont and Lévi, 1997; Mormont et al., 2002).

Exogenous corticosteroids have been shown to synchronise and enhance the circadian rhythm in 5-FU disposition (Joly et al., 2001).

Rhythms in cellular determinants of 5-FU toxicity have been re-evaluated recently (Lincoln et al., 2000; Lévi, 2002; Milano, 2002). Interestingly, peaks in 5-FU concentrations and in DPD activity occurred 12 h apart in the study with low-dose infusional 5-FU, while they coincided in studies with high infusional doses (Lévi, 2002; Milano, 2002). Clearance

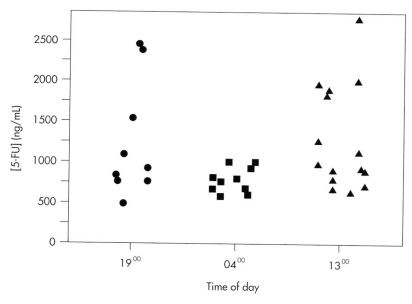

Figure 10.5 Role of peak time of 5-FU chronomodulated delivery on 5-FU C_{max} in patients with metastatic colorectal cancer. Eighteen patients received chronomodulated 5-FU–LV–oxaliplatin, with peak rate of 5-FU–LV infusion scheduled at 19.00, 04.00 or 13.00. Blood was sampled at time of peak delivery on the first and fourth infusional days. Interpatient and intrapatient variability were significantly less in the patients receiving chronomodulated 5-FU with peak delivery at 04.00. 5-FU, fluorouracil; LV, folinic acid (leucovorin). Courtesy of F. Lévi.

of 5-FU was correlated with liver blood flow (Hori et al., 1996). This suggests, according to Lévi (2002), that the main mechanism of the circadian rhythms in 5-FU plasma levels is the rhythm in DPD activity when low doses are used, while the rhythm in liver blood flow, which exhibits a 50% increase in the second part of the night, might be the most relevant mechanism when high daily doses are considered (Lévi, 2002).

Infusional chronomodulation of 5-FU has been implemented widely since the early 1990s, i.e. from the time when new technological development allowed the safe and ambulatory delivery of drugs by portable programmable-in-time pumps (e.g. Intelliject, Melodie, Z pumps).

Numerous trials were performed in phases I to III with 5-FU in monotherapy (Lévi et al., 1995), 5-FU associated with its biochemical modulator (folinic acid (leucovorin; LV)) and 5-FU–LV combined with other drugs, i.e. platinum derivatives (oxali-, carbo-, cisplatin), irinotecan (CPT-11) and mitozantrone (mitoxantrone) (Focan et al., 1995, 1999; Giacchetti et al., 2001; Lévi, 1999, 2002; Lévi et al., 2000, 2001a,b) (Figure 10.6). Initially tested in colorectal cancer, the chronotherapeutic concepts were further developed in other digestive or nondigestive tumours (Focan, 1995b). All these data have been thoroughly reviewed elsewhere, so we will focus here only on the most relevant recent achievements.

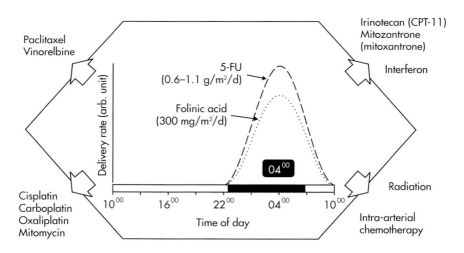

Figure 10.6 Chronomodulated scheduling of 5-FU ± folinic acid (leucovorin; LV), automatically delivered with a programmable pump, without hospitalisation. This schedule has been the basis for subsequent combination chronotherapy regimens involving chronomodulated or standard administrations of platinum complexes, antimitotic drugs or other agents, including radiation therapy.

For colorectal cancer, studies of the European group for cancer chronotherapy (initially activated in Villejuif and Liège) can be summarised as follows. Initial scheduling of theoretical times of minimum toxicities in humans was derived from animal data (Figure 10.3) (hour of peak of tolerance in animal ± 12 h).

1. Chronotolerance of 5-FU in monotherapy was confirmed and allowed the dose to be increased by 40% when the peak of administration was scheduled at 04.00 (Lévi *et al.*, 1995).
2. Chronotolerance of a complex infusional treatment with 5-FU–LV and oxaliplatin has also been assessed in two consecutive randomised multicentre trials (IOCC, 1997). Patients received the drugs 5 days every 3 weeks either by continuous infusion at constant rate or by chronomodulated sinusoidally varying infusion with peaks of administration at 04.00 for 5-FU–LV and at 16.00 for oxaliplatin. The chronomodulated infusion produced substantially reduced toxicity (5 times less mucosal toxicity, hospitalisations reduced by one-third) and allowed significantly increased overall drug doses without any further toxicity.
3. Circadian chronotolerance of the same combined treatment has been prospectively assessed by treating patients with peak decays (Gruia *et al.*, 1996). The reference peaks (04.00 for 5-FU–LV; 16.00 for oxaliplatin) were displaced in the nycthemeral period by ±3 h (thus, respective peaks at 07.00 and 19.00 or at 01.00 and 13.00); ±6 h (peaks at 10.00 and 22.00); ±9 h (peaks at 13.00 and 01.00); and ±12 h (peaks at 16.00 and 04.00). The preliminary analysis of this study tended to confirm that minimum toxicity occurred at times predicted from animal data (peak of 5-FU (and LV) at 04.00; peak of oxaliplatin at 16.00; also in the group treated at −03.00) (Gruia *et al.*, 1996).
4. Combinations of irinotecan (CPT-11) either with 5-FU–LV or with 5-FU–LV and oxaliplatin have been tried recently (Giacchetti *et al.*, 2001; Lévi *et al.*, 2001b). The administration of irinotecan around 05.00 seemed to reduce asthenia, diarrhoea and haematological toxicity (Giacchetti *et al.*, 2001; Mormont *et al.*, 2001).

Combined infusional chronotherapy has also been developed in other solid tumours, such as upper respiratory–digestive tract (Focan, 1995b, 2002b), stomach (Focan *et al.*, 2000a), pancreatic cancers (Focan *et al.*, 2000b) and non-small-cell lung cancer (NSCLC) (Focan, 2003). In all these indications a generally excellent tolerance of the combined medications was reported (Focan, 1995b; Focan *et al.*, 1995, 1999, 2000a,b) (see Tables 10.1–10.5). Focan *et al.* prospectively evaluated the tolerance to infusional chronotherapy with 5-FU–LV and carboplatin given in advanced NSCLC (Focan *et al.*, 2001). Patients were randomised to receive either the reference B protocol (with peaks of 5-FU at 04.00 and carboplatin at 16.00) or a protocol with peak shifts of ±8 h (A, −8 h; C, + 8 h) (Figure 10.7). Despite a limited number of patients, a clear difference in toxicity in favour of the reference schedule (B) was evident (Figure 10.8). Those results prompted the initiation of phase II trials in lung (68 cases) and head and neck (48 cases) cancer with the use of protocol B (Focan, 2002a,b; Tables 10.1–10.5). Also, a more intense version of protocol B (4 days every 2 weeks) was applied to metastatic colorectal cancer with excellent overall tolerance (Focan *et al.*, 1999).

Table 10.1 Infusional fluorouracil, folinic acid and carboplatin chronotherapy for advanced non-small-cell lung cancer: patient characteristics

Number	68
Age (years)	
Median	62
Range	40–77
Sex (M/F)	49/19
Histology	
Squamous cell	26
Adenocarcinoma	36
Anaplastic	3
Other	1
Performance status	
0	19
1	22
2	18
3	6
4	1
X	2
Stage	
II	2
IIIA	18
IIIB	9
IV	39
Previous chemotherapy courses[a]	
Number	369
Median	4
Range	1–11
Assessable patients	66

[a]Doses of carboplatin were increased from 40 (11 cases) to 50–55 mg/m^2/day (26 cases) and 60 mg/m^2/day (31 cases).

Table 10.2 Infusional fluorouracil, folinic acid and carboplatin chronotherapy for advanced non-small-cell lung cancer: Grades 3–4 toxicities (%)

	Courses (359)	Patients (68)
Nausea–vomiting	5.8	17.6
Diarrhoea	1.7	7.4
Stomatitis	2.8	13.2
Skin	0.03	1.5
Alopecia	9.9	0.0
Haemoglobin	3.9	13.2
Leukocytes	2.8	10.3
Granulocytes	2.5	11.8
Platelets	5.8	14.7
Others	1.7	5.9

Table 10.3 Infusional fluorouracil, folinic acid and carboplatin chronotherapy for advanced non-small-cell lung cancer: response rate (%); 68 cases

CR	2.8
PR	35.3
Overall	38.2
Stage II and IIIA[a]	60

CR, complete response; PR, partial response; Overall, overall response.
[a]7/20 CR after surgery ± radiotherapy.

Early studies with the combination of 5-FU–LV and CPT-11 have suggested a possible improvement in tolerance of CPT-11 when its administration is scheduled around 05.00 (less asthenia, diarrhoea, haematological toxicity) (Lévi et al., 2001; Mormont et al., 2001). Those results prompted the European Organisation for Research and Treatment of Cancer (EORTC) chronotherapy group to launch a study intended to determine the best time of tolerance of CPT-11 when combined with chrono-5-FU–LV and oxaliplatin (Table 10.6). Such a study is currently running under the auspices of the same group looking for the best tolerance schedule of vinorelbine (combined with chrono-5-FU) in second- or third-line treatment for advanced breast cancer (Coudert et al., 2002).

Hepatic artery infusion (HAI) has been used to increase the selective exposure to chemotherapy of neoplastic cells in liver. This strategy was mostly applied to the treatment of liver metastases from colorectal cancer using protracted (14/28 days) floxuridine (FUDR) arterial infusion,

Table 10.4 Infusional fluorouracil, folinic acid and carboplatin chronotherapy for advanced head and neck cancer: patient characteristics

Number	48
Age (years)	
Median	60.1
Range	38.9–83.5
Sex (M/F)	31/17
Performance status (0/1/2)	29/17/2
Histology	
(EPI/ADE/x)	43/4/1
Grade (1/2/3/4/x)	22/12/8/1/5
Stage	
II	1
III	8
IV	39
Distant metastases (O/1/x)	41/6/1
Radiotherapy (yes/no)	30/18
Concomitant radiotherapy (yes/no)	8/40
Surgery (yes/no)	19/29

Table 10.5 Infusional fluorouracil, folinic acid and carboplatin chronotherapy for advanced head and neck cancer: Grades 3–4 toxicities (%)[a]

	Courses (208)	Patients (48)
Nausea–vomiting	0.5	2.1
Diarrhoea	0.5	2.1
Mucositis	5.3	18.8
Skin	0.03	1.5
Haemoglobin	0.5	2.1
Leukocytes	3.4	14.6
Granulocytes	10.0	35.4
Platelets	7.2	22.9

[a]Mucous and granulocyte toxicity enhanced by concomitant radiotherapy (one toxic death in this group).

although direct hepatic toxicity, biliary sclerosis or gastroduodenal irritation or ulceration often compromise the therapeutic index of HAI of FUDR. Experimental data have indicated that the toxicities of FUDR, like those of 5-FU, varied by 50% or more according to the time of administration during the 24 h period (Figure 10.3) (Focan, 1995b; Lévi, 1999, 2001). Tolerability of continuous 14-day infusion of FUDR, given via the intravenous or the intra-arterial route, was compared according to whether the delivery rate was constant or chronomodulated. In the

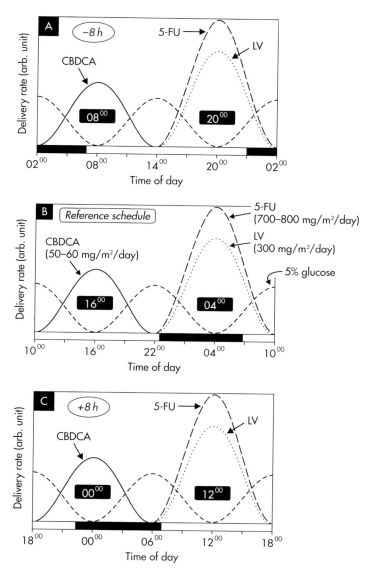

Figure 10.7 Schedules of chronomodulated intravenous infusion of 5-FU, folinic acid (leucovorin; LV) and carboplatin (CBDCA) for advanced non-small-cell lung cancer. Local time in hours is plotted against the drug delivery rate. This cycle was repeated automatically for five consecutive days. Groups differed according to the timing of delivery peaks: reference schedule (B) with peak of CBDCA at 16.00 and 5-FU (+ LV) at 04.00; schedule −8 h (A) with respective peaks at 08.00 and 20.00; schedule +8 h (C) with respective peaks at 00.00 and 12:00. Modified with permission from Focan (2002a), courtesy of Marcel Dekker, Inc.

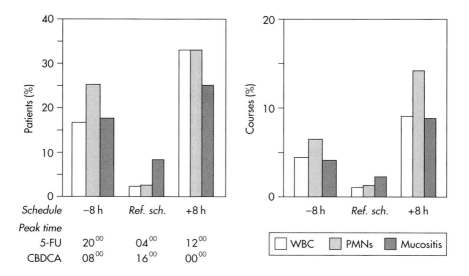

Figure 10.8 Grades 3–4 toxicity per course (right) and per patient (left) according to the circadian schedule (reference; ±8 h) in advanced non-small-cell lung cancer. PMNs, polymorphonucleocytes. Reproduced with permission from Focan (2002a), courtesy of Marcel Dekker Inc.

latter schedule, two-thirds of the daily FUDR dose was given between 15.00 and 21.00. Patients on the chronomodulated schedule could receive a 30–45% higher dose but experienced reduced overall toxicities as compared to those receiving flat infusion (Hrushesky and Bjarnason, 1993; Kemeny et al., 1994).

High-dose chemotherapy combining regional hepatic artery infusion of FUDR and systemic venous infusion of 5-FU was delivered against liver metastases from colorectal cancer (Focan et al., 1999). The hypothesis was tested that chronomodulation of delivery rate along the 24 h time scale would increase the tolerable doses of both drugs. Combined HAI FUDR and systemic venous infusion of 5-FU were administered for 5 consecutive days every 3 weeks, either as a constant-rate infusion or as chronotherapy. The latter regimen consisted of a sinusoidal modulation of the delivery rate over 24 h with a maximum at 16.00 for FUDR and 04.00 for 5-FU. Intrapatient dose escalation up to the individual MTD was planned for both drugs in the absence of any previous grade 3 or 4 toxicity. Severe stomatitis occurred in 71% of the patients and was dose-limiting. No hepatic toxicity was encountered. Dose reductions of 5-FU and/or FUDR were required more frequently in the flat schedule after reaching the individual MTD ($p < 0.05$). Granulopenia

Table 10.6 Multicentre trials led by the EORTC Chronotherapy Group

Trial	Tumour and treatment	Schedule	End point	Number of patients	Status
05962	Pancreatic cancer 5-FU ± cisplatin	Chrono versus flat	Survival	200	Ongoing
05963	Colorectal cancer 5-FU–LV–oxaliplatin	Chrono versus flat	Survival	554	Closed[a]
05971	Breast cancer Vinorelbine (VNR)–5-FU	8 chrono schedules	Chrono-toxicity of VNR	80	Ongoing
05991	Biliary cancer Chrono 5-FU + RT	Chrono (phase II)	Efficacy	50	Ongoing
05011	Colorectal cancer CPT-11–5-FU–LV–oxaliplatin	6 chrono schedules	Chrono-toxicity of CPT-11	180	Ongoing
	Lung cancer Gemcitabine–carboplatin	6 chrono schedules	Chrono-toxicity of gemcitabine	130	In preparation

CPT-11, irinotecan; 5-FU, fluorouracil; LV, folinic acid (leucovorin); RT, radiotherapy.
[a]Accrual attained (03.2002).

was also more common in the flat infusion group ($p = 0.002$). Over the first six cycles, patients on chronotherapy could receive higher doses ($p = 0.009$) and higher dose intensities ($p < 0.04$) of both drugs. These results support the need for further exploration of chronotherapy of liver metastases arising from colorectal cancer with combined arterial and intravenous fluoropyrimidine chemotherapy.

This type of combined treatment was recently compared as adjuvant treatment to systemic 5-FU treatment alone after curative resections of hepatic metastases from colorectal cancer (Kemeny *et al.*, 1999). The combined (arterial + intravenous) adjuvant therapy significantly improved the outcome of patients at 2 years (Kemeny *et al.*, 1999).

10.6.2 Other drugs

Platinum derivatives (oxaliplatin, carboplatin) and irinotecan (CPT-11) have been assessed regarding their circadian tolerance (Lévi *et al.*, 2000, 2001; Focan *et al.*, 2001). Most of the results have been discussed above.

The importance of the circadian timing of cisplatin administration was re-evaluated prospectively by Kobayashi *et al.* (2001). The clock time of cisplatin administration had an impact on the frequency of emesis, which was always greater after morning (versus evening) dosing; prophylactic administration of ondansetron diminished the time-of-day dependency of cisplatin-induced vomiting. Also, administration of cisplatin in the morning rather than in the evening appears to cause less renal damage. These last results contrasted with previous evaluations in ovarian cancer (Focan, 1995a; Lévi, 1997).

Chronopharmacology of high-dose oral busulfan was studied in 21 children with malignant solid tumours (Vassal *et al.*, 1993). Plasma levels of busulfan exhibited significant circadian variation with the acrophase of the rhythm appearing at around 06.00. Furthermore, urinary excretion of busulfan exhibited a significant circadian rhythm apparently linked to the physiological circadian rhythm in urinary output (Vassal *et al.*, 1993).

Ando *et al.* (1996) failed to observe diurnal variations in the plasma titres of low-dose etoposide given in continuous intravenous infusion, while Yamamoto *et al.* (1997) found higher plasma concentrations at 09.00 than at 21.00. Anecdotally, a circadian distribution of febrile episodes in neutropenic patients was observed with a maximum between 17.00 and midnight (acrophase 21.30) (Maher *et al.*, 1993).

10.7 Disruption of spontaneous rhythmicity of the host with tumour development or tumour treatment

The spontaneous circadian rhythmicity of the host systems may be altered in the presence of a tumour. Moreover, anticancer treatment has been shown capable of producing alterations of circadian structure.

10.7.1 Influence of tumour development on host circadian rhythmicity

The influence of tumour development on actively dividing healthy tissues was studied by the group of Scheving *et al.* (1992). They studied rhythms of DNA synthesis in selected organs of mice bearing a transplanted Lewis lung carcinoma. The 24 h changes were more pronounced in mice with 10- or 14-day-old tumours than in those with 6-day-old tumours. Moreover, in bone marrow, a second peak of DNA synthesis appeared in mice with 6-day carcinoma but not in controls. The peak in DNA synthesis was sharper and occurred earlier in animals with a 10-day-old tumour. Finally, ultradian rhythms with a period shorter than 20 h were seen in DNA synthesis of animals bearing 14-day-old tumours (Scheving et al., 1992; Mormont and Lévi, 1997).

More recently, E. Filipski and colleagues (unpublished observation) have assessed individual alteration of the temperature or rest–activity rhythms in relation to tumour growth. A damping of the circadian amplitude of both rhythms with shifting of diurnal peaks of plasma corticosterone levels and the number of lymphocytes was observed, correlated with tumour weight.

Reyna et al. (1997) have studied the evolution of the rate of mitoses after administration of colchicine, which induced temporary blockade of metaphases, in duodenal crypt enterocytes of ES 12A hepatocarcinoma-bearing mice. In the presence of the tumour, mitotic activity was inhibited, accompanied by an apparent temporal shift in the circadian curve for this growth parameter. However, other authors have failed to demonstrate any difference in the circadian evolution of bone marrow proliferation in the presence of tumour (Blank et al., 1995; Granda et al., 2001a).

Serum tumour markers (CEA, α-fetoprotein, CA 125, CA 15.3, thymidine kinase) have been evaluated in relation to cancer development with regard to their circadian profiles in humans (Touitou and Focan, 1992; Touitou et al., 1995, 1996; Hallek et al., 1997). Generally, in healthy subjects, grouped data of serum titres of tumour markers present a clear circadian pattern (Touitou and Focan, 1992; Touitou et al., 1995, 1996; Hallek et al., 1997). Conversely, in cancer patients, no group phenomenon could be observed despite, in some individual patients, the presence of circadian patterns, but generally with varying amplitudes and acrophases. Also in those patients, as discussed earlier, frequent alterations (with damped amplitude, shifted acrophases) of circadian structure were reported in relation to tumour burden, liver metastases and/or poor performance status (Touitou and Focan, 1992; Hallek et al., 1997; Touitou et al., 1995, 1996). A decrease of nocturnal excretion of melatonin was also observed (Taverna et al., 1997).

Circadian variations of cytokine levels (i.e. basic fibroblast growth factor, EGF, insulin-like growth factor 1) or cytokine receptor numbers (insulin-like growth factor binding protein) were studied in 28 hospitalised women with metastatic breast cancer (Haus et al., 2001). Circadian rhythms with acrophases between 11.00 and 11.20 were measured, sometimes with a superimposed 12 h periodicity. However, other groups failed to find reproducible patterns; Akbulut et al. (1999)

even reported an abolition of the GM-CSF circadian rhythmicity in early breast cancer.

The importance of the hypothalamo-hypophyseal region for the control of endocrine circadian rhythms is illustrated by Veldman *et al.* (2000), who showed the possibility of a complete restoration of normal daily adenocorticotrophin, cortisol, growth hormone and prolactin secretory dynamics in adults with Cushing's disease successfully treated by transsphenoidal microadenomectomy (Veldman *et al.*, 2000).

A comprehensive review of circadian system alterations during cancer processes has been published by Mormont and Lévi (1997). More recently, the same authors have studied the relative variation of four rhythms known to reflect circadian clock function (rest–activity as gauged by actimetry, plasma concentration of melatonin and cortisol, and lymphocyte counts) in patients with metastatic colorectal cancer and excellent performance status (Minors *et al.*, 1996; Mormont *et al.*, 1998, 2000).

10.7.2 Influence of anticancer treatment on host circadian rhythmicity

Li *et al.* studied the pharmacological effect of vinorelbine on body temperature and locomotor activity circadian rhythms in B6D2 F_1 mice (Li *et al.*, 2002). There was a coincidence between the peak of these two output rhythms of SCN function and the best tolerability for vinorelbine. It was observed that vinorelbine profoundly altered both circadian clock outputs in a dose-dependent and dosing-time-dependent manner (higher alterations at the times of least tolerance). Such results could suggest that chemotherapy-induced circadian dysfunction may contribute to the toxicokinetics of vinorelbine and possibly other anticancer agents (Li *et al.*, 2002).

Combination chemotherapy with 5-FU and irinotecan (CPT-11) appears promising for treating human colorectal cancer. As discussed above, the circadian-dependent toxicity – and efficacy? – of 5-FU is related to the circadian variation in activity of DPD, which is a rate-limiting enzyme in the pyrimidine catabolic pathway. In control mice, the DPD mRNA level was significantly higher at 14.00 than at 02.00 (Shimuzu *et al.*, 2001). Intravenous administration of CPT-11 at 20.00 suppressed the circadian rhythm of the DPD mRNA level for a dose-dependent period (Shimuzu *et al.*, 2001). This phenomenon was not observed after administration of CPT-11 at other times of day (Shimuzu *et al.*, 2001). These results suggest that CPT-11 may influence the circadian rhythm of DPD at the transcriptional level. This modulation of the circadian rhythm of DPD by CPT-11 may be a critical factor in optimising the combination of 5-FU and CPT-11.

The profound disruptive effect of interferon α (IFN-α) on the rhythms of locomotor activity, body temperature and clock gene mRNA was studied in peripheral organs of mice (liver, adrenals) and in SCN, the primary circadian pacemaker (Ohdo et al., 2001). The rhythmicity of clock genes (mBmal1, mClock, mPer1, mPer2) was disturbed in SCN only by the repetitive administration of IFN-α in a time dependent manner (no alteration when given at 0 HALO; major alteration when given at 12 HALO). Interestingly, a constant subcutaneous infusion of corticosterone induced alteration of the rest–activity cycle but had no impact on the circadian expression of clock genes in the SCN. Thus, alteration of clock function, a new concept in adverse reactions to drugs, may be overcome by optimising the dosing schedule (Ohdo et al., 2001).

The influence of treatment with recombinant TNF-α on the circadian rhythm of cortisol secretion in patients with advanced cancer was evaluated by Zubelewicz et al. (1995, 2001). A decreased cortisol secretion by the adrenals was observed with maintenance of a normal circadian rhythmicity.

Finally, Mormont et al. have recently suggested that the patient's rest–activity cycle may be maintained, restored or altered according to the delivery pattern of an infusional chemotherapy with irinotecan (CPT-11): maintenance of circadian rest–activity pattern with chronomodulated infusion; alteration of the same pattern with standard infusion (Mormont et al., 2001).

10.8 Chronoefficacy

10.8.1 Animal data

This section summarises some recent experimental data emphasising the importance of the time of administration of monochemotherapy or combination chemotherapy in relation to tumour outcome.

Balb/c mice bearing murine colon carcinoma 26 were treated with 5-FU every 6 h over a 24 h period (Kojima et al., 1999). As in a previous assessment (Focan, 1995a), the antitumour effect of treatment at 0 HALO was significantly better (with lower toxicity) than at other times of day, resulting in a significant improvement of survival (Kojima et al., 1999).

Kemeny et al. (1994) compared the effects of floxuridine (FUDR) given by hepatic artery infusion (HAI) in rats with hepatic adenocarcinoma metastases either with a 'day-cycled' or a constant-rate administration for 14 days (see Section 10.5.1.1). A 90% objective response rate (RR) was observed, but 88% of rats died from toxicity with the flat infusion. In contrast, with the 'day-cycled' HAI with 60% of FUDR delivered between 15.00 and 21.00, no mortality was recorded with the same objective RR.

Tumour growth of Walker-256 carcinoma was faster during night-time than during daytime (Sato, 1994). Treatment with 5-FU was more efficient when the drug was given at night than during the day: the tumour volume was significantly diminished, while body weight was conserved (Sato, 1994). Other Japanese authors (Hori *et al.*, 1996) observed that, in normal tissues (subcutaneous tissue, liver, kidney cortex, bone marrow), there were no apparent differences between average tissue blood flow at two different times, while tumour tissue blood flow increased significantly at night. Based on this functional characteristic of tumour microcirculation, antitumour effects were compared between a group in which doxorubicin was administered at 4 HALO and a group in which it was administered at 16 HALO. The therapeutic effect of doxorubicin administered at 16 HALO was significantly greater, particularly in large rat tumours, than when administered at 4 HALO. The main reason for the therapeutic improvement may be the selective increase in delivery of anticancer drugs to tumour tissues brought about by a circadian increase in tumour tissue blood flow (Hori *et al.*, 1996).

The greater tolerance to carboplatin in the early sleep phase of the mice (Lu and Yin, 1994) was associated with improved antitumour activity (greatest survival) and lowest marrow toxicity when the drug was also given at the beginning of the sleep phase.

The relevance of dosing time for antitumour efficacy of vinorelbine was assessed in P-388 leukaemia-bearing mice by Filipski *et al.* (1999). Vinorelbine was injected as a single or weekly dose at one of six circadian times, 4 h apart. A significant correlation between single dose and median survival time was limited to vinorelbine administration at 19 or 23 HALO. An increase in the vinorelbine weekly dose shortened median survival time in the mice treated at 7 HALO but significantly improved it in those treated at 19 HALO. The study demonstrates the circadian rhythm dependence of MTD and the need to deliver the MTD when it is least toxic in order to achieve improvement of survival through chronotherapy (Filipski *et al.*, 1999).

The circadian time dependency of docetaxel and doxorubicin as single agents or in combination was studied by Granda *et al.* (2001b) in

MA 13/C-bearing mice. As a single drug, each agent was most effective at the time of its best tolerance, at 7 HALO, irrespective of dose. Similarly, docetaxel–doxorubicin combinations were most effective (in terms of rate of complete tumour responses and long-term survival rate) following dosing at the beginning of the rest period or shortly after the onset of the activity period.

Granda *et al*. (2002b) also investigated the relevance of circadian rhythms in irinotecan (CPT-11) and oxaliplatin tolerability with regard to antitumour activity in mice bearing Glasgow osteosarcoma. CPT-11 and oxaliplatin were given 1 min apart at 7 or 15 HALO or at either their respective times of best tolerability (7 HALO for CPT-11 and 15 HALO for oxaliplatin) or worst tolerability (15 HALO for CPT-11 and 7 HALO for oxaliplatin). Tumour growth rate was nearly halved and life period was almost doubled in the mice receiving CPT-11 at 7 HALO compared to 15 HALO. Results of similar magnitude were obtained with oxaliplatin for both end points, but with 7 HALO corresponding to least efficacy and 15 HALO to greatest efficacy. Combination of CPT-11 with oxaliplatin produced therapeutic benefit only if the schedule consisted of CPT-11 administration at 7 HALO and oxaliplatin delivery at 15 HALO, i.e. when both drugs were given near their respective 'best' circadian times (Granda *et al*., 2002b).

Thus, if the work from Lévi's group is translated to the human situation, it may be hypothesised that better therapeutic indices could be obtained by treating cancer patients in late afternoon with mitozantrone (mitoxantrone), in the evening with vinorelbine, near the middle of the night with CPT-11 or docetaxel or the middle of the day with oxaliplatin (Granda and Lévi, 2002; Lévi, 2002; Figure 10.3).

In 1993, Koren *et al*. studied the circadian dependence of the antitumour activity of recombinant IFN-α and IFN-γ in B16 melanoma-bearing C57 B16 mice (Koren *et al*., 1993). The oncostatic activity of both interferons depends largely on the time of their administration. As examples, to generate maximum antitumour effectiveness, approximately 5-fold higher amounts of IFN-α were required at 12 HALO than at 4 HALO, and at least 8.5-fold greater doses of IFN-γ at 8 HALO than at 16 HALO (Koren *et al*., 1993). In this murine model, maximum antitumour activity was obtained with administration of IFN-α at 4 HALO and of IFN-γ at 16 HALO. More recently, Takane *et al*. (2000) investigated further the mechanisms underlying the dosing-time-dependent change in the antitumour efficacy of IFN-β. The greatest efficacy was observed in early light phase. More precisely, the higher antitumour effect of IFN-β was observed when specific binding to the IFN receptor and DNA synthesis in tumour cells increased, and the lowest efficacy was observed when these levels decreased. The dosing-time-dependent effect of IFN-β was supported by the time-dependent expression of transcription factor (signal transducers and activators of transcription 1) and cell proliferation inhibitor (p21 wild-type p53-activated fragment 1) proteins induced by IFN-β. There was a significant dosing-time-dependent change in IFN-β concentration in the tumour, with a higher level in early light phase and a lower level in early dark phase. The dosing-time-dependent change of IFN-β concentration in tumour was associated with that of IFN-β-induced antitumour effect (Takane *et al.*, 2000).

Blumenthal *et al.* (2001) were able to correlate the chronotoxicity and the chronoefficacy of the COX-2 inhibitor celecoxib in athymic nude mice bearing human breast cancer xenografts. Interestingly, tumour growth response to the MTD at each HALO revealed that there was no clear relationship between dose administered and therapeutic response. Indeed, the optimised MTD was found to be 25 mg/kg at 7–10 HALO but only 10 mg/kg at 17–20 HALO, while optimal tumour regression was observed when dosing was done at 23 to 7 HALO (05.00–15.00); no therapeutic response was observed when dosing was done at 10–13 HALO (16.00–19.00) and rapid tumour growth was reported after dosing at 17 HALO (23.00). Furthermore, COX-2 expression did not account for either the chronotherapy or the chronotoxicity results. As an example, MCF-7 tumour COX-2 is expressed most strongly at the gene level at 20 HALO and to a lesser degree at 13 HALO; however, COX-2-protein was not found in MCF-7 tumour at any time, although celecoxib was shown to be therapeutic in these tumours between 20 HALO and 7 HALO (Blumenthal *et al.*, 2001).

10.8.2 Human data

Previous observations (Rivard *et al.*, 1985) on the impact of morning versus evening schedule for oral methotrexate (MTX) and mercaptopurine on relapse risk for children with acute lymphoblastic leukaemia (ALL) were recently confirmed by the Nordic Society for Paediatric Haematology and Oncology (Schmiegelow *et al.*, 1997). In a study on 294 children, the authors confirmed the superior outcome with the evening schedule; they also found a relationship between tumour outcome and erythrocyte levels of intracellular metabolites of absorbed drugs (MTX polyglutamates and 6-thioguanine nucleotides), which deserves further evaluation in prospective trials (Schmiegelow *et al.*, 1997).

Taking tumour outcome or therapeutic index (i.e. a positive ratio between tumour efficacy and clinical toxicity) as end points, clinically effective drug delivery according to chronobiological concepts has been developed in digestive, lung and upper aerodigestive tract carcinomas. Up to now, with the exception of a pioneering study (Focan, 1979), it has only been postulated in phase I–III clinical trials that reduced clinical toxicities at specific times will promote antitumour activity by allowing increased drug levels (Focan, 1995b; Coudert *et al.*, 2002; Lévi, 1999, 2000, 2002).

The pivotal chronotherapy protocol is based on the chronomodulation of 5-FU ±folinic acid (leucovorin; LV) with a 12 h infusion exhibiting a sinusoidal variation of delivery rate with a peak at 04.00 (Figure 10.6). Circadian tolerance of this schedule in combination with oxaliplatin or carboplatin has been validated in randomised studies performed in colorectal cancer for oxaliplatin (IOCC, 1997; Lévi *et al.*, 2001) and non-small-cell lung carcinoma (Focan *et al.*, 2001; Focan, 2002a). In colorectal cancer, the Lévi group also observed a better antitumour activity of the combined treatment when the peaks of delivery were close to that of the best circadian tolerance of the drugs (Gruia *et al.*, 1996; F. Lévi, personal communication).

In previously untreated metastatic colorectal cancer, this three-drug schedule was compared by the European Group of Chronotherapy (International Organisation for Cancer Chronotherapy (IOCC) now acting as the Chronotherapy Group (CTG) of the EORTC) to the same combination given in flat infusion (IOCC, 1997). The chronomodulated chemotherapy was not only clearly better tolerated (as discussed above) but it could be given in higher doses. Also, the response rate was significantly enhanced (52% versus 30%) as well as the median time after which tumours regrew (time to progression). The studies of the European Group emphasised the potential curability of the disease thanks to further surgical resection of residual metastases (Misset and Lévi, 2000; Giacchetti, 2002). Thus, the opportunity for a curative resection of metastases was directly related to the response rate and therefore to the dose-intensity of oxaliplatin and 5-FU (Giacchetti, 2002). The median survival of patients curatively resected was estimated to be around 6 years. Comprehensive reviews on the influence of chronotherapy on colorectal cancer have been published (Giachetti, 2002; Lévi, 2001, 2002).

In non-small-cell lung carcinoma (NSCLC) and upper respiratory/digestive tract cancers, Focan (2002a, 2002b) has reviewed experimental and clinical data providing the basis for a chronobiological approach to medical treatment. In a pioneering study based upon chronoefficacy of a sequential chemotherapy (with the hope of inducing cell synchronisation and recruitment phenomena), Focan (1979) observed an enhancement of antitumour activity (i.e. in lung cancer) with the best-timed chemotherapy, with higher response rate and longer durations of responses and survival. It is to be noted that these results were observed despite the higher haematological toxicity of the most active schedule (Focan, 2002a).

In a prospective study performed in previously untreated advanced NSCLC patients, Focan *et al.* (2001) confirmed the chronotolerance of an infusional combination chemotherapy with 5-FU, folinic acid (leucovorin; LV) and carboplatin. The same authors then delivered the best schedule (Figure 10.7; reference schedule B) (which was also the one foreseen from experimental data) in a large phase II study to 68 patients with similar disease (Tables 10.1–10.3). The remarkable therapeutic indices – adequate tumour control and good host tolerance to treatments allowing sustained quality of life – prompted further trials aimed at defining the role of this protocol in the multidisciplinary management of NSCLC, sometimes with addition of newer drugs such as gemcitabine and concomitant radiotherapy (unpublished data). In the near future, the EORTC Chronotherapy Study Group hopes to evaluate the chronotolerance of gemcitabine in association with carboplatin as a first-line treatment of advanced NSCLC (Table 10.6). Studies on the clinical timing of radiotherapy are also warranted.

Similar infusional chronotherapy was proposed for treatment of upper respiratory/digestive tract cancers (head and neck and oesophageal cancers). Focan also reviewed the chronobiological concepts justifying the application of chronotherapy for head and neck cancer (Focan, 2003). A phase II trial was performed with the same chronotherapy protocol (Figure 10.7; reference schedule B) in 48 patients suffering from an advanced head and neck squamous cell carcinoma (Table 10.4). Again, the excellent tolerance was obvious despite the frequent radiotherapy (Table 10.5). An overall 63% response (19% CR, 44% PR), higher in earlier stages (89% versus 56% for stage IV) was observed; long-term survival was 47%, more frequent when radiotherapy was delivered (61% versus 31%) and in non-stage-IV patients (71% versus 40%) (Focan, 2003).

The results obtained in oesophageal and gastric cancers are presently under evaluation.

Chronobiological considerations for pancreatic cancer management have also been emphasised (Focan *et al.*, 2000b). The effectiveness of a chronomodulated infusional chemotherapy (with 5-FU ± cisplatin) versus a conventional one is being prospectively evaluated in a trial launched by the EORTC Chronotherapy Group (trial 05962) (Table 10.6).

10.8.3 Organisation of chronotherapy centres within the EORTC Chronotherapy Group

Since the early 1990s, thanks to the marketing of programmable-in-time portable pumps, the progressive development of ambulatory chronotherapy cancer units within day-hospitals of haemato-oncology departments has been possible.

The various aspects (maintenance, nursing, pharmacy, computer programming of pumps, data handling) have been progressively controlled and refined (Figure 10.9) with cooperation of nursing and medical staff of hospitals and/or of home-care organisations (Mohnen *et al.*, 1995). The first commercially available pumps (Intelliject) have been abandoned in favour of higher-performance pumps with an internal clock (Melodie, Aguettant Lab., Lyon, France, with four channels; or Z pump, Inpharzam, with two channels). In our hospital, the research staff programs and manages a set of 35 Melodie pumps (and three 3 Z pumps) with an 85% occupation rate. All patients treated are fully ambulatory (Mohnen *et al.*, 1995). The delivery of drugs is individually scheduled according to clinical need, using the software associated with

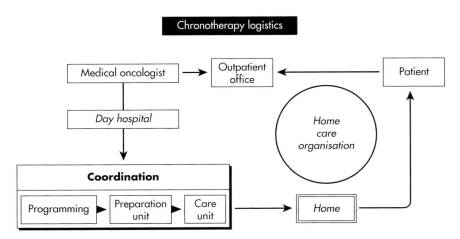

Figure 10.9 Chronotherapy logistics with multichannel programmable-in-time pumps. Modified with permission from Levi (2002).

the pumps. The preparation of drugs supervised by the pharmacist is performed under laminar flow by the oncology nurses within the day-hospital. The reservoirs of pumps are then filled, pumps are installed and patients are released. In case of any problem, the patients may come back to the hospital haemato-oncology unit at any time.

Because of the additional cost to the hospitals of chronotherapy delivery – in terms of specific training of nursing and medical staff, initial cost of the pumps, programming and maintenance and/or cost of disposable materials – results of recent trials comparing combination chemotherapy consisting of 5-FU, folinic acid (leucovorin; LV) and oxaliplatin, given either as flat or chronomodulated infusion for metastatic colorectal cancer, were subjected to pharmacoeconomic evaluation (Focan, 2002b). The overall costs of treatment with the flat and chronomodulated protocols appeared equivalent. The expense of the delivery of medication with the chronotherapeutic arm was greater than with the standard arm because it was feasible to administer more courses (requiring more frequent doctor visits) and higher doses (higher cost of medications) within the limitations of toxicity. However, chronomodulation was definitively more cost-effective than standard treatment because the outcome of treatment was more effective: there was greater tumour response rate and slower progression with less treatment-associated toxicity. Finally, selection of the Melodie brand infusion pump resulted in a further 18% reduction of overall costs and made it possible for patients to enjoy increased autonomy and improved quality of life (Focan, 2002b).

Most of the large chronotherapy centres now participate in the trials of the EORTC CTG. The strategies developed with this cooperative group (more than 45 centres with centres, also in the USA, Canada, Israel and China) have been published recently (Coudert *et al.*, 2002; Lévi, 2002). The circadian rhythm in anticancer treatment tolerability results in a predictable variation in MTD, this end point being determined in phase I trials. If a standard dose is given in the best circadian schedule, tolerability and quality of life are expected to improve while efficacy will remain similar to that of conventional treatment modalities. Conversely, if the dose delivered in the 'best' schedule is increased to MTD, efficacy should be enhanced while tolerability remains similar to that of conventional treatments. Of course, it is also possible that, owing to circadian variation in the sensitivity of the tumour target to cancer therapy, elective tumour responses at conventional or low doses could be observed (Focan, 1995a). Systematic evaluations of items of quality of life, psychological issues and circadian profiles of patients are carried out daily (Bertolini *et al.*, 1995; Pugliese *et al.*, 2002; Mormont and Waterhouse, 2002). Two trials of the EORTC CGT are evaluating before and after treatment, not only quality of life but also the circadian structure (by rest–activity actimetry and gradient 8/16 h titrations of plasma cortisol and circulating granulocytes). Relevant titres of cortisol or melatonin may also be obtained more easily from saliva samples (Mormont *et al.*, 1998). In the near future, it is hoped also to evaluate the diurnal variation in some easily accessible clock genes, e.g. in lymphocyte nuclei.

10.9 Conclusion

Since our last contribution to reviews on cancer chronotherapeutics (Focan, 1995a), major discoveries in biology have opened the door to fascinating prospects. Molecular and genetic biology have demonstrated the universal presence, in all normal and cancer cells, of clock genes. Therapeutic agents may act on some components of the circadian system or on various clock outputs; the potential usefulness to chronotherapeutics of the study of the genetic components of the circadian system is thus emphasised.

In oncology, the development of chronotherapy has strictly followed the usual steps of clinical pharmacology (experimental and clinical prerequisites; phase I, II and III trials). With this methodology, a chronotherapy schedule has been shown to be beneficial and a pertinent clinical benefit has been demonstrated in populations of cancer patients with definite characteristics (metastatic colorectal cancer).

Our efforts are now directed to the following goals:

1. The demonstration of definitive survival benefits in patients adequately treated with chronotherapy (EORTC trial 05963, now completed with 564 cases of metastatic colorectal cancer).
2. The refinement and further understanding of molecular mechanisms governing given outputs of circadian clocks. In this domain, interest will be focused on various mechanisms of rhythm alteration in cancer patients (alteration sometimes differing according to the rhythm considered; for example, cortisol versus rest–activity).
3. Progressively implementing individualised chronotherapy schedules according to the characteristics of the circadian structure of individual patients.
4. Development of active tools (e.g. cortisone, adrenocorticotrophic hormone, melatonin, bright light) aimed at resynchronising patients who have lost their basic circadian profile.

From all these refinements and developments, we hope ultimately to improve overall well-being, quality of life, psychological issues and tumour outcome of cancer patients.

References

Akbulut H, Icli F, Buyukcelik A, et al. (1999). The role of granulocyte-macrophage-colony stimulating factor, cortisol, and melatonin in the regulation of the circadian rhythms of peripheral blood cells in healthy volunteers and patients with breast cancer. *J Pineal Res* 26: 1–8.

Anderson L E, Morris J E, Sasser L B, Stevens R G (2000). Effect of constant light on DMBA mammary tumorigenesis in rats. *Cancer Lett* 148: 121–126.

Ando Y, Minami H, Sakai S, Shimokata K (1996). Is there a circadian variation in plasma concentrations of etoposide given by prolonged continuous infusion? *Cancer Chemother Pharmacol* 37: 616–618.

Ballard R G, Lagorio S, De Angelis G, Verdechia A (2000). Cancer incidence and mortality among flight personnel: a meta-analysis. *Aviat Space Environ Med* 71: 216–224.

Balsalobre A, Damiola F, Schibler U (1998). A serum shock induces circadian gene expression in mammalian tissue culture cells. *Cell* 93: 929–937.

Barrat M A, Mormont M C, Renée N, et al. (2002). Circadian changes in dehydropyrimidine dehydrogenase activity, cortisol and melatonin in previously-treated patients with very advanced metastatic colorectal cancer. *Proc American Society of Clinical Oncology* 21: 88b (2163).

Bartsch C, Bartsch H, Karenovics A, et al. (1997). Nocturnal urinary 6-sulphatoxymelatonin excretion is decreased in primary breast cancer patients compared to age-matched controls and shows negative correlation with tumor-size. *J Pineal Res* 23: 53–58.

Bertolini R, Focan C, Bartholome F, Baro V (1995). Comparative psychological aspects of two different types of chemotherapy administration (chronotherapy vs traditional chemotherapy) on quality of life of cancer patients at advanced stage. *In Vivo* 9: 583–588.

Bjarnason G A, Jordan R C K (2000). Circadian variation of cell proliferation and cell cycle protein expression in man: clinical implications. *Prog Cell Cycle Res* 4: 193–206.

Bjarnason G A, Jordan R C K (2002). Rhythms in human gastrointestinal mucosa and skin. *Chronobiol Int* 19: 129–140.

Bjarnason G A, Jordan R C K, Sothern R B (1999). Circadian variation in the expression of cell-cycle proteins in human oral epithelium. *Am J Pathol* 154: 613–622.

Bjarnason G A, Jordan R C K, Wood P A, et al. (2001). Circadian expression of clock genes in human oral mucosa and skin. *Am J Pathol* 158: 1793–1801.

Blank M A, Gushchin V A, Halberg F, et al. (1995). X-irradiation chronosensitivity and circadian rhythmic proliferation in healthy and sarcoma-carrying rats' bone marrow. *In Vivo* 9: 395–400.

Blumenthal R D, Waskewich C, Goldenberg D M, et al. (2001). Chronotherapy and chronotoxicity of the cyclooxygenase-2-inhibitor, celecoxib in athymic mice bearing human breast cancer xenografts. *Clin Cancer Res* 7: 3178–3185.

Boughattas N, Lévi F, Fournier C, et al. (1989). Circadian rhythm in toxicities and tissue uptake of 1,2-diamminocyclohexane(*trans*-1)oxalatoplatinum (II) in mice. *Cancer Res* 49: 3362–3368.

Boughattas N, Lévi F, Fournier C, et al. (1990). Stable circadian mechanisms of toxicity of two platinum analogs (cisplatin and carboplatin) despite repeated dosages in mice. *J Pharmacol Exp Ther* 255: 672–679.

Boughattas N A, Li X M, Filipski J, et al. (1996). Modulation of cisplatin chronotoxicity related to reduced glutathione in mice. *Hum Exp Toxicol* 15: 563–572.

Bressolle F, Joudia J M, Pinguet F, et al. (1999). Circadian rhythm of 5-fluorouracil population pharmacokinetics in patients with metastatic colorectal cancer. *Cancer Chemother Pharmacol* 44: 295–302.

Buchi K N, Moore J G, Hrushesky W J, et al. (1991). Circadian rhythm of cellular proliferation in the human rectal mucosa. *Gastroenterology* 101: 410–415.

Buemi M, Allegra A, Anastasi G, et al. (1997). Loss of circadian rhythm in erythropoietin production in a patient with renal erythropoietin secreting neoplasia. *Clin Nephrol* 47: 134–135.

Chevalier V, Papon J, Pezet D, et al. (2002). Blood concentrations of interleukin-15 (IL-15) and interleukin-2 (IL-2) in colorectal cancer patients: preliminary results. *Proc Am Assoc Cancer Res* 43: 448 (2226).

Codacci-Pisanelli G, van der Wilt C L, Pinedo H M, et al. (1995). Antitumour activity, toxicity and inhibition of thymidylate synthase on prolonged administration of 5-fluorouracil in mice. *Eur J Cancer* 31A: 1517–1525

Colombo L L, Mazzoni E O, Meiss R P (2000). The time of tumor cell division and death depends on the site of growth. *Oncol Rep* 7(6): 1363–1366.

Coudert B, Focan C, Donato di Paola E, Lévi F (2002). It's time for chronotherapy! *Eur J Cancer* 38: S50–S53.

Davis S, Mirick D K, Stevens R G (2001). Night shift work, light at night and risk of breast cancer. *J Natl Cancer Inst* 93: 1557–1562.

Delaunay F, Grechez A, Grundshober C, et al. (2002). Analyse globale de l'expression génique circadienne. *Proc 12ème colloque de l'Association de Chronobiologie Médicale*, Aussois, France (abstract).

Diasio R B, Harris B E (1989). Clinical pharmacology of 5-fluorouracil. *Clin Pharmacokinet* 16: 215–237.

Filipski E, Lévi F, Vadrot N, et al. (1997). Circadian changes in irinotecan toxicity in mice. *Proc Am Assoc Cancer Res*, 38: 305 (2048).

Filipski E, Amat S, Lemaigre G, et al. (1999). Relationship between circadian rhythm of vinorelbine toxicity and efficacy in P388-bearing mice. *J Pharmacol Exp Ther* 289: 231–235.

Filipski E, King V M, Li X, et al. (2002a). Host circadian clock as a control point in tumor progression. *J Natl Cancer Inst* 94: 690–697.

Filipski E, King V D, Sun J, et al. (2002b). Experimental jet-lag accelerates malignant growth and shortens survival in tumor-bearing mice. *Proc Am Assoc Cancer Res* 430: 862 (4270).

Fleming G F, Schilsky R L, Mick R, et al. (1994). Circadian variation of 5-fluorouracil (5-FU) and cortisol plasma levels during continuous-infusion 5-FU and leucovorin (LV) in patients with hepatic or renal dysfunction. *Proc Am Soc Clin Oncol* 13: (Abstr.) 352.

Focan C (1979). Sequential chemotherapy and circadian rhythm in human solid tumours: randomized trial. *Cancer Chemother Pharmacol* 3: 197–200.
Focan C (1995a). Circadian rhythms and cancer chemotherapy. *Pharmacol Ther* 67: 1–52.
Focan C (1995b). Chronotherapy for human solid tumors other than colorectal. *In Vivo* 9: 549–554.
Focan C (2002a). Chronobiological concepts underlying the chronotherapy of human lung cancer. *Chronobiol Int* 19: 253–273.
Focan C (2002b). Pharmaco-economic comparative evaluation of combination chronotherapy vs standard chemotherapy for colorectal cancer. *Chronobiol Int* 19: 289–297.
Focan C (2003). Considérations chronobiologiques pour la prise en charge des tumeurs de la tête et du cou chez l'homme. In press.
Focan C, Denis B, Kreutz F, et al. (1995). Ambulatory chronotherapy with 5-fluorouracil, folinic acid, and carboplatin for advanced non-small cell lung cancer. A phase II feasibility trial. *J Infus Chemother* 5(3 Suppl 1): 148–152.
Focan C, Lévi F, Kreutz F, et al. (1999). Continuous delivery of venous 5-fluorouracil and arterial 5-fluorodeoxyuridine for hepatic metastases from colorectal cancer: feasibility and tolerance in a randomized phase II trial comparing flat versus chronomodulated infusion. *Anti-Cancer Drugs* 10: 385–392.
Focan C, Roland S, Kreutz F, et al. (2000a). Chronotherapy with 5-fluorouracil, folinic acid and carboplatin for stomach cancer. A pilot study. In: *Abstracts of 2nd Annual Symposium of the Belgian Society of Medical Oncology*.
Focan C, Kreutz F, Focan-Henrard D, et al. (2000b). Approche chronopharmacologique de l'administration des agents anticancéreux dans l'adénocarcinome pancréatique. Expérience pilote. *Rev Med Liege* 55: 3: 149–153.
Focan C, Kreutz F, Moeneclaye N, et al. (2003). Étude randomisée évaluant la chronotolérance d'une association de 5-fluorouracile, acide folinique et carboplatine chez des patients porteurs d'un cancer pulmonaire non à petites cellules (CPNPC). In press.
Giacchetti S (2002). Chronotherapy of colorectal cancer. *Chronobiol Int* 19: 207–219.
Giacchetti S, Curé H, Adenis A, et al. (2001). Chronomodulated irinotecan (CPT 11) versus standard infusion in patients with metastatic colorectal cancer, a randomized multicenter trial. *Eur J Cancer* 37(Suppl 6): S309 (1144).
Granda T G, Lévi F (2002). Tumor-based rhythms of anticancer efficacy in experimental models. *Chronobiol Int.* 19: 21–42.
Granda T G, Liu X H, Filipski E, et al. (2001a). Different rhythms in cell cycle distribution in bone marrow and in MA13/C mammary tumor of C3H/HeN mice. *Proc Am Assoc Cancer Res* 42: 912 (4897).
Granda T G, D'Attino R M, Filipski E, et al. (2001b). Experimental chronotherapy of mouse mammary adenocarcinoma MA 13/C with docetaxel and doxorubicin as single agents and in combination. *Cancer Res* 61: 1996–2001.
Granda T G, Filipski E, Lévi F (2002a). Expression des protéines BCL-2 et P53 dans l'adénocarcinome mammaire MA13/C. Effet du docetaxel. *Proc 12ème colloque de l'Association de Chronobiologie Médicale*, Aussois, France (abstract).
Granda T G, D'Attino R M, Filipski E, et al. (2002b). Circadian optimisation of irinotecan and oxaliplatin efficacy in mice with Glasgow osteosarcoma. *Br J Cancer* 86: 999–1005.

Gruia G, Giacchetti S, Depres P, *et al.* (1996). Role of time of peak delivery of chronomodulated 5-fluorouracil (5-FU), L-leucovorin (LV) and oxaliplatin (L-OHP) in patient (PTS) with metastatic colorectal cancer (MCC). *Proc Am Soc Clin Oncol* 15: 180 (372).

Grundschober C, Delaunay F, Puhlhofer A, *et al.* (2001). Circadian regulation of diverse gene products revealed by mRNA expression profiling of synchronized fibroblasts. *J Biol Chem* 276: 46751–46758.

Halberg F, Cornelissen G, Sothern R B, *et al.* (1981). International geographic studies of oncological interest on chronobiological variables. In: Kaiser, H E, ed. *Neoplasm – Comparative Pathology of Growth in Animals, Plants and Man.* Baltimore: Williams and Wilkins, 553–596.

Hallek M, Touitou Y, Lévi F, *et al.* (1997). Serum thymidine kinase levels are elevated and exhibit diurnal variations in patients with advanced ovarian cancer. *Clin Chim Acta* 267: 155–166.

Harris B, Song R, Soong S, *et al.* (1990). Relationship between dihydropyrimidine dehydrogenase activity and plasma 5-fluorouracil levels: evidence for circadian variation of plasma drug levels in cancer patients receiving 5-fluorouracil by protracted continuous infusion. *Cancer Res* 50: 197–201.

Haus E, Dumitriu L, Nicolan G Y, *et al.* (2001). Circadian rhythms of basic fibroblast growth factor (bFGF), epidermal growth factor (EGF), insulin-like growth factor-1 (IGF-1), insulin-like growth factor binding protein-3 (IGFBP-3), cortisol, and melatonin in women with breast cancer. *Chronobiol Int* 18: 709–727.

Hori K, Zhang Q H, Li H C, *et al.* (1996). Timing of cancer chemotherapy based on circadian variations in tumor tissue blood flow. *Int J Cancer* 65: 360–364.

Hrushesky W J, Bjarnason G A (1993). Circadian cancer therapy. *J Clin Oncol* 11: 1403–1417.

Hrushesky W J, Langevin T, Kim Y J, Wood P A (1994). Circadian dynamics of tumor necrosis factor alpha (cachectin) lethality. *J Exp Med* 180: 1059–1065.

Ikeda M (2000). BMALI and circadian rhythm. *Nihon Shinkei Seishin Yakurigaku Zasshi* 20: 203–212.

IOCC, International Organization for Cancer Chronotherapy; Lévi F, Zidani R, Misset J L (1997). Randomized multicentre trial of chronotherapy with oxaliplatin, fluorouracil and folinic acid in metastatic colorectal cancer. *Lancet* 350: 681–686.

Izumi M, Ohwada S, Morishita Y (1995). Evaluation of intravenous infusion with 5-fluorouracil for advanced or recurrent gastric or colorectal cancer. *J Jpn Soc Cancer Ther* 30: 24–29.

Joly A C (1998). Perfusion continue de 5-FU et cancers bronchiques non à petites cellules: rythmes circadien et influence d'une corticothérapie (Thèse de DES (Diplôme d'Études Spécialisées) de Pharmacie Spécialisée, Université Paris XI, Faculté de Chatenay-Malabry).

Joly A C, Monnet J U, Chouaid C, *et al.* (2001). Influence d'une corticothérapie sur les concentrations plasmatiques de 5-FU et sur les rythmes circadiens de patients atteints de cancer bronchique non petite cellule (CBNPC). *Proc 11ème colloque de l'Association de Chronobiologie Médicale*, Aussois, France (abstract).

Kemeny M M, Alava G, Oliver J M (1994). The effects on liver metastases of circadian patterned continuous hepatic arterial infusion of FUDR. *HPB Surg* 7: 219–224.

Kemeny N, Huang Y, Cohen A M, et al. (1999). Hepatic arterial infusion of chemotherapy after resection of hepatic metastases from colorectal cancer. *N Engl J Med* 341: 2039–2048.

Kobayashi M, To H, Tokue A, et al. (2001). Cisplatin-induced vomiting depends on circadian timing. *Chronobiol Int* 18: 851–863.

Kojima H, Sakamoto J, Yasue M (1999). Circadian rhythm-modulated chemotherapy with high dose 5-fluorouracil against gastrointestinal cancers: evaluation and case report. *Nagoya J Med Sci* 62: 29–38.

Koren S, Whorton E B Jr, Fleischmann W R Jr (1993). Circadian dependence of interferon antitumor activity in mice. *J Natl Cancer Inst* 85: 1927–1932.

Kramer A, Yang F C, Snodgrass P, et al. (2001). Regulation of daily locomotor activity and sleep by hypothalamic EGF receptor signaling. *Science* 294: 2511–2515.

Kwiatkowski F, Chevalier V, Cure H, et al. (2001). Méthode d'optimisation des chimiothérapies par patient dans les traitements conventionnels et en chronothérapie. *Proc 11ème colloque de l'Association de Chronobiologie Médicale*, Aussois, France (abstract).

Lévi F (1997). Chronopharmacology of anticancer agents. In: Redfern P H, Lemmer B, eds. *Physiology and Pharmacology of Biological Rhythms, Handbook of Experimental Pharmacology*. Berlin: Springer-Verlag, 299–331.

Lévi F (1999). Cancer chronotherapy. *J Pharm Pharmacol* 51: 891–898.

Lévi F (2000). Therapeutic implications of circadian rhythms in cancer patients. *Novartis Found Symp* 227: 119–136.

Lévi F (2001). Cancer Chronotherapy. *Lancet Oncol* 2: 307–315.

Lévi F (2002). From circadian rhythms to cancer chronotherapeutics. *Chronobiol Int* 19: 1–19.

Lévi F, Tampellini M, Metzger G, et al. (1994). Circadian changes in mitoxantrone toxicity in mice: relationship with plasma pharmacokinetics. *Int J Cancer* 59: 543–547.

Lévi F, Soussan A, Adam R, et al. (1995). A phase I–II trial of five-day continuous intravenous infusion of 5-fluorouracil delivered at circadian rhythm modulated rate in patients with metastatic colorectal cancer. *J Infus Chemother* 5(3 Suppl 1): 153–158.

Lévi F, Metzger G, Massari C, Milano G (2000). Oxaliplatin: pharmacokinetics and chronopharmacological aspects. *Clin Pharmacokinet* 38: 1–21.

Lévi F, Zidani R, Coudert B, et al. (2001). Chronomodulated irinotecan (I)–fluorouracil (F)–leucovorin (L)–oxaliplatin (O) (CHRONO IFLO) as salvage therapy in patients with heavily pretreated metastatic colorectal cancer (MCC). *Proc Am Soc Clin Oncol* 20: 139 (552).

Li J C, Xu F (1997). Influences of light–dark shifting on the immune system, tumor growth and life span of rats, mice and fruit flies as well as on the counteraction of melatonin. *Biol Signals* 6: 77–89.

Li X M, Metzger G, Filipski E, et al. (1997). Pharmacologic modulation of reduced glutathione circadian rhythms with buthionine sulfoximine: relationship with cisplatin toxicity in mice. *Toxicol Appl Pharmacol* 143: 281–290.

Li X M, Filipski E, Lévi F (1998). Pharmacological modulation of cisplatin toxicity rhythms with buthionine sulfoximine in mice bearing pancreatic adenocarcinoma (P03). *Chronobiol Int* 15: 323–335.

Li X M, Vincenti M, Lévi F (2002). Pharmacological effects of vinorelbine on body temperature and locomotor activity circadian rhythms in mice. *Chronobiol Int* 19: 43–55.

Lincoln D W, Hrushesky W J, Wood P A (2000). Circadian organization of thymidylate synthase activity in normal tissues: a possible basis for 5-fluorouracil chronotherapeutic advantage. *Int J Cancer* 88: 479–485.

Liu X H, Metzger G, Lévi F (2001a). Circadian rhythms in Bcl-2 and bax protein expressions in mouse bone marrow. *Proc Am Assoc Cancer Res* 42: 305 (1646).

Liu X H, Cermakian N, Sassone-Corsi P, Lévi F (2001b). Rythmic circadian expression of BCL-2 and bax proteins in relation with clock genes transcription in mouse bone marrow. *Proc 11ème colloque de l'Association de Chronobiologie Médicale*, Aussois, France (abstract).

Lu X H, Yin L J (1994). Circadian rhythm in susceptibility of mice to the anti-tumor drug carboplatin. *Zhonghua FU Chan Ke Za Zhi* 29: 729–731.

Maher J, Browne P, Daly L, *et al.* (1993). A circadian distribution to febrile episodes in neutropenic patients. *Support Care Cancer* 1: 98–100.

Marchenay C, Cellarier E, Lévi F, *et al.* (2001). Circadian variation in O^6-alkylguanine-DNA alkyltransferase activity in circulating blood mononuclear cells of healthy human subjects. *Int J Cancer* 91: 60–66.

Marra G, Anti M, Percesepe A, *et al.* (1994). Circadian variations of epithelial cell proliferation in human rectal crypts. *Gastroenterology* 106: 982–987.

Martineau-Pivoteau N, Lévi F, Rolhion C, *et al.* (1996). Circadian rhythm in toxic effects of cystemustine in mice: relevance for chronomodulated delivery. *Int J Cancer* 68: 669–674.

Metzger G, Massari C, Etienne M C, *et al.* (1994). Spontaneous or imposed circadian changes in plasma concentrations of 5-fluorouracil coadministered with folinic acid and oxaliplatin: relationship with mucosal toxicity in cancer patients. *Clin Pharm Ther* 56: 190–201.

Micke O, Schafer U, Wormann B, *et al.* (1997). Circadian variations of interleukin-2 receptors, serum thymidine kinase and beta-2-microglobulin in patients with non-Hodgkin's lymphoma and normal controls. *Anticancer Res* 17: 3007–3010.

Milano G (2002). Clinical pharmacokinetics of 5-fluorouracil with considerations on chronopharmacokinetics. *Chronobiol Int* 19: 177–189.

Minors D, Akerstedt T, Atkinson G, *et al.* (1996). The difference between activity when in bed and out of bed. I. Healthy subjects and selected patients. *Chronobiol Int* 13: 27–34.

Misset J L, Lévi F (2000). Chronotherapy with 5-fluorouracil and other drugs in gastrointestinal malignancies. *Semin Oncol* 27 (5 Suppl 10): 78–82.

Mohnen L, Heedfeld T, Focan-Henrard D, *et al.* (1995). Ambulatory chrono-chemotherapy by portable pumps: feasibility and compliance. Nursing aspects. *In Vivo* 9: 565–572.

Mormont M C, Lévi F (1997). Circadian-system alterations during cancer processes: a review. *Int J Cancer* 70: 241–247.

Mormont M C, Waterhouse J (2002). Contribution of the rest–activity circadian rhythm to quality of life in cancer patients. *Chronobiol Int* 19: 313–323.

Mormont M C, Hecquet B, Bogdan A, *et al.* (1998). Non-invasive estimation of the circadian rhythm in serum cortisol in patients with ovarian or colorectal cancer. *Int J Cancer* 78: 421–424.

Mormont M C, Waterhouse J, Bleuzen P, et al. (2000). Marked 24-h rest/activity rhythms are associated with better quality of life, better response, and longer survival in patients with metastatic colorectal cancer and good performance status. *Clin Cancer Res* 6: 3038–3045.

Mormont M C, Chedouba-Messali L, Chevalier V, et al. (2001). Chronomodulated versus standard infusion of irinotecan maintains or alters circadian rest/activity rhythm in patients with metastatic colorectal cancer. *Proc Am Soc Clin Oncol* 20: 112 (2200).

Mormont M C, Langouët A M, Claustrat B, et al. (2002). Marker rhythms of circadian system function: a study of patients with metastatic colorectal cancer and good performance status. *Chronobiol Int* 19: 141–155.

Nowakowska-Dulawa E (1990). Circadian rhythm of 5-fluorouracil (FU) pharmacokinetics and tolerance. *Chronobiologia* 17: 27–35.

Ohdo S, Makinosumi T, Ishizaki T, et al. (1997). Cell-cycle-dependent chronotoxicity of irinotecan hydrochloride in mice. *J Pharmacol Exp Ther* 283: 1383–1388.

Ohdo S, Koyanagi S, Suyama H, et al. (2001). Changing the dosing schedule minimizes the disruptive effects of interferon on clock function. *Nat Med* 7: 356–360.

Pasqualetti P, Collacciani A, Casale R (1996). Circadian rhythm of serum erythropoietin in multiple myeloma. *Am J Hematol* 53: 40–42.

Petit E, Milano G, Lévi F, et al. (1998). Circadian varying plasma concentration of 5-FU during 5-day continuous venous infusion at constant rate in cancer patients. *Cancer Res* 48: 1676–1679.

Pugliese P, Garufi C, Perrone M, et al. (2002). Quality of life and chronotherapy. *Chronobiol Int* 19: 299–312.

Rafnsson V, Hrafnkelsson J, Tulinius H (2000). Incidence of cancer among commercial airline pilots. *Occup Environ Med* 57: 175–179.

Reyna J C, Barbeito C G, Badran A F, Moreno F R (1997). Mitotic activity of duodenal-crypt enterocytes in mice with hepatocarcinoma. *Medicina (B Aires)* 57: 708–712.

Rivard G, Infante-Rivard C, Hoyeux C, Champagne J (1985). Maintenance chemotherapy for childhood acute lymphoblastic leukemia: better in the evening. *Lancet* ii: 1264–1266.

Sato T (1994). Circadian chronotherapy for Walker-256 of Wistar rat. *Gan to Kagaku Ryoho* 21: 2287–2290.

Schernhammer E S, Laden F, Speizer F E, et al. (2001). Rotating night shifts and risk of breast cancer in women participating in the nurses' health study. *J Natl Cancer Inst* 93: 1563–1568.

Scheving L E, Tsai T H, Scheving L A, et al. (1992). Normal and abnormal cell proliferation in mice especially as it relates to cancer. In: Touitou Y, Haus E, eds. *Biologic Rhythms in Clinical and Laboratory Medicine*. Berlin: Springer-Verlag, 566–599.

Schmiegelow K, Glomstein A, Kristinsson J, et al. (1997). Impact of morning versus evening schedule for oral methotrexate and 6-mercaptopurine on relapse risk for children with acute lymphoblastic leukemia. Nordic Society for Pediatric Hematology and Oncology (NOPHO). *J Pediatr Hematol Oncol* 19: 102–109.

Sephton S E, Sapolsky R M, Kraemer H C, Spiegel D (2000). Diurnal cortisol rhythm as a predictor of breast cancer survival. *J Natl Cancer Inst* 92: 994–1000.

Shakhar G, Bar-Ziv I, Ben-Eliyahu S (2001). Diurnal changes in lung tumor clearance and their relation to NK cell cytotoxicity in the blood and spleen. *Int J Cancer* 94: 401–406.

Shakil A, Hirabayashi N, Toge T (1993). Circadian variation of 5-fluorouracil and *cis*-platinum toxicity in mice. *Hiroshima J Med Sci* 42: 147–154.

Shimizu M, Tamura T, Yamada Y, *et al.* (2001). CPT-11 alters the circadian rhythm of dihydropyrimidine dehydrogenase mRNA in mouse liver. *Jpn J Cancer Res* 92: 554–561.

Sothern R B, Smaaland R, Moore J G (1995). Circannual rhythm in DNA synthesis (S-phase) in healthy human bone marrow and rectal mucosa. *FASEB J* 9: 397–403.

Sparano J A, Wadler S, Diasio R B, *et al.* (1993). Phase I trial of low-dose, prolonged continuous infusion fluorouracil plus interferon-alfa: evidence for enhanced fluorouracil toxicity without pharmacokinetic perturbation. *J Clin Oncol* 11: 1609–1617.

Stevens R G, Rea M S (2001). Light in the built environment: potential role of circadian disruption in endocrine disruption and breast cancer. *Cancer Causes Control* 12: 279–287.

Storch K F, Lipan O, Leykin I, *et al.* (2002). Extensive and divergent circadian gene expression in liver and heart. *Nature* 417: 78–83.

Takane H, Ohdo S, Yamada T, *et al.* (2000). Chronopharmacology of antitumor effect induced by interferon-beta in tumor bearing mice. *J Pharmacol Exp Ther* 294: 746–752.

Takane H, Ohdo S, Yamada T, *et al.* (2001). Relationship between diurnal rhythm of cell cycle and interferon receptor expression in implanted-tumor cells. *Life Sci* 68: 1449–1455.

Takimoto C H, Yee L K, Venzon D J, *et al.* (1999). High inter- and intrapatient variation in 5-fluorouracil plasma concentrations during a prolonged drug infusion. *Clin Cancer Res* 5: 1347–1352.

Tampellini M, Filipski E, Liu X H, *et al.* (1998). Docetaxel chronopharmacology in mice. *Cancer Res* 58: 3896–3904.

Taverna G, Trinchieri A, Mandressi A, *et al.* (1997). Variation in nocturnal urinary excretion of melatonin in a group of patients older than 55 years suffering from urogenital tract disorders. *Arch Ital Urol Androl* 69: 293–297.

Thiberville L, Compagnon C, Moore N, *et al.* (1994). Plasma 5-fluorouracil and alpha-fluoro-beta-alanin accumulation in lung cancer patients treated with continuous infusion of cisplatin and 5-fluorouracil. *Cancer Chemother Pharmacol* 35: 64–70.

Ticher A, Haus E, Ron I G, *et al.* (1996). The patterns of hormonal circadian time structure (acrophase) as an assessor of breast-cancer risk. *Int J Cancer* 65: 591–593.

Touitou Y, Focan C (1992). Rhythms in tumor markers. In: Touitou Y, Haus E, eds. *Biologic Rhythms in Clinical and Laboratory Medicine*. Berlin: Springer-Verlag, 648–657.

Touitou Y, Lévi F, Bogdan A, *et al.* (1995). Rhythm alteration in patients with metastatic breast cancer and poor prognostic factors. *J Cancer Res Clin Oncol* 121: 181–188.

Touitou Y, Bogdan A, Levi F, *et al.* (1996). Disruption of the circadian patterns of serum cortisol in breast and ovarian cancer patients: relationships with tumour marker antigens. *Br J Cancer* 74: 1248–1252.

Travlos G S, Wilson R E, Murrell *et al.* (2001). The effect of short intermittent light exposures on the melatonin circadian rhythm and NMU-induced breast cancer in female F344/N rats. *Toxicol Pathol* 29: 126–136.

Van den Heiligenberg S, Depres-Brummer P, Barbason H, *et al.* (1999). The tumor promoting effect of constant light exposure on diethylnitrosamine-induced hepatocarcinogenesis in rats. *Life Sci* 64: 2523–2534.

Vassal G, Challine D, Koscielny S, *et al.* (1993). Chronopharmacology of high-dose busulfan in children. *Cancer Res* 53: 1534–1537.

Veldman R G, Frolich M, Pincus S M, *et al.* (2000). Apparently complete restoration of normal daily adrenocorticotropin, cortisol, growth hormone, and prolactin secretory dynamics in adults with Cushing's disease after clinically successful transsphenoidal adenomectomy. *J Clin Endocrinol Metab* 85: 4039–4046.

Weinberg R A (1996). The molecular basis of carcinogenesis: understanding the cell cycle clock. *Cytokines Mol Ther* 2: 105–110.

Wood P A, Hrushesky W J (1996). Circadian rhythms and cancer chemotherapy. *Crit Rev Eukaryot Gene Expr* 6: 299–343.

Xian L J, Jian S, Cao Q-Y, *et al.* (2002). Circadian rhythms of DNA synthesis in nasopharyngeal carcinoma cells. *Chronobiol Int* 19: 69–76.

Yamamoto N, Tamura T, Ohe Y, *et al.* (1997). Chronopharmacology of etoposide given by low dose prolonged infusion in lung cancer patients. *Anticancer Res* 17: 669–672.

Zubelewicz B, Braczkowski R, Romanowski W, Grzeszczak W (1995). Influence of treatment using recombinant tumor necrosis factor alpha (hrec TNF alpha) on circadian rhythm of cortisol secretion in patients with advanced neoplastic disease. *Pol Arch Med Wewn* 94: 506–511.

Zubelewicz-Szkodzinska B, Muc-Wierzgon M, Wierzgon J, Brodziak A (2001). Dynamics of circadian flutuations in serum concentration of cortisol and TNF-alpha soluble receptors in gastrointestinal cancer patients. *Oncol Rep* 8: 207–212.

11

Chronopharmaceutical drug delivery

Howard N E Stevens

11.1 Introduction

Oral drug delivery remains the most popular and convenient form of administration of drugs to patients. Over the latter half of the 20th century, oral formulations increased in complexity from formulations that released drug as a bolus soon after ingestion through to a range of more complex formulation technologies capable of affording controlled release. During the late 1980s and the 1990s, pharmaceutical scientists were inspired by the challenge of developing novel kinds of drug delivery systems that undergo a defined lag period following ingestion by the patient and either release drug as a bolus or provide sustained release following this delay period (Figure 11.1).

The rationale of this time-delayed delivery approach, referred to as 'chronopharmaceutical drug delivery' (Stevens, 1998), is to deliver drugs in accordance with the circadian rhythms of the disease. As will be evident from much of the information presented in previous chapters, circadian variation in physiological and pathological processes is commonly gradual. When sampled with sufficient frequency over 24 h, the physiological and pathological rhythms tend towards a sigmoidal curve. Delivery of drug so that the plasma level – and by extrapolation the concentration of drug at the 'receptor' – exactly parallels this sigmoidal pattern can be viewed as the ultimate goal of the chronopharmaceutical scientist. More realistically in the shorter term, pharmaceutical formulators have concentrated on methods of timing the pulsatile delivery of drug to coincide with the peak time of clinical need or of receptor sensitivity. The significant challenge facing the pharmaceutical formulator is to identify a particular trigger that can be exploited to provoke drug release from the formulation in a time-dependent manner.

The formulation strategies to be discussed in this chapter involve the incorporation of a lag phase using a wide variety of different formulation strategies. The objective is that no drug will be released from the formulation during the lag phase and that drug is only released following

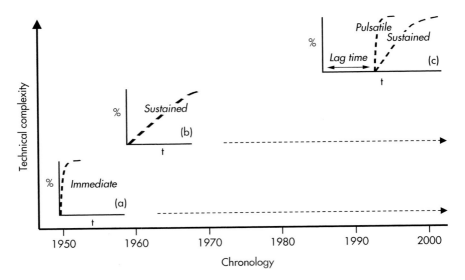

Figure 11.1 Evolution of drug dissolution profiles from oral drug delivery systems: (a) immediate release; (b) sustained release; (c) pulsatile or sustained release following a lag period.

elapse of this lag time. In this context, pulsatile release is described as taking the form of a bolus comprising all the drug content, released after the defined lag period.

Innovative time-dependent mechanisms have been described in the literature for tablet, pellet and capsule formulations utilising a wide range of physicochemical and physicomechanical strategies and many of these will be discussed in this chapter. The common feature of all such formulations is that they are activated by contact with fluids following ingestion by the patient and the drug is released at a predetermined time after administration. Only after the formulations come into contact with gastric fluids does the 'clock' start. Using such formulation strategies, it will be possible to design delivery systems capable of releasing drugs according to chronotherapeutic principles and targeting release to the circadian rhythm of disease states (Stevens, 1998). However, although the time of drug release can be controlled in such systems, one consequence of time-delayed formulation strategies is that drug will be released at a range of sites in the gastrointestinal (GI) tract. Where it is known that the drug is effectively absorbed from the entire length of the GI tract, this will not to be a constraint. However, where a drug is known to be variably absorbed as a function of GI tract site, such a drug may not be a good candidate for chronopharmaceutical drug delivery. The case for conducting scintigraphic/pharmacokinetic studies to identify any regional variation in absorption from different GI tract sites for a particular drug molecule is particularly appropriate before considering the development of chronopharmaceutical formulations.

11.2 Time-delayed tablet formulations

11.2.1 Osmotic control

Two technologically independent modifications of the Oros tablet technology have been reported by scientists from the ALZA Corporation that are capable of providing delayed release. Magruder *et al.* (1988) described a time-delayed formulation for the release of salbutamol in which the rate of release varied in proportion to the changing content of sodium chloride, included as a functional excipient in the tablet formulation (Figure 11.2).

In common with other Oros formulations, fluids diffuse into the tablet through the semipermeable film coat and interact with the osmotic core contents. Internal pressure develops, which then provides the driving force for pumping dissolved drug out through the laser-drilled hole. At saturation concentrations of sodium chloride, salbutamol solubility was depressed to 11 mg/mL; however, following depletion of sodium chloride, salbutamol solubility progressively increased to >300 mg/mL. However, since salbutamol has a limited solubility even at high sodium chloride concentrations, some drug will be released in a continuous manner, giving rise to a modulated release profile rather than true pulsatile release (Figure 11.3).

Development of the Oros technology was extended by Ayer *et al.* (1989), who incorporated a hydrophilic barrier layer under the

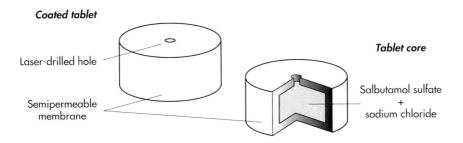

Figure 11.2 Osmotic delivery system for providing modulated drug delivery of salbutamol. Redrawn from Magruder *et al.* (1988).

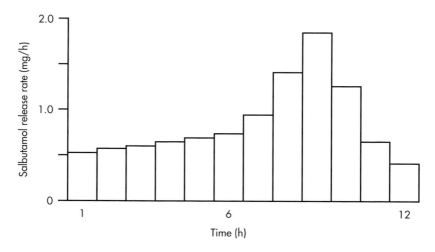

Figure 11.3 Modulated release of salbutamol as a function of diminishing sodium chloride content in Oros tablet. Redrawn from Magruder *et al.* (1988).

semipermeable tablet coating (Figure 11.4). Fluid penetration into the core was delayed by the hydrophilic barrier layer, which must be hydrated before fluid can diffuse across and activate the osmotic core.

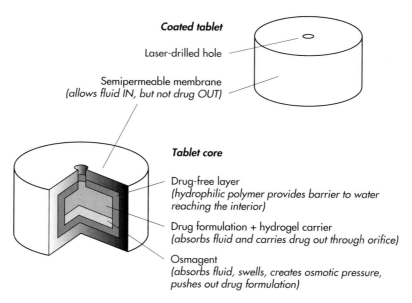

Figure 11.4 Oros tablet with barrier layer to provide delayed release of drug. Redrawn from Ayer *et al.* (1989).

Gupta *et al.* (1996) demonstrated that the calcium channel blocker verapamil was delivered in a sustained-release profile from such a preparation following a 4 h delay. An 'onset-controlled' formulation containing verapamil based on this delivery system has been commercialised (Covera HS, Searle, USA).

11.2.2 Gellable barrier layers using hydrophilic polymers

The principle underlying the use of gellable barrier layers around a core tablet is that polymer hydration is time-dependent. The hydration rate of the barrier layer will depend on both the nature of the polymer and the thickness of the coating layer; only after complete hydration of the barrier layer can fluid diffuse into the core and initiate drug release. Since in most cases the gel layer remains intact, drug release will occur by diffusion through this layer, resulting in a lag period followed by a sustained-release profile (Figure 11.5).

Gazzaniga and colleagues (Gazzaniga and Giordano, 1993; Gazzaniga *et al.*, 1994, 1995, 1996, 1997) and Sangalli *et al.* (1998,

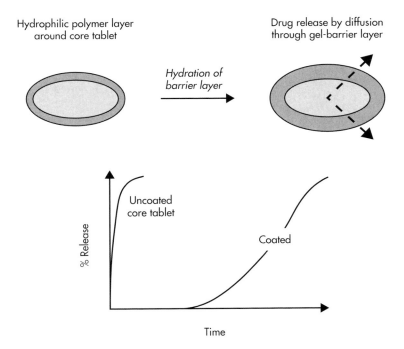

Figure 11.5 Time-delayed drug release from tablets coated with a swellable barrier layer.

1999) have extensively studied the use of spray coating technology to apply time-dependent hydrophilic barriers to tablet formulations. With HPMC (hydroxypropyl methylcellulose; hypromellose) barrier layers it was demonstrated that the thickness of the barrier provided control of lag time, following which a sustained-release profile was obtained over several hours as drug diffused through the swollen gel layer. The technology is being developed as Chronotopic (Poli Industria Chimica, Italy). The use of compression coating techniques to apply a barrier layer around core tablets was earlier studied by Conte *et al.* (1992, 1993). In this the barrier layer formed a gellable layer; drug release occurred by diffusion across this layer and was again slower than from the uncoated core. However, if the barrier layer eroded, the drug release profile exhibited the same release rate as the original uncoated core. Control of the lag time was achieved by variation of the composition or thickness of the barrier layer.

In an earlier report, Conte *et al.* (1989) described a biphasic release formulation of the NSAID ibuprofen (Figure 11.6). This consisted of an immediately available drug layer, which was separated from a second drug compartment by means of a barrier formed from hydrophilic polymeric material. The second layer of drug and the barrier layer were encased in an impermeable layer that prevented fluid access to the second drug layer until the removal of the first drug layer and the barrier material. A report in the patent literature described a preparation in which a removable top portion of the coated tablet detached to expose the first drug layer (Conte *et al.*, 1997). Maggi *et al.* (1999) employed

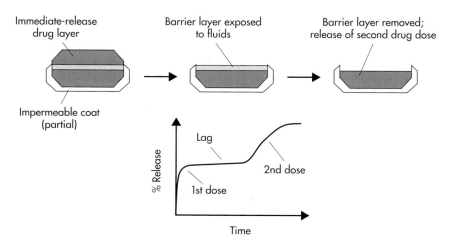

Figure 11.6 Biphasic release from a partially coated tablet.

similar partial coating techniques around tablets possessing internal barrier layers to achieve immediate release followed by time-delayed release of the NSAIDs diclofenac and ibuprofen.

Press coating technology was also utilised by Conte and Maggi (1996, 1997) to apply barrier layer coatings to one or both sides of an active core, although this afforded a continuous sustained-release rather than a time-delayed release profile (commercialised as Geomatrix, SkyePharma).

11.2.3 Erodible barrier layers

The principle underlying the use of erodible barrier layers around a core tablet is to use the disintegration/dissolution of the barrier components as a time-dependent process. The erosion rate of the barrier layer will depend on both the nature of the excipients employed and the thickness of the coating layer; only after erosion of the barrier layer can fluid diffuse into the core and initiate drug release. Since the barrier layer is removed during this process, drug release occurs following disintegration of the core, resulting in a lag period followed by a pulsatile release profile (Figure 11.7).

Pozzi *et al.* (1994) employed erodible barrier layers for the Time-Clock formulation (Zambon, Italy). In this technology, a core formulation was surrounded by a barrier layer comprising hydrophobic and surfactant components that eroded in a time-dependent manner. The components were applied as an aqueous dispersion containing a hydrosoluble polymer to aid adhesion to the core. Control of time delay was by variation of thickness of the coat. In scintigraphic and pharmacokinetic studies, the β-adrenoreceptor agonist salbutamol was released following 3 or 6 h delay periods. In a further gammascintigraphic study, an enteric-coated Time-Clock formulation demonstrated release in the colon that was independent of residence time in the stomach. It was reported that the rate of disintegration of the core was protracted in the large bowel (Wilding *et al.*, 1994) compared to when the core disintegration occurred in the upper GI tract (Pozzi *et al.*, 1994).

A core tablet enclosed in a hollow cylindrical polymer matrix was described by Vandelli *et al.* (1996) in which the upper and lower surfaces were then coated with an impermeable film. The gel layer formed by hydration of the polymer matrix dissolved and eroded in a time-dependent manner and afforded release of isosorbide mononitrate after 6 to 12 h delay periods.

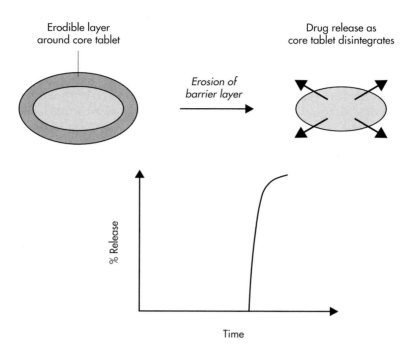

Figure 11.7 Time-delayed drug release from a tablet coated with an erodible barrier layer.

11.2.4 Tablets that rupture

The principle underlying bursting tablets is that a coating is applied around a core tablet that contains disintegrant. As water enters the core at a controlled rate dependent on the nature and thickness of the coating, internal pressure is generated by the swelling action of the disintegrant material in the core. When the internal pressure exceeds the strength of the coating, the coating ruptures and exposes the disintegrated core to fluids. A formulation that burst after a predetermined lag time was reported by Held *et al.* (1990) for the release of levodopa. The authors described a core tablet coated with a protective layer followed by a semipermeable membrane. The tablets were reported to take up water progressively by diffusion through the semipermeable membrane, resulting in increasing internal pressure. This caused the membrane to burst and the contents to be expelled (Figure 11.8). Studies in humans and dogs showed that maximum plasma concentrations of levodopa were obtained between 5 and 7 h after administration, demonstrating good correlation with *in vitro* performance.

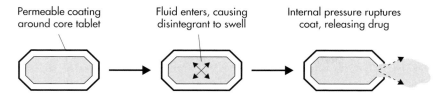

Figure 11.8 Time-delayed drug release from a bursting tablet.

Ishino *et al.* (1992a) described a poorly permeable wax compression coating applied around a core tablet containing disintegrant. A lag time was obtained while water slowly penetrated the waxy outer shell. On reaching the core, the water caused the disintegrant to swell and rupture the waxy coat. The lag time could be controlled by variation of the composition or thickness of the coat. Absorption of the calcium channel blocker diltiazem from a formulation containing hydrogenated castor oil and polyvinyl chloride as coating was studied in three beagle dogs (Ishino *et al.*, 1992b). A good correlation was obtained between *in vitro* and *in vivo* results in fed dogs, but large variability in lag times was reported in the fasted state.

11.2.5 Three-dimensional printing techniques

A novel approach to the development of programmable release has been described by Kastra *et al.* (1998, 1999), who utilised a three-dimensional printing technology (Therics Inc., USA). Liquid binder and drug were deposited onto excipient material using a computer-controlled printer to build up two-dimensional layers, one at a time. Excipient layers were built up using mechanical techniques until a three-dimensional structure was achieved. *In vitro* release profiles showing successive release of different drug layers were reported.

11.3 Time-delayed pellet formulations

11.3.1 Pellet formulation with a sigmoidal release profile

Stevens *et al.* (1992) used extrusion/spheronisation technology to produce a novel pellet formulation containing diltiazem that was coated with a mixed film coat comprising ethylcellulose and Eudragit RS polymers (Figure 11.9). While the ethylcellulose component acted as a diffusion barrier retarding release of diltiazem, the permeability of the Eudragit RS increased progressively. The overall effect was a sigmoidal release profile

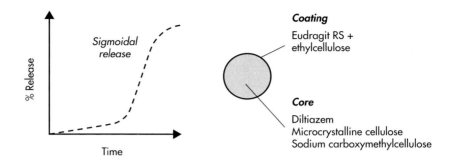

Figure 11.9 Coated pellet delivery system affording a sigmoidal dissolution profile.

with an increasing rate of release with time, providing an alternative to the immediate- or sustained-release profiles available at that time. The formulation was commercialised by Synthélabo in France as Tildiem LP.

Other workers (Narisawa *et al.*, 1994; Kao *et al.*, 1997; Beckert *et al.*, 1999) have subsequently employed similar formulation techniques to afford sigmoidal release profiles.

11.3.2 Pellets that rupture

The preparation of bursting pellets was first described by Ueda *et al.* (1989, 1994) with a formulation they aptly named the Time-Controlled Explosion System (TES). Four-layered pellets comprised drug coated around nonpareil seeds followed by a layer of a swelling agent, low-substituted hydroxypropylcellulose. The composition was completed with a final coat of a water-insoluble polymer, preferably ethylcellulose (Figure 11.10). Contact with fluid resulted in ingress of water into the core and contact with the swelling agent, leading to development of internal pressure. When the coat was brittle (e.g. ethylcellulose) it ruptured; however, if a flexible Eudragit RL film was used, the more elastic nature allowed the pellet to swell rather than burst. TES was shown to be capable of providing a programmable lag time controlled by the thickness of the outer membrane, with drug release being independent of pH of the dissolution medium.

An alternative pulsatile pellet formulation proposed by Chen (1993) was based on a drug-containing core that also contained a swelling agent that was coated with a more complex film than proposed by Ueda, comprising an insoluble permeable film incorporating a permeability-reducing agent and a water-soluble component (Figure 11.11).

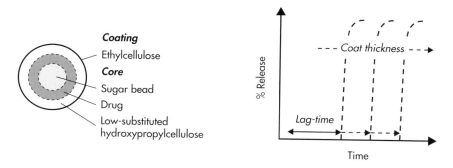

Figure 11.10 Pulsatile dissolution profiles from the TES pellet delivery system showing influence of coating thickness on burst time. Redrawn from Ueda *et al.* (1994).

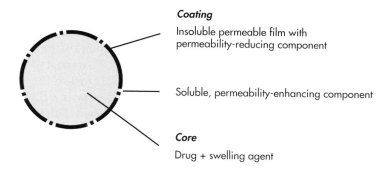

Figure 11.11 Pulsatile pellet system (Chen, 1993).

Similarly to TES, Chen's particles ruptured following ingress of water and development of internal pressure that eventually exceeded the strength of the coating layer. The lag time could be controlled by manipulation of the proportions of the three coating ingredients as well as the coating thickness.

11.4 Time-delayed capsule formulations

11.4.1 Plug separation from permeable capsules

Rashid (1990) first described a water-permeable capsule device (Figure 11.12) constructed from a water-swellable hydrogel polymer prepared from crosslinked polyethylene glycol (PEG). The internal cavity of the moulded capsule body was filled with a swelling agent (high-swelling

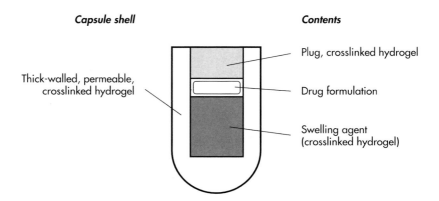

Figure 11.12 Pulsatile hydrogel capsule system developed by Rashid (1990).

hydrogel polymer) mixed with drug, and the contents were sealed into the capsule using a hydrogel plug. The hydrogel composition and wall thickness controlled the rate at which water diffused through the capsule wall. Following a delay period caused by water diffusion through the wall, the contents swelled and generated internal pressure that caused the plug to be expelled, affording pulsatile drug release. Time delay was controlled by optimisation of wall composition and thickness, strength of the osmotic contents and length of the plug sealing the drug formulation inside the capsule body.

This early formulation was evaluated in the human in a gammascintigraphy study in which capsules with a 5 h lag time were designed to target release of the ACE (angiotesin-converting enzyme) inhibitor captopril to the colon in fasted volunteers. Although pharmacokinetic analysis demonstrated minimal absorption from the colon, scintigraphic observations confirmed that pulsatile release had taken place at the desired site (Wilding *et al.*, 1992).

More recently, the concept of water diffusing through the capsule wall to cause generation of internal pressure and ejection of a plug has been re-examined by Crison *et al.* (1995), who described the PORT system. This capsule device comprises a water-permeable, coated gelatin capsule containing a swellable osmotic charge sealed into the capsule by an insoluble wax plug (Figure 11.13).

The capability of the PORT delivery system to provide *in vivo* release following a lag time has been confirmed in studies in dogs with paracetamol (acetaminophen) (Lipka *et al.*, 1996) and the β-adrenoreceptor antagonist metoprolol (Loebenberg *et al.*, 1999).

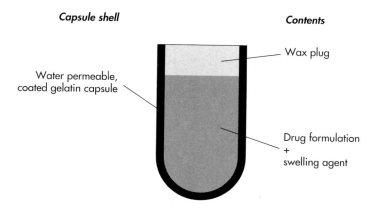

Figure 11.13 PORT capsule system developed by Crison *et al.* (1995).

11.4.2 Impermeable capsules with separable components

In the late 1980s, Polysystems Ltd, a small Scottish company, filed patent applications for the Pulsincap formulation (McNeill *et al.*, 1993). The company was subsequently acquired by RP Scherer Corporation and the technology was then developed by Scherer DDS Ltd. Pulsincap represented a simplification of the earlier capsule described by Rashid (1990) in that the capsule body of the Pulsincap device was a gelatin capsule, coated with ethylcellulose to render it impermeable, or alternatively a moulded plastic capsule (Figure 11.14). The contents of the capsule were sealed into the body by a hydrogel plug prepared from crosslinked PEG. In the presence of fluid, the hydrogel plug swelled at a controlled rate independent of the nature or pH of the medium (Binns *et al.*, 1993). In the presence of fluids, the plug developed a frustroconical shape as it swelled and slowly pulled itself out of the capsule. The lag time was controlled by the length of the plug and its insertion distance into the capsule. A longer plug ejected after a more prolonged lag time compared to a short plug and this provided a reliable mechanism for controlling the time delay.

The performance of Pulsincap has been investigated in numerous human studies (Bakhshaee *et al.*, 1992, Hebden *et al.*, 1999) and was shown to be well tolerated in humans (Binns *et al.*, 1996). The importance of an active expulsion system to ensure that the contents were rapidly and completely expelled from the capsule once the plug had ejected was demonstrated in human studies of salbutamol (Stevens *et al.*, 1995a, 1999) (Figure 11.15).

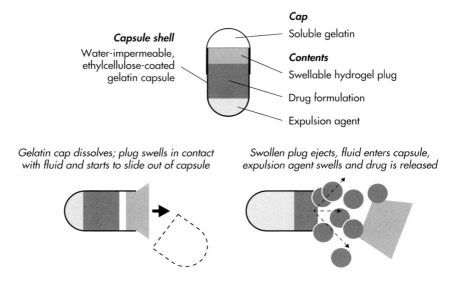

Figure 11.14 Pulsincap capsule system developed by Scherer DDS Ltd.

The tightness of fit of the plug in the capsule was found to have particular significance; if it fitted too tightly, drug was released erratically; if the plug was too slack, it ejected prematurely (Hegarty and Atkins, 1995). A diagnostic application of Pulsincap based on plastic capsules releasing nutrient components into a microbial test medium after a 6 h lag time was commercialised in 1997 (SprintSalmonella, Oxoid Ltd, Basingstoke, UK).

Scientists at the Alza Corporation (Wong et al., 1994) achieved a similar result for the Chronset technology using a sliding mechanism to remove the cap from an impermeable capsule (Figure 11.16). The driving force for separation of capsule components was an osmotically active compartment in the semipermeable cap. In contact with fluids, this osmotic fill swells and pushes against the rigid barrier component, causing the cap to slide off the capsule body after a predetermined time. Following cap removal, fluid enters the capsule and causes drug release.

In a study investigating the oral absorption of insulin in humans, insulin (25 U) released from a 6 h Chronset caused a fall in blood glucose when co-administered with permeation enhancers but not when enhancers were not included. The Alza Corporation also described the absorption of paracetamol in 18 human subjects following 2 h and 6 h lag times (Wong et al., 1998). Wong and colleagues (1995) have also

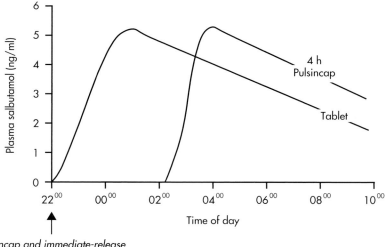

Figure 11.15 Pharmacokinetics of salbutamol when given at 22.00 as tablet and 4 h delayed Pulsincap.

described more complex sliding telescopic arrangements in the patent literature.

11.4.3 Impermeable capsules fitted with erodible plugs

In an attempt to overcome the technical constraints imposed by dimensional tolerances of sliding components, Stevens *et al.* (1995b) and Ross *et al.* (2000a) simplified the Pulsincap technology by replacing the complex swelling hydrogel plug with a simple erodible compressed tablet (Figure 11.17). This erodible tablet did not move relative to the capsule body and overcame the requirement for the precise dimensional tolerances between plug and capsule demanded by the Pulsincap formulation.

Using a variety of erodible tablet formulations and ethylcellulose-coated gelatin capsules, it was demonstrated that pulsatile delivery of β-adrenoreceptor antagonist propranolol was obtained following a controllable lag time that was determined by either plug composition or plug thickness. Ross *et al.* (2000a) utilised low-substituted hydroxypropylcellulose (LH21, Shin-Etsu) as a swellable expulsion system and demonstrated release of propranolol over a controllable 2–10 h range using erodible tablet plugs compressed from mixed lactose and HPMC

298 Chronopharmaceutical drug delivery

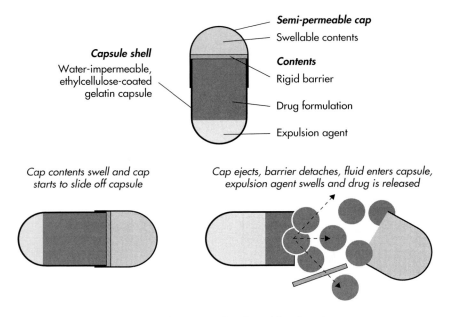

Figure 11.16 Chronset capsule system developed by the Alza Corporation.

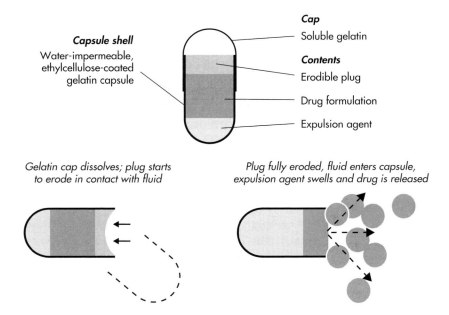

Figure 11.17 Erodible plug, time-delayed capsule system developed at Strathclyde University.

excipients (Figure 11.18). It was found that release time was dependent on erodible tablet weight (i.e. thickness of the erodible tablet) as well as on the composition (e.g. percentage of HPMC in erodible tablets also containing lactose) (Figure 11.19).

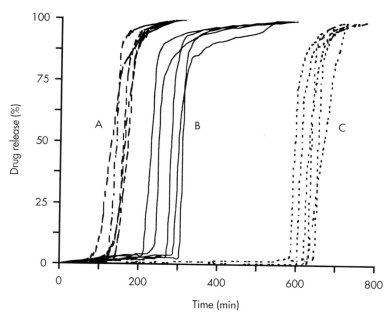

Figure 11.18 Time-delayed dissolution profiles of propranolol (50 mg) for fast (A) → slow (C) eroding tablets (Ross *et al.*, 2000a).

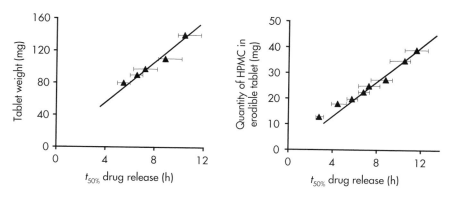

Figure 11.19 Influence of tablet weight (thickness) and composition (HPMC content) on time-delayed release of propranolol (50 mg) (Ross *et al.*, 2000a).

Drug delivery formulations based on erodible plug concepts were also studied by Krögel and Bodmeier (1998) in studies using plastic capsules and erodible formulations based on either compressed tablets or congealed semi-solid materials. Control of lag time of chlorphenamine (chlorpheniramine) release was obtained by varying either plug composition or its weight. These authors employed mildly effervescent fills to ensure that contents were expelled efficiently following erosion of the plug. In later studies, the same authors investigated plugs that eroded owing to the enzymatic action of pectinolytic enzyme in the plug formulation (Krögel and Bodmeier, 1999). The pectin:enzyme ratio controlled the time of release of chlorphenamine from the formulations.

The concept of erodible plugs was also exploited by Bar-Shalom and Kindt-Larsen (1993), who used injection moulding technology to prepare a multilayered dosage formulation in the form of a tube with erodible plugs sealing the ends. They employed selected grades of PEG together with surfactants and waxes to produce formulations such that the end plugs eroded faster than the tube walls. A drug contained in the tube was released after controllable lag times (Bar-Shalom *et al.*, 1991).

11.4.4 Pressure-controlled systems

Niwa *et al.* (1995) prepared a capsule perforated with micropores at one end. The capsule contained a high swelling excipient that swelled following ingress of fluids via the micropores. Internal pressure was generated, causing disruption of the capsule cap and provoking drug release. The same researchers later described a thin-walled ethylcellulose capsule. It had sufficient integrity to survive stomach and intestinal transit but on passage into the colon it ruptured because of the pressure of colonic peristalsis (Takaya *et al.*, 1996) (Figure 11.20).

11.4.5 Hydrophilic sandwich capsule

Ross *et al.* (2000b) devised a manually assembled time-delayed probe capsule based on a capsule within a capsule, with the intercapsular space filled with a layer of hydrophilic polymer (HPMC). The time delay was controlled by the molecular weight of the polymer and by the inclusion of lactose as a soluble filler in the hydrophilic layer (Figure 11.21).

Soutar *et al.* (2001) used a gastroresistant version of the hydrophilic sandwich capsule to deliver 500 mg paracetamol to a group of 13 volunteers. Using salivary determination of absorbed drug, a mean t_{max} value of 7.9 h (SD 0.96) was observed (Figure 11.22).

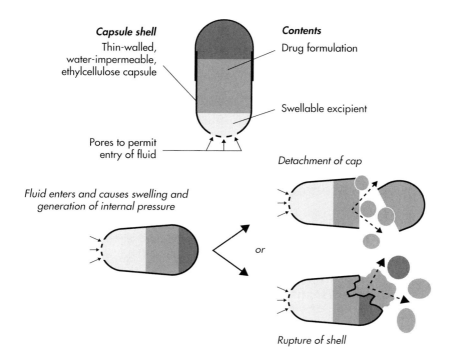

Figure 11.20 Time-delayed delivery from an insoluble capsule with pores and swellable contents (Niwa *et al.*, 1995).

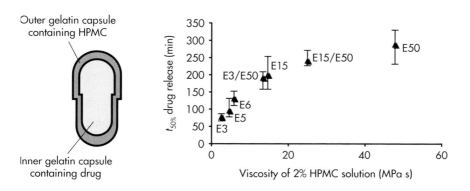

Figure 11.21 The hydrophilic sandwich capsule and the influence of HPMC viscosity on $t_{50\%}$ drug release (Stevens *et al.*, 2000).

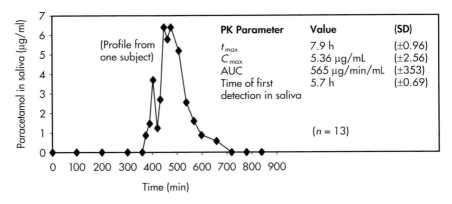

Figure 11.22 Absorption of paracetamol from the hydrophilic sandwich capsule (Soutar et al., 2001).

11.5 The potential use of pH-based systems using gastroresistant coatings

While the main focus of this chapter has been on true time-delayed drug delivery, it is appropriate to review briefly the use of enteric coatings as a means of inducing a time delay. Enteric coating technology applied to an oral dosage form can give rise to a time delay; however, the result is unpredictable in terms of the timing of drug release.

While insoluble at the acidic pH values typically found in the stomach, enteric coating polymers have been designed to disintegrate at the neutral or slightly alkaline pH encountered at the ileum (Rhodes and Porter, 1998). Thus, while an enteric-coated dosage form remains in the stomach, drug release will be inhibited; but once gastric emptying occurs and the dosage form encounters the neutral pH environment of the small intestine, dissolution of the coat can occur and then drug release. The trigger for release of drug is not elapsed time but the rise in pH on moving from acidic to neutral pH environment in the GI tract (Evans *et al.*, 1988).

In vivo sampling techniques (Lindahl *et al.*, 1997) have further demonstrated the wide extent of pH variability that exists even in the stomach, adding further doubt to the capability of gastroresistant coatings to perform predictably. Ashford *et al.* (1992, 1993) seriously questioned the appropriateness of using pH-dependent triggers as mechanisms for time-dependent delivery in a series of *in vitro* and *in vivo* studies using Eudragit S-coated tablets. *In vitro* dissolution was found to be inconsistent and depended strongly on coating thickness, dissolution pH and buffer composition, and considerable *in vivo* variability was demonstrated in a series of human gammascintigraphy studies. Whereas the enteric coat was capable of protecting the tablets from disintegration in the stomach and upper small intestine, disintegration following stomach emptying was extremely variable in terms of time of release (5–15 h) as well as intestinal site (ranging from ileum to splenic flexure in the colon). From the work of Ashford *et al.* (1993), it can be concluded that enteric coating is not a sufficiently reliable mechanism for affording reproducible time-delayed drug delivery.

In addition, earlier scintigraphic studies clearly demonstrated the variability of gastric emptying times. This is substantially determined by the fed/fasted state of the subject (Davis *et al.*, 1986, 1988) and will further contribute to the time-dependency of drug release from such enteric-coated formulations.

11.6 Conclusion

Interest in time-delayed oral drug delivery systems has developed within the past 10 to 15 years. Prior to that, enteric coats had been used to achieve lag times, but their use for this purpose has subsequently been discredited.

Considerable ingenuity has been demonstrated by pharmaceutical scientists over the past decade and a range of innovative techniques have been reported that appear capable of affording true time-delayed drug delivery. Some of these have now been commercialised. Over the next decade we shall undoubtedly see further instances in which the emergence of novel oral drug delivery systems can play a crucial role in making chronotherapeutics a reality. So, although the development of oral pulsatile drug delivery systems will more appropriately synchronise drug release with the time of clinical need, much remains to be achieved in terms of true chronotherapeutic drug delivery. This will only be realised when formulators can devise technologies capable of recognising 'biofeedback' from measured rhythms and then converting that signal into mechanism for provoking drug release from the formulation. The field of chronopharmaceutics is in its infancy. We look forward with great anticipation to the coming decade when chronopharmaceutical drug delivery moves through adolescence into adulthood.

References

Ashford M., Fell J T, Attwood D, Woodhead P J (1992). An in-vitro investigation into the suitability of pH-dependent polymers for colon targeting. *Int J Pharm* 94: 241–245.

Ashford M, Fell J T, Attwood D, et al. (1993). An in-vivo investigation into the suitability of pH dependent polymers for colonic targeting. *Int J Pharm* 95: 193–199.

Ayer A D, Theeuwes F, Wong P S (1989). Pulsed drug delivery. *US patent* 4 842 867.

Bakhshaee M, Binns J S, Stevens H N E, Miller C J (1992). Pulsatile drug delivery to the colon monitored by gamma scintigraphy. *Pharm Res* 9(Suppl): F230.

Bar-Shalom D, Kindt-Larsen T (1993). Controlled release composition. *US patent* 5 213 808.

Bar-Shalom D, Bukh N, Kindt-Larsen T (1991). Eagalet®, a novel controlled release system. *Ann NY Acad Sci* 618: 578–580.

Beckert T E, Pogarell K, Hack I, Petereit H-U (1999). Pulsed drug release with film coatings of Eudragit RS30D. *Proc Int Symp Controlled Release of Bioactive Materials* 26: 533–534.

Binns J S, Bakhshaee M, Miller C J, Stevens H N E (1993). Application of a pH independent PEG based hydrogel to afford pulsatile delivery. *Proc Int Symp Controlled Release of Bioactive Materials* 20: 226–227.

Binns J S, Stevens H N E, McEwen J, et al. (1996). The tolerability of multiple oral doses of Pulsincap capsules in healthy volunteers. *J Control Release* 38: 151–158.

Chen C-M (1993). Pulsatile particles drug delivery system. *US patent* 5 260 069.

Conte U, Maggi L (1996). Modulation of the dissolution profiles from Geomatrix® multi-layer matrix tablets containing drugs of different solubility. *Biomaterials* 17: 889–896.

Conte U, Maggi L (1997). Geomatrix® tablets for the pulsatile release of drugs. *Proc Int Symp Controlled Release of Bioactive Materials* 24: 291–292.

Conte U, Colombo P, La Manna A, et al. (1989). A new ibuprofen pulsed release oral dosage form. *Drug Dev Ind Pharm* 15: 2583–2596.

Conte U, Maggi L, Giunchedi A, La Manna A (1992). New oral system for timing-release of drugs. *Boll Chim Farm* 131: 199–204.

Conte U, Maggi L, Torre M L, et al. (1993). Press-coated tablets for time-programmed release of drugs. *Biomaterials*, 14: 1017–1023.

Conte U, La Manna A, Maggi L (1997). Pharmaceutical tablet capable of releasing the active ingredients contained therein at subsequent times. *US patent* 5 650 169.

Crison J R, Siersma P R, Taylor M D, Amidon G L (1995). Programmable oral release technology PORT System®: a novel dosage form for time and site specific oral drug delivery. *Proc Int Symp Controlled Release of Bioactive Materials* 22: 278–279.

Davis S S, Hardy J G, Fara J W (1986). The transit of pharmaceutical dosage forms through the small intestine. *Gut* 27: 886–892.

Davis S S, Norring-Christensen F, Khosla R, Feely L C (1988). Gastric emptying of large single unit dosage forms. *J Pharm Pharmacol* 40: 205–207.

Evans D F, Pye G, Bramley R, et al. (1988). Measurement of gastrointestinal pH in normal ambulatory human subjects. *Gut* 29: 1035–1041.

Gazzaniga A, Giordano F (1993). Delayed release systems (DRS) based on retarding swellable hydrophilic coating. *Proc 12th Pharmaceutical International Technology Conference* 1: 400–408.

Gazzaniga A, Iamartino P, Maffione G, Sangalli M E (1994). Oral delayed-release system for colonic specific delivery. *Int J Pharm* 108: 77–83.

Gazzaniga A, Busetti C, Moro L, *et al*. (1995). Time-dependent oral delivery systems for colon targeting. *STP Pharma Sci*, 1: 83–88.

Gazzaniga A, Busetti C, Sangalli M E, *et al*. (1996). *In vivo* evaluation of oral Chronotopic® system for time and site specific delivery in the GI tract. *Proc Int Symp Controlled Release of Bioactive Materials* 23: 571–572.

Gazzaniga A, Brambini L, Busetti C, *et al*. (1997). HPMC aqueous coating formulations for delayed drug delivery: an investigation on process parameters and *in vitro* drug release. *Proc Int Symp Controlled Release of Bioactive Materials* 24: 341–342.

Gupta S K, Atkinson L, Theeuwes F, *et al*. (1996). Pharmacokinetics of verapamil from an osmotic system with delayed onset. *Eur J Pharm Biopharm* 42: 74–81.

Hebden J M, Wilson C G, Spiller R C, *et al*. (1999). Regional differences in quinine absorption from the undisturbed human colon assessed using a timed release delivery system. *Pharm Res* 16: 1087–1092.

Hegarty M, Atkins G (1995). Controlled release device. *International patent application* WO 95/10263A1.

Held K, Marolf T, Erni W (1990). Principle and development of a tablet formulation with pulsatile drug delivery after a controlled and pH independent latency period. *Acta Pharm Technol* 36: 38s.

Ishino R, Yoshino H, Hirakawa Y, Noda K (1992a). Design and preparation of a pulsatile release tablet as a new oral drug delivery system. *Chem Pharm Bull* 40: 3036–3041.

Ishino R, Yoshino H, Hirakawa Y, Noda K (1992b). Absorption of diltiazem in beagle dog from pulsatile release tablet. *Chem Pharm Bull* 40: 3094–3096.

Kao C-C, Chen S-C, Sheu M-T (1997). Lag time method to delay drug release to various sites in the gastrointestinal tract. *J Control Release*, 44: 263–270.

Katstra W, Cima M J, Palazzolo B, Rowe B (1998). Oral dosage forms by three-dimensional printing. *Proc Int Symp Controlled Release of Bioactive Materials* 25: 760–761.

Katstra W E, Rowe C W, Teung P, Cima M J (1999). Pulsatory oral drug delivery devices fabricated by three-dimensional printing. *Proc Int Symp Controlled Release of Bioactive Materials* 26: 569–570.

Krögel I, Bodmeier R (1998). Pulsatile drug release from an insoluble capsule body controlled by an insoluble plug. *Pharm Res* 15: 474–481.

Krögel I, Bodmeier R (1999). Evaluation of an enzyme-containing capsular shaped pulsatile delivery system. *Pharm Res* 16: 1424–1429.

Lindahl A, Ungell A-L, Knutson L, Lennernäs H (1997). Characterisation of fluids from the stomach and proximal jejunum in men and women. *Pharm Res* 14: 497–502.

Lipka E, Kim J S, Siersma C A, *et al*. (1996). Drug delivery utilizing a novel programmable oral release technology: in-vivo–in-vitro correlation of release times. *Proc Int Symp Controlled Release of Bioactive Materials* 23: 575–576.

Loebenberg R, Kim J S, Crison J, Amidon G L (1999). Effect of food on pulsatile delivery of metoprolol. *AAPS PharmSci* 1: s-448 (3312).

Maggi L, Conte U, Bruni R (1999). Delivery device for the release of the active ingredient in subsequent times. *Proc Int Symp Controlled Release of Bioactive Materials* 26: 741–742.

Magruder P R, Barclay B, Wong P S L, Theeuwes F (1988). Composition comprising salbutamol. *US patent* 4 751 071.

McNeill M E, Rashid A, Stevens H N E (1993). Drug dispensing device. *GB patent* 2 230 442.

Narisawa S, Nagata M, Danyoshi C, *et al*. (1994). An organic acid-induced sigmoidal release system for oral controlled-release preparations. *Pharm Res* 11: 111–116.

Niwa K, Takaya T, Morimoto T, Takada K (1995). Preparation and evaluation of a time-controlled release capsule made of ethylcellulose for colon delivery of drugs. *J Drug Target* 3: 83–89.

Pozzi F, Furlani P, Gazzaniga A, *et al*. (1994). The Time Clock™ system: a new oral dosage form for fast and complete release of drug after a predetermined lag time. *J Control Release* 31: 99–108.

Rashid A (1990). Dispensing device. *European patent* EP 0 384 642.

Rhodes C T, Porter S C (1998). Coatings for controlled-release drug delivery systems. *Drug Dev Ind Pharm* 24: 1139–1154.

Ross A C, MacRae R J, Walther M, Stevens H N E (2000a). Chronopharmaceutical drug delivery from a pulsatile capsule device based on programmable erosion. *J Pharm Pharmacol* 52: 903–909.

Ross A C, Stevens H N E, Johnson J R, (2000b). A novel oral probe formulation: The hydrophilic sandwich (HS) capsule. *AAPS PharmSci* 2(4): s3136.

Sangalli M E, Busetti C, Maroni A, *et al*. (1998). Chronotopic® system for colon specific drug delivery: pharmacokinetic and scintigraphic evaluation. *Proc Int Symp Controlled Release of Bioactive Materials* 25: 858–859.

Sangalli M E, Maroni A, Busetti C, *et al*. (1999). *In vitro* and *in vivo* evaluation of oral systems for time and site specific delivery of drugs (Chronotopic® technology). *Boll Chim Farm* 138: 68–73.

Soutar S, O'Mahony B, Perkins A C, *et al*. (2001). Time-delayed release of paracetamol in man from the hydrophilic sandwich (HS) capsule. *Proc Int Symp Controlled Release of Bioactive Materials* 28: 790–791.

Stevens H N E (1998). Chronopharmaceutical drug delivery. *J Pharm Pharmacol* 50: s5.

Stevens H N E, Chariot M, Arnold F, Lewis G A (1992). Sustained release pharmaceutical compositions of diltiazem. *US patent* 5 112 621.

Stevens H N E, Binns J S, Guy M I (1995a). Drug expelled from oral delivery device by gas. *International patent application* WO 95/17173.

Stevens H N E, Rashid A, Bakhshaee M (1995b). Drug dispensing device. *US patent* 5 474 784.

Stevens H N E, Rashid A, Bakhshaee M, *et al*. (1999). Expulsion of material from a delivery device. *US patent* 5 897 874.

Stevens H N E, Ross A C, Johnson J R, (2000). The hydrophilic sandwich (HS) capsule: a convenient time-delayed oral probe device. *J Pharm Pharmacol* 52: s41.

Takaya T, Niwa K, Matsuda K, *et al*. (1996). Evaluation of pressure controlled colon delivery capsule made of ethylcellulose. *Proc Int Symp Controlled Release of Bioactive Materials* 23: 603–604.

Ueda S, Hata T, Asakura S, *et al*. (1994). Development of a novel drug release system, time-controlled explosion system (TES). I Concept and design. *J Drug Target* 2: 35–44.

Ueda Y, Hata T, Yamaguchi H, *et al*. (1989). Time-controlled explosion systems. *US patent* 4 871 549.

Vandelli M A, Leo E, Formi F, Bernabei M T (1996). *In vitro* evaluation of a potential colonic delivery system that releases drug after a controllable lag time. *Eur J Pharm Biopharm* 43: 148–151.

Wilding I R, Davis S S, Bakhshaee M, *et al*. (1992). Gastrointestinal transit and systemic absorption of captopril from a pulsed-release formulation. *Pharm Res* 9: 654–657.

Wilding I, Davis S S, Pozzi F, *et al*. (1994). Enteric coated timed release systems for colonic targeting. *Int J Pharm* 111: 99–102.

Wong P S L, Theeuwes F, Larsen S D (1994). Osmotic device for delayed delivery of agent. *US patent* 5 312 388.

Wong P S L, Theeuwes F, Larsen S D, Dong L C (1995). Osmotic device for delayed delivery of agent. *US patent* 5 443 459.

Wong P S L, Dong L C, Ferrari V J, *et al*. (1998). Osmotic device with high drug loading and delayed activation of drug delivery. *US patent* 5 817 335.

12

Circadian rhythm abnormalities

James M Waterhouse and David S Minors

12.1 Normal variations in the phase of circadian rhythms

As was described in Chapter 1, the 'body clock' (endogenous circadian oscillator) has a period of about 24.3 h when it is studied in the absence of time cues from the environment and the individual's sleep–activity cycle. In subjects living normally, however, it is adjusted by zeitgebers, which normally adjust the body clock to an average period of exactly 24 h, equal to that of the zeitgebers.

Evidence supporting this can be obtained from cross-sectional studies of the normal population, which indicate that there is a fairly tight coupling between the timing of the zeitgebers and that of various circadian rhythms (see, for example, Chapter 1, Figure 1.1). Evidence can also be obtained by longitudinal studies of individuals, particularly if the rhythm is not being masked by the daily sleep–activity cycle. Thus, estimates of the phase of the rectal temperature rhythm, measured in subjects otherwise living normally in two constant routines undertaken a week apart, indicated a mean phase difference that was not significantly different from zero, as would be expected in subjects entrained to the solar day (Lushington *et al.*, 1989).

However, other studies of the phase of the body clock have indicated that there is some day-by-day intra-individual variation as well as inter-individual variation. Evidence about the amount of inter-individual variation can be obtained from cross-sectional studies of individuals living their self-selected pattern of sleep and activity. Considering the healthy population as a whole, there is a Gaussian distribution of the times of peak of the core temperature rhythm, with a mean at about 18.00, and about two-thirds of the population fall within an hour of this mean. The 5% 'tails' of the population are about 2 h earlier or later than this; these subgroups are called 'larks' (morning types) or 'owls' (evening types), respectively (Kerkhof, 1985).

In assessing the intra-individual variation in timing of circadian rhythms, it is necessary to measure their phase on several successive days while the subject is living normally. The demands of 'constant routines' (see Chapter 1) preclude the use of this method. The only method used so far has been to measure the daily timing of the circadian rhythm of rectal temperature after 'purification' of the raw data for the direct effects of the sleep–activity cycle. Results indicated that the phase of the temperature rhythm ranged over a band about 3–4 h wide (Waterhouse et al., 2000), but this might have been an overestimate of the true variability since the process of purification would have resulted in an accentuation of the effect of residual 'noise' in the data.

Studies of the sleep–activity cycles of students (Elliott et al., 1970) have indicated that there is a considerable daily variation, with times of going to bed varying over a range of 4 h being quite common (Figure 12.1). There is also anecdotal evidence that some retire and rise progressively later until they miss a night's sleep altogether. Presumably, this is a reflection of the fact that, in the absence of meetings and other commitments, and with the opportunity to plan one's own habits, there is an opportunity to ignore conventional zeitgebers, as a result of which the sleep–activity cycle can 'free-run' with a period slightly greater than 24 h.

'Larks' and 'owls' have been defined above; their circadian rhythms and sleep–wake cycles are phased considerably earlier or later, respectively, than the norm. These differences could be due to differences in the phasing of the body clock and/or to differences in chosen lifestyle. Differences in lifestyle will play a role, but the observation that the differences in phase remain in subjects who keep the same sleep–wake cycle points to at least part of the difference being endogenous. The role of this endogenous component can also be inferred from the observation that differences remain after purification of temperature data obtained from successive normal sleep–wake cycles, and when assessments are made of the phase of rectal temperature during constant routines (Kerkhof and Van Dongen, 1996).

There are also considerable variations in the timing and amplitude of circadian rhythms with age, particularly in infants and the elderly. In infants, circadian rhythms are poorly developed at birth, though a low-amplitude circadian rhythm of core temperature has been found in several studies (see, for example, Weinert et al., 1997). The rhythms increase in amplitude and synchrony with each other and the environment over the first months of life (Mirmiran and Kok, 1991). This is believed to reflect an increasingly robust circadian output from the

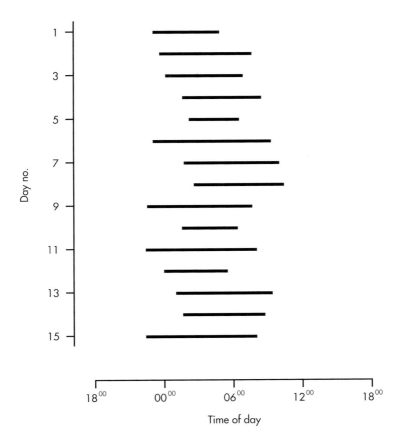

Figure 12.1 The sleep–wake cycle of a student. Bars indicate times of sleep. Reproduced from Minors and Waterhouse (1981).

suprachiasmatic nuclei (SCN), which, in turn, is attributed to increasing synaptic connections between the individual cells of these nuclei (Honma et al., 2000). However, the infant's awareness of the environment, sleep profiles, activity when awake and feeding patterns are all developing at this time. These factors also will contribute to the development of circadian rhythms, some of them acting as zeitgebers and promoting entrainment of the body clock, and all of them increasing the exogenous component of a circadian rhythm.

In old age, too, circadian rhythms are less marked, the system as a whole showing changes that mirror, to some extent, those observed in infancy (Van Cauter et al., 1998). There is a decrease in circadian output from the body clock in at least some aged subjects, as judged by a fall in amplitude of core temperature and an increased phase variability in circadian rhythms in the population as a whole. These changes have also been observed when using a 'constant routine' protocol (Monk et al., 1995). Degeneration of the SCN themselves might contribute to this (Zhou and Swaab, 1999; Hofman, 2000). There is also a decreasing ability to maintain consolidated sleep, which is coupled with an increasing need to micturate at night as the renal circadian rhythms deteriorate and urine production is not decreased as much at this time. Napping in the daytime might reduce even further the dichotomy between daytime activity and nocturnal sleep, as do declining daytime physical activities (due to retirement from employment or disability) and even declining social and mental activities. The result of all these factors is to lessen the strength of the normal zeitgebers, though this might be alleviated somewhat by an increased rigidity of habits (Minors et al., 1998). Clinical aspects of rhythms in ageing, particularly with regard to the role of melatonin, are considered in Chapter 14.

12.2 Abnormal variations in the phase of circadian rhythms

The above variations in phase and differences in mean phase between individuals often do not cause any problems to the individual, reflecting instead that individual's choice of lifestyle. Even in those cases (infants and the aged) where difficulties can arise, they seem to be no more than reflections of a more general development and deterioration of the body at different stages of life, and could hardly be classed as 'abnormal'.

Other examples exist among the young and the aged where differences from the norm are even more marked and the term 'abnormal' seems more appropriate. Some children, and even some adults, suffer from nocturnal enuresis. This is often due to an inadequate development of the circadian rhythm of concentration of antidiuretic hormone in the blood, levels of which normally rise at night so reducing urine production at this time (Rittig et al., 1989). In general, urine production at night does not fall as much in the elderly as in the rest of the adult population, and so, as mentioned above, waking up to micturate can become another reason for broken sleep.

Just as waking at night is one consequence of an increased rate of urine production, so too, in other cases where there are altered circadian rhythms, it is often a disturbed sleep pattern that is noticed by the patient. It is now considered that one subgroup of the insomnias is due to disorders of the circadian system (Dagan and Eisenstein, 1999). The most common disorder is delayed sleep phase syndrome (DSPS); associated with it, though this is less common, is advanced sleep phase syndrome (ASPS). In both disorders, the phases of all circadian rhythms seem to be affected equally (Ozaki *et al.*, 1996; Shibui *et al.*, 1999). However, because of the link between sleep propensity and the phases of the circadian rhythms of temperature and melatonin, it is the problem with sleep that causes inconvenience. Individuals wish to sleep at hours that are too late (about 04.00–12.00 in the case of DSPS sufferers) or too early (20.00–04.00 in the case of ASPS) to be compatible with a conventional lifestyle and hours of work. When allowed to live as they wish, such individuals appear to have no problems with sleeping normally (provided that their bedrooms are isolated from external noise and lighting), and show a normal phase relationship between their rhythms. The abnormality seems to derive from an abnormal phase relationship between the body clock and environmental zeitgebers, though the reason for this is unknown.

For some individuals, there is evidence that the body clock is unable to adjust to a 24 h day (Nakagawa *et al.*, 1992). This maladjustment is found occasionally in sighted people (McArthur *et al.*, 1996), but is quite common in the blind, a result that supports the view that the light–dark cycle is normally an important zeitgeber. In this disorder, all circadian rhythms continue, including those that affect the abilities to get to sleep and to stay asleep, but they show a period greater than 24 h (Figure 12.2). As the body rhythms drift out of phase with the requirements of a normal lifestyle, the problems of sleep loss at night and fatigue during the daytime worsen. They are at their most marked when the body rhythms and lifestyles differ by about 12 h (Lockley *et al.*, 1999), when body temperature peaks during the night and melatonin secretion starts at about 09.00 (if the individual is not in bright light). At this time, subjects are most likely to nap in the daytime. With a continuation of this drift, the body clock and lifestyle come back into synchrony, bringing better sleep at night and more alertness in the daytime – though only temporarily, as the drifting continues.

Altered times of sleep and activity can be marked in patients suffering from Alzheimer's disease and multiple infarct dementia, much of the normal dichotomy between daytime activity and nocturnal sleep being lost (Mishima *et al.*, 2000). The extent to which this disorder can be linked to a specific abnormality of the circadian timing system and, if so, whether it is the zeitgeber input, the clock itself or the output from the clock have not been firmly established.

314 Circadian rhythm abnormalities

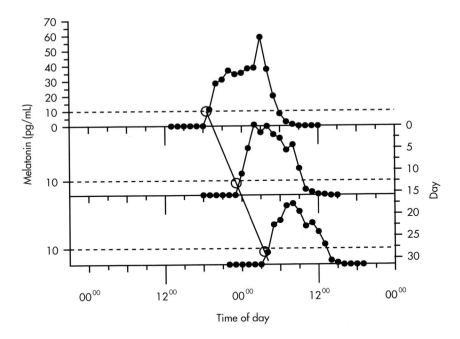

Figure 12.2 Free-running rhythm of plasma melatonin in a blind subject who was studied on three occasions. Reproduced from Sack et al. (2000).

12.3 Occupational health and long-haul travellers; jet lag

After a rapid flight across several time zones, a general malaise, commonly known as 'jet lag', appears transiently (Waterhouse et al., 1997). The exact symptoms, their severity and their duration depend upon the individual, the direction of flight (they are absent after flights to the north or south, present after ones to the west, and most marked after flights to the east), and the number of time zones crossed (generally increasing with the number of time zones crossed in excess of 2–3). The symptoms include some or all of the following: fatigue during the new daytime and yet inability to sleep satisfactorily at night time; loss of appetite, and indigestion and feeling bloated after food intake; decreased ability to concentrate and loss of motivation; increases in irritability and the frequency of headaches.

The symptoms are due to the sudden change in timing of the individual's lifestyle, caused by travelling to a new time zone, not to any abnormality of the circadian system. One of the properties of the body clock (see Chapter 1) is that it is robust and little affected by transient changes in habits – a daytime nap or a brief nocturnal awakening, for example. This stability becomes a disadvantage when, for some days after arrival in the new time zone, the body clock has not adjusted its phase in accord with the new local time.

Before adjustment has taken place, there is a loss of the normal synchrony between the individual's body clock and the new environment (Figure 12.3). For example, after a westward flight across eight time zones, the individual feels tired at 16.00 local time (equivalent to 24.00 in the time zone just left, the time zone to which the body clock is still adjusted) and begins to wake up at midnight (as this corresponds to 08.00 by 'body time'). In the same way, after an eastward flight across eight time zones, the individual does not feel tired at midnight by local time, but is ready to sleep as the new day dawns at 08.00! (16.00 and 24.00, respectively, by body time). The symptoms alleviate as the body clock adjusts to the new time zone. Consequently, there is an obvious premium on promoting such adjustment.

Any long-haul flight is associated with 'travel fatigue', possibly compounded by loss of baggage, a general feeling of being hassled, and 'culture shock', but these difficulties can be separated from jet lag (Reilly *et al.*, 1997). Travel fatigue, the general hassles and problems associated with travel, applies whatever the direction of the flight. On the day of arrival, the subject feels tired and travel-worn owing to sleep loss, a change in feeding habits, and dehydration from the dry air in the plane. These problems wear off rapidly after the traveller takes a shower, replaces fluid loss, settles down and has a good sleep. However, while a good sleep at night is possible if no time-zone transition has been involved (since body time and local time are coincident), it is difficult after long-distance flights to the west or east because body and local times no longer coincide. Culture shock also does not depend upon the direction or distance of flight, being minimal after travel from the UK to as far away as Australia and New Zealand, for example.

12.3.1 Differences between individuals

Some individuals suffer more from jet lag than do others, but there has been little success so far in predicting those who will suffer most. It has been argued that it will be less in the young, the fit and those whose sleeping habits are flexible, but the supporting evidence is weak. It has also been argued that 'larks' will have less trouble with eastward flights than do 'owls' (because an advance of the body clock is required), and

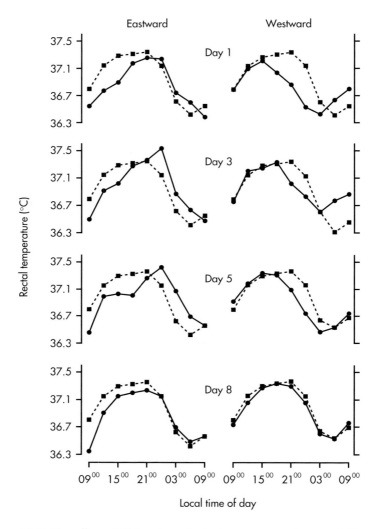

Figure 12.3 The effect of flights through six time zones to the east (left) and west (right) upon rectal temperature on (top downwards) days 1, 3, 5 and 8 after the flights. Full line: post-flight data; dashed lines, pre-flight data expressed on the new local time. Reproduced from Minors and Waterhouse (1981); data of Klein *et al.* (1972).

the opposite will be the case with westward flights (where a phase delay is required); again, the evidence is weak. A recent study (Waterhouse *et al.*, 2002) has summarised the relevant evidence but indicated also that the amount of jet lag experienced might depend more upon some aspects of the individuals' travel arrangements than upon physiological and psychological differences between them. In this study, jet lag was assessed for the first 6 days after a flight from the UK to Australia (an eastward time-zone transition of 10 h, the flights taking about 24 h). All subjects experienced jet lag and a disruption of circadian rhythms. However, jet lag over the course of the 6 days of measurement was felt less by those subjects whose travel arrangements were such that they had left the UK in the morning (UK time) and arrived in Australia in the late afternoon (local time in Australia). This meant that they could attempt a full sleep (during the new local night-time) about 32 h after waking from their last nocturnal sleep in the UK. By contrast, more jet lag was felt by those who had left the UK in the evening (UK time) and arrived in Australia in the morning (local time in Australia). Even though this latter group had slept more than the other group during the flights (since the flights took place mainly at night), and they were allowed a 2 h nap on arrival in Australia, by the time they attempted a nocturnal sleep they had been out of bed for more than 50 h.

For aircrew, there is the possibility of long-term effects of continual time-zone transitions. There is evidence for a greater sleep loss in older crew members, and for secondary amenorrhoea and alterations of melatonin and reproductive hormone profiles in female cabin staff (Harma *et al.*, 1994). Many aircrew return home after only a short stay in a different time zone (as a result of which there is insufficient time for adjustment of the body clock to take place), and some retain home time during the course of their tour of duty (Lowden and Akerstedt, 1998). In addition, long-haul flights often take place during the 'night' (whether this is defined by the home or destination time), and opportunities for naps are shared during the course of the flight; times of sleep in general appear to be determined by duty hours and opportunity rather than biological pressure. The result of these complexities is that it is difficult to ascertain to what extent any difficulties in aircrew can be assigned to the time-zone transitions per se; instead, the problem could be viewed as one due to shift work.

12.4 Occupational health and night work; shift-workers' malaise

We live in a society in which it is expected that emergency services, the supplies of electricity gas and water, and the service industries will all be available round the clock. The hours during which shops are open and banking facilities are available have also increased, and this, coupled with the time of preparation that is needed to make this possible, extends the hours when the workforce has to work. Many industrial processes must be run continuously, and the news-gathering media need to work round the clock. About 20% of the working population is involved in shift work.

Do these demands of a '24 h society' negatively affect the workforce involved, particularly as working at night and sleeping in the daytime will conflict with circadian rhythms (unless adjustment takes place)? From a biological viewpoint, concern has been expressed with regard to the frequency of accidents in the workplace and to the short-term and long-term effects of work patterns upon the health and general well-being of the individuals concerned.

12.4.1 Accidents and short-term effects

Accidents and errors in the workplace are comparatively rare (but see Dinges, 1995), so it is more common to assess mental and physical performance and subjective mood (Folkard, 1990). These measurements indicate that human error occurs in the workplace, this error arising from several interacting causes. These include environmental conditions (lighting, temperature, noise, weather), time of day (circadian factors), time awake (particularly time on duty), work load (and other aspects of 'stress'), sleep loss (particularly due to less daytime sleep), and sleeping in the workplace. Three of these – circadian factors, sleep loss and time awake – apply directly to the night worker.

For those working at night (Akerstedt, 1995), the night shift coincides with the time of worst performance (coincident with the temperature and adrenaline (epinephrine) secretion minima), daytime sleep is shorter and is broken (owing to environmentally mediated disturbances and attempting sleep when temperature and adrenaline secretion rhythms are rising), and subjects tend to work the night shift after spending leisure time with the family and colleagues (on day work, work is performed soon after sleep when the subject is less fatigued). Even if the circadian rhythms were to adjust to night work (which is unlikely, see below), disturbances due to the environment and working at the end of the waking span would be unlikely to prevent completely the decline in performance.

There is evidence that unintentional sleeping at work, often for only a few moments, occurs (Wedderburn, 1987). 'Night-shift paralysis' can also occur, in which the individual, though apparently awake, becomes immobile and unable to respond to the environment (Folkard and Condon, 1987). Those who work at night generally complain of a lack of sleep, a fall of motivation, and an increase in fatigue at work.

Much of their time away from the workplace, including rest days, seems to be associated with catching up on lost sleep and domestic duties rather than relaxation and being able to pursue hobbies and interests. Fatigue is associated with increased irritability with workmates, with friends in general and, particularly, with the partner and family.

Models have been developed to predict alertness when the waking time is extended, as is frequently the case, for example, with a first night shift (Akerstedt and Folkard, 1995). The combination of a long time awake and the circadian trough in the night produces a change in alertness during the night that falls below a 'risk level', i.e. below a level where alertness is adequate for responding effectively to the environment. Such models also predict that alertness benefits from short sleeps or naps, and from an adjustment in the phase of the circadian component of the performance rhythm.

Loss of appetite is common, and the food that is eaten seems to cause indigestion and a feeling of being 'bloated'. Night workers tend to change their eating habits in two main ways compared with when they work on the day shifts. First, they tend to have snacks throughout the night shift, rather than a single substantial 'lunch' at about 01.00. Second, and linked to this, the kind of food eaten during the night shifts, and even during the daytime if the individual does not bother to cook anything, tends to be chocolate bars, crisps and convenience foods (Lennernas *et al.*, 1995). Such a change of eating habits might reflect the unavailability of suitable food in the middle of the night, but habit, the loss of motivation when tired, and the inappropriate phasing of the body clock are all also implicated.

12.4.2 Long-term effects

Apart from the possible long-term consequences of difficulties with interpersonal relationships with friends, partner and family, there are also, with continued night work, significant increases in the incidence of gastrointestinal disturbance, particularly ulcers, and cardiovascular morbidity (Boggild and Knutsson, 1999).

Recent work offers a possible link between night work and cardiovascular morbidity, based on food intake patterns (Lennernas *et al.*, 1994). Normal eating habits result in glucose being the main metabolite during the daytime, with the excess being stored as fat in adipose tissue. At night, the different pattern of hormone secretion and decrease in tissue sensitivity to insulin result in the mobilisation and metabolism of fat. However, it has been shown that the level of plasma LDL ('bad') cholesterol – a predictor of cardiovascular morbidity – is directly related to the proportion of the daily glucose intake that occurs at night. Normally, this proportion is low (since we are fasting then), but it will be considerably higher in a night worker who snacks his or her way through the night shift.

12.4.3 Differences between individuals

Individuals differ in their susceptibilities to the factors discussed above. Many night workers can shrug off feelings of fatigue and not all the workforce suffers from ulcers or cardiovascular disorders. In part, this might indicate that individuals differ intrinsically in their ability to adjust to night work (Harma, 1995). A summary of these differences follows.

12.4.3.1 General sleeping habits

The advantage of being able to nap has already been mentioned. More generally, it is disadvantageous if an individual has rigid sleeping habits and so cannot take the opportunity to sleep whenever it is presented. Also, an inability to throw off feelings of fatigue during a work period is an obvious disadvantage. Although there are few reports describing how workers combat such feelings of fatigue, methods used in one study included taking a brief break or nap, opening windows, scratching or slapping oneself, and eating sweets or slices of lemon (Wedderburn, 1987). A brief bout of exercise can also be successful but, as with all these methods, it appears that the effects are rather short-lived.

12.4.3.2 'Larks' versus 'owls'

'Larks' (morning types) are better able to wake up and perform mental tasks in the early morning, and to go to sleep early in the evening. This is probably associated with the fact that they show an earlier phasing of the circadian rhythm of core temperature (Kerkhof and Van Dongen, 1996). While all of this supports the intuition that 'owls' (evening types) will find night work less disturbing than will larks, some recent work indicates that there is a greater incidence of self-reported psychological disturbances in owls, regardless of their work pattern. It has also been reported that owls suffer more from health disorders (mood disturbance and indigestion, for example) when on night shifts. It must be remembered, however, that most members of the population are 'intermediate' in type, rather than larks or owls.

12.4.3.3 Age, coping mechanisms and commitment.

Ageing is associated with a decreasing tolerance to shift work; decreasing flexibility in sleep times and physical fitness are probably contributory factors. As in the case of aircrew (see above), workers tend to opt out of night work as they grow older. Against this must be set the value of increased experience, particularly if coping strategies are considered (Van Reeth, 1998). These strategies, which have also been called 'commitment', would include the following:

- An acceptance of a nonconventional social and family life
- The adoption of methods for gaining enough sleep
- The use of methods for overcoming tiredness

A worker who initially possessed the attributes outlined above and/or developed the necessary coping mechanisms would be predicted to be the most tolerant of night work and to be the safest and the most productive; however, such attributes and coping mechanisms might also exact a cost, in the form of a restricted lifestyle. Conversely, those who cannot (or do not wish to) develop such coping mechanisms are likely to leave night work at the first opportunity. Those who continue in night work tend to be 'survivors', chosen by a self-selection process.

12.4.4 Adjustment of the body clock

Many of the problems confronting night workers, as with those who travel across time zones, stem from a lifestyle that is at variance with the timing of the body clock. Adjustment of circadian rhythms to night work and daytime sleep would appear to be advantageous, therefore. However, it has been shown on several occasions that this process is much slower than occurs after time-zone transitions (Figure 12.4) because, unlike the case of the traveller across time zones, the night worker faces a conflict between the timing of zeitgebers.

For the traveller, all zeitgebers change with arrival in the new time zone, unless a conscious effort is made to retain home time (as is the case for some aircrew, see above), a strategy that normally requires the individual to stay indoors and live independently of the population as a whole. By contrast, trying to retain home time with regard to meals and the sleep–wake cycle, and yet being exposed to the natural light–dark cycle and participating in some activities in the new time zone, would not be expected to be successful, owing to the conflict in timing between zeitgebers that this would produce. It is exactly such a conflict that the night

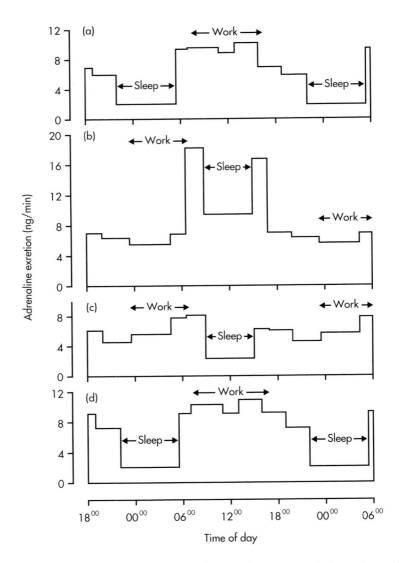

Figure 12.4 Daily excretion of adrenaline in the urine in shift workers during (a) first week of day work; (b) first week of night work; (c) third week of night work; (d) first week of returning to day work. Reproduced from Minors and Waterhouse (1981); data of Akerstedt and Froberg (1975).

worker is normally exposed to; even if meal times and the sleep–wake cycle were adjusted to fit in with the demands of night work, special care would be required to avoid exposure to a day-orientated light–dark cycle. Moreover, on rest days, when there is social and personal pressure to revert to a 'normal' diurnal lifestyle, all zeitgebers will tend to cause any adjustment to night work to be lost rapidly. Those who can avoid this dilemma, by virtue of living an isolated existence or in an environment where there is equal provision for work, meals, leisure or sleep at all times – submariners during a tour of duty, for example – are likely to be able to adjust fully to night work (or, indeed to any other shift). However, they must be considered to be living in exceptional circumstances, and such a lifestyle, if prolonged, might well result in the individual having an impoverished social life.

Because the process of adjustment of the body clock to night work is so difficult to achieve, it is common for night work to be performed for only a small number of consecutive days. As a result, hardly any adjustment to night work can take place (with all the undesirable effects implied by this, see Akerstedt, 1995), but the negative consequences of an accumulating sleep loss can be minimised. When the number of consecutive nights that is worked is larger, it becomes important to attempt to promote adjustment of the body clock. Since night work might be performed for years, and might involve a substantial proportion of an individual's time spent at work, any method used, as with reducing the effects of jet lag in long-haul aircrew, needs to be safe in the long term.

12.5 Treatment of circadian rhythm disorders

A summary of the main differences between normal subjects and patients, with regard to some of the problems affecting the circadian timing system and the main methods for dealing with them, is given in Table 12.1.

Jet lag (Section 12.3) and shift-workers' malaise (Section 12.4) reflect the fact that a normal circadian system has not adjusted to a change in the timing of an individual's sleep–wake cycle. Consequently, treatment of these problems can consist of using putative zeitgebers (see Chapter 1) to promote adjustment of the body clock to the new schedule, or to prevent adjustment in the case of rapidly rotating shift systems where adjustment is not required (Redfern, 1989; Dawson and Armstrong, 1996). By contrast, with the problems that appear to arise as a result of some error in the circadian system (Section 12.2), the position is more complex and will require further comment below (Section 12.5.3).

12.5.1 Treatment of jet lag

Even if the advice given below is followed, adjustment is not immediate but takes up to about 10 days after an eastward transition of 10 time

Table 12.1 A comparison of some aspects of normal and abnormal circadian rhythms

Subjects or patients	Entrained?	Phasing of rhythms	Can advance?	Can delay?
1. Normal subjects				
Living normally	Yes	Stable and normal	Yes	Yes
Without commitments	?	Irregular	Yes	Yes
After time-zone transitions	Not initially[a]	'Jet lag'	Eastward shifts	Westward shifts
During night work	Incompletely[a]	'Shift-workers' malaise'	Rare	Yes
'Larks' or 'owls'	Yes	Rather early or late	Yes	Yes
2. Delayed sleep phase syndrome	Yes	Inconveniently late	Difficult	Unusually easy
3. Advanced sleep phase syndrome	Yes	Inconveniently early	Unusually easy	Difficult
4. 'Free-running' rhythms				
Sighted patient	No[a]	Variable	N/A	N/A
Blind patient	No[b]	Variable	N/A	N/A

[a]Can be adjusted by behaviour cycle, bright light and melatonin.
[b]Can be adjusted by behaviour cycle and melatonin.
N/A, not applicable.

zones, slightly less after a similar transition to the west (Reilly *et al.*, 1997). Accordingly, for stays in the new time zone of only a few days before returning to the 'home' time zone, adjustment is not possible and the traveller is advised to time appointments in the new time zone to coincide with daytime on home time. The following methods have been used to promote adjustment of the body clock to time-zone transitions.

12.5.1.1 Meal times

It has been proposed that the appropriate timing and composition of meals acts as a zeitgeber in humans. The rationale of the method is that a high-protein breakfast will raise the concentration of tyrosine in the plasma and so promote its uptake into the brain, where it acts as a precursor for the formation of the neurotransmitters noradrenaline (norepinephrine) and dopamine. These transmitters are released during the active phase of the sleep–wake cycle. By contrast, a high-carbohydrate evening meal will raise the concentration of tryptophan in the plasma and promote its uptake into the brain, where it acts as a precursor for the neurotransmitters serotonin and melatonin (Leathwood, 1989). At one time, kits were produced that required the traveller to take one pill (containing, *inter alia*, tyrosine) in the morning in the new time zone, and another (containing, *inter alia*, tryptophan) in the evening. Tests of this 'feeding hypothesis' have suffered from methodological problems and any effects have been found to be only marginal, probably because, although *de novo* synthesis of neurotransmitter follows precursor loading, little 'functional' transmitter reaches appropriate postsynaptic receptors.

12.5.1.2 Physical activity

This method is based upon experiments on hamsters and have shown that a new running wheel, certain drugs and the presence of a potential mate all promote an extensive bout of physical exercise (Mrosovsky and Salmon, 1987). This can promote adjustment of the body clock, possibly through the excitement engendered, the direction of which depends upon the time at which the exercise takes place. The mechanisms responsible for this effect are discussed in more detail in Chapter 13. Adapting this protocol to humans – that is, requiring them to undergo a bout of exercise – has not produced convincing results; unlike in hamsters, only small phase shifts have been produced in humans, even when substantial amounts of exercise (70% of maximum oxygen uptake for 30 min) have been involved (Edwards *et al.*, 2002). The cause of this apparent species difference is not known, but the amount of exercise might be relevant. Thus, a new wheel can induce the hamster to run more than 5 km in a period of 3 h; it is unclear what this translates into when the amount of exercise required in the case of a human is considered, but it might well be more than many travellers are prepared (or even able) to take.

12.5.1.3 Bright light

The effects of light of different intensity, but particularly bright light, in shifting the body clock in humans have been well documented in laboratory-based experiments in which a phase–response curve (PRC) to light has been established (see Chapter 1). Even though there are far fewer data from field studies, this PRC can be used for promoting adjustment of the body clock in travellers who have crossed time zones.

In principle, after an eastward flight of up to nine time zones, the aim is to promote an advance of the body clock. This requires exposure to light in the 6 h after the core temperature minimum. (On the day immediately after the flight, this temperature minimum is at about 04.00 on 'home' time, since the body clock is still adjusted to the home time zone.) Also, avoidance of light in the 6 h before the temperature minimum is necessary, since this would produce a phase delay. With both light exposure and light avoidance, these times must be converted to the new local time. For flights through more than nine time zones to the east, and for all westward flights, the aim is to promote a delay of the clock by exposure to light in the 6 h before the temperature minimum and avoidance of light in the 6 h after it (Waterhouse *et al.*, 1997). On subsequent days, the local times of light exposure and avoidance can be advanced (when advancing the body clock) or delayed (when delaying the body clock) by 1–2 h each day until they coincide with the late afternoon and evening (when delaying the body clock) or with morning hours immediately after waking up (when advancing the body clock). Table 12.2 translates this protocol into times of exposure to light and avoidance of it in terms of local times in the new time zone.

It will be noted that living in accord with the habits of the inhabitants in the new time zone ('When in Rome, do as the Romans do') produces the right times of exposure to, and avoidance of, light after a flight to the west, but that such a behaviour pattern is often inappropriate after time-zone transitions to the east. For example, after a flight

Table 12.2 The use of bright light to adjust the body clock after time-zone transitions

	Bad local times for exposure to bright light	Good local times for exposure to bright light
Time zones to the west		
4 h	01.00–07.00[a]	17.00–23.00[b]
8 h	21.00–03.00[a]	13.00–19.00[b]
12 h	17.00–23.00[a]	09.00–15.00[b]
Time zones to the east		
4 h	01.00–07.00[b]	09.00–15.00[a]
8 h	05.00–11.00[b]	13.00–19.00[a]
10–12 h	Treat this as 12–14 h to the west[c]	

From Waterhouse *et al.* (1997).
[a]This will advance the body clock.
[b]This will delay the body clock.
[c]This is because the body clock adjusts to large delays more easily than to large advances.

from the UK to Los Angeles (eight time zones to the west), light is initially required at 13.00–19.00 local time (equivalent to 21.00–03.00 on 'body time') and should be avoided at 21.00–03.00 local time (05.00–11.00 on 'body time'). This requirement fits in easily with the natural lighting in Los Angeles. By contrast, after a flight from the UK to Hong Kong (eight time zones to the east), light is required at 13.00–19.00 on local time (equivalent to 05.00–11.00 on 'body time') and should be avoided at 05.00–11.00 local time (equivalent to 21.00–03.00 on 'body time'). Unfortunately, therefore, the traveller to the east who tries to promote adjustment by 'getting out and about' at the first opportunity might actually be causing a delay of the body clock. It seems likely that this is one reason why jet lag after an eastward flight can last longer than after one to the west.

Portable light sources, in the form of battery-operated visors, are available so that the traveller can ensure that light is available when required, even during the flight, but few detailed reports of their efficacy have been published. Since, when using the light protocol just described, it is also important to avoid light at other times, a pair of strong sunglasses or, preferably, staying indoors and away from windows at certain times becomes an equally important part of the regimen.

Bright light (more than about 2500 lux) also activates the individual, and dim light (less than about 100 lux) acts as a soporific; these effects probably occur at whatever time they are given. However, unless their timing is adjusted as described above, the effect would not be through adjustment of the body clock but rather a direct one that masks the underlying problem of an inappropriate phasing of the body clock.

12.5.1.4 Melatonin

In normal circumstances, melatonin is secreted from the pineal gland into the bloodstream between about 21.00 and 07.00 and can be regarded as a 'dark pulse' or 'internal zeitgeber' for the body clock (for more details, see Chapter 14). Melatonin capsules taken in the evening by local time in the new time zone have been found in several (see Haimov and Arendt, 1999, for review; Takahashi *et al.*, 2000) but not in all (Spitzer *et al.*, 1999) studies to reduce the individual's perception of the amount of jet lag suffered. Beneficial effects apply to both sexes, after flights in both directions and at whatever time the flight itself takes place.

There are some caveats with regard to the general advisability of taking melatonin capsules (Guardiola-Lemaître, 1997). First, even though melatonin reduces jet lag and improves sleep, it is not clear whether there are comparable improvements in mental and physical performance, as would be predicted if melatonin were shifting the body clock. This doubt reflects the fact that melatonin exerts two effects. It has a hypnotic action, possibly because it lowers body temperature via cutaneous vasodilation (Cagnacci, 1997). Also, laboratory studies indicate that melatonin can adjust the body clock, ingestion in the hours around and after the minimum of the circadian rhythm of core temperature causing a phase delay, and in the hours around and after the maximum causing a phase advance (see Chapter 1). There are no studies to indicate which of these effects is responsible for the effectiveness of melatonin in the field, though only an appropriate adjustment of the body clock would remove the underlying cause of jet lag.

The second caveat relates to the availability of melatonin. It does not have a licence in the UK and many other countries, and so is generally available only by prescription. It can be obtained commercially in the USA, but as a food supplement, often mixed with other substances, rather than in a pure form. No long-term clinical trials appear to have been performed, if only because melatonin itself cannot be patented. However, melatonin analogues have been produced and are currently undergoing trials.

For these reasons, whatever its value, the use of melatonin, particularly in the long term, cannot be recommended unless under medical supervision.

12.5.1.5 Drugs

The careful use of hypnotics can to some extent offset the difficulties with sleep in the new time zone. Benzodiazepines such as temazepam and benzodiazepine-like drugs such as zolpidem have been used with some success, but it is possible that they have residual effects upon mental and physical performance the next day, which would negate their value (Buxton *et al.*, 2000; Daurat *et al.*, 2000). Drugs that promote and maintain alertness have also been investigated (Akerstedt and Ficca, 1997); such drugs include amphetamines, caffeine and pemoline. Although these drugs improve performance in several mental tasks, they adversely affect the ability to initiate and sustain sleep, and so their use might be counterproductive.

With neither group of drugs does it appear that the underlying cause of jet lag is being addressed. Consequently, drugs do not provide a remedy for the syndrome.

12.5.2 Treatment of shift-workers' malaise

It must be remembered that night work generally applies to individuals for a significant fraction of their working time, often extending over a period of years; a comparable frequency of changes of the sleep–wake pattern rarely applies to anyone but aircrew regularly engaged in long-haul flights or the most ardent 'jet-setter'. Therefore, any advice must bear in mind the long-term implications. Moreover, with rapidly rotating shift systems, it is a stable, day-orientated phasing of circadian rhythms that should be promoted, adjustment of circadian rhythms being desirable only if the night shift is being worked for extended periods (at least a week).

12.5.2.1 Drugs and melatonin

Some reports (Folkard *et al.*, 1993; Walsh *et al.*, 1995) indicate that melatonin and other short-acting hypnotics can improve daytime sleep and some aspects of mood. However, lack of data regarding the consequences of the long-term use of these substances means that their use cannot be recommended unless it is under medical supervision. It appears that many night workers use caffeine and cigarettes to promote wakefulness during the night shift and, perhaps, an alcoholic drink to promote daytime sleep. There is laboratory-based evidence supporting the effectiveness of caffeine (Walsh *et al.*, 1995), but the combination of this with nicotine and alcohol might exacerbate any problems with indigestion and gastrointestinal ulcers.

12.5.2.2 Meals

The use of the food content of a meal to promote adjustment or stability of the body clock in night workers, a variant of the 'feeding hypothesis', see Section 12.5.1.1, does not seem to have been investigated. However, the social role of communal mealtimes, both at work and at home, might act to reinforce the changed sleep–wake schedule and/or to maintain family contacts.

12.5.2.3 Exercise

Here also the use of timed periods of exercise to promote clock adjustment or stability has not been tested. There is some research that indicates that female shift workers benefitted from a general exercise programme (Harma *et al.*, 1988), but whether this was due to an increase of fitness or to any effects upon the body clock is not known. When exercise was combined with light treatment (see below), no effect that could be attributed to exercise was found (Baehr *et al.*, 1999).

12.5.2.4 Light exposure

Several reports exist of attempts to promote adjustment of the body clock by exposure to bright light, often in studies on volunteers in laboratory-based simulations of night work (Czeisler *et al.*, 1990; Eastman and Martin, 1999). Subjects have been exposed to bright light during the first part of the first night shift, with the time of exposure becoming progressively later on each subsequent night shift (Figure 12.5). A more complete and rapid delay of the body clock, as assessed from the core temperature rhythm, was produced by this method. This shift of the temperature rhythm not only moves its time of the minimum away from the night shift, so promoting performance during the shift, but also moves it towards the hours before noon, as a result of which sleep at home should be facilitated.

As with the use of light after time-zone transitions, if a phase delay of the rhythms is required, care must be taken not to expose the worker

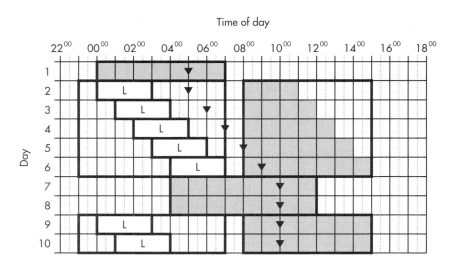

Figure 12.5 A schedule designed to promote adjustment of the circadian rhythms to night work by appropriate timing of light exposure and avoidance. Blocks show 8 h night shifts. Shaded areas show times for light avoidance. Times of giving 3 h of bright light are labelled L. Triangles indicate typical times of core temperature minimum, this becoming delayed by the protocol. Reproduced from Eastman and Martin (1999).

to too much light after the minimum of the core temperature rhythm (which would tend to cause a phase advance). This requires that light be dimmed towards the end of the first night shift and, if it is light outdoors when the subject returns home, that dark goggles should be worn. By contrast, if it is desired to prevent a shift of the body clock, then bright light throughout the shift and, if possible, on the way home after the shift is appropriate.

Two field studies have been published on nurses working at night in rapidly rotating shift systems. The first used light boxes in a room that was used by the nurses at the start and end of the night shift as well as during rest breaks (Costa *et al.*, 1994). The second used light visors worn throughout the shift (Costa *et al.*, 1997). The phase-shifting effects in both studies were small, which is desirable in a rapidly rotating shift system. The lack of phase-shifting effect can be attributed to the light being available throughout the shift (so combining possible phase-advancing and phase-delaying effects) as well as to its being used for only two nights. There was also a positive effect upon mood, which might have been a direct effect of the light, regardless of any effect upon the body clock. Such an effect might benefit all night workers, regardless of attempts to promote adjustment or stability of the phase of the body clock.

12.5.3 Treatment of clinical problems associated with circadian rhythm disorders

A priori, an observed disorder could originate from a fault in one or more of the following: the clock itself; its output to the different systems of the body; the zeitgebers present; or the transmission of zeitgeber information to the body clock. In a clinical setting, it is rare that the exact cause of an abnormality is known, and it might, anyway, be multifactorial. The deterioration of circadian rhythms with old age is likely to be an example of this, with decreases in the efficiency of many aspects of the circadian system – from poorer zeitgeber inputs, through a deteriorating body clock, to a decreased ability to implement the clock output – all being implicated (see Chapter 14).

There are some exceptions to this ignorance of causes. For example, in many cases of nocturnal enuresis it appears to be the nocturnal secretion of antidiuretic hormone that is abnormal, though the reason for this is unknown (Rittig *et al.*, 1989). An injection of an analogue of arginine vasopressin each evening is effective (Robertson *et al.*, 1999). Similarly, for the abnormalities associated with blind people, it is the inability to use the light–dark cycle as a zeitgeber that is almost certainly responsible, and so treatment can concentrate on the use of other zeitgebers (see below).

The observation that the circadian system is acting abnormally need not mean that this is the primary cause of the disorder; the primary defect might be elsewhere (in a neurotransmitter system, for example) and the effects upon the circadian system may be only one of several abnormalities. Treating the circadian system might improve the individual's sleep at night and activity during the daytime, but it is not treating the underlying cause of the problem. Similarly, a treatment might be effective because it combats some of the symptoms of the disorder rather than the disorder itself. This applies also to treatments for jet lag and shift-workers' malaise (see Sections 12.5.1 and 12.5.2). Thus, light exposure can promote a general activation of the central nervous system; the ingestion of melatonin and other hypnotics can promote sleep; and daytime activity can promote nocturnal fatigue. One particularly clear example of treating the symptoms rather than the cause is the use of 'strategic napping' by night workers (Rosekind et al., 1995). Moreover, if the primary error is not in the circadian system, then the treatment's effectiveness might arise because it strengthens patient's dichotomy between daytime activity and nocturnal rest and sleep to such an extent that this obscures the underlying problem.

These caveats do not present a problem for those who wish to help patients; but they do for those who wish to understand a disorder and to attempt to develop a rationale for its treatment. A comprehensive review of the diagnosis and treatment of clinical disorders of the circadian system has been published (Campbell et al., 1999). The three main areas of treatment are behavioural therapy, light therapy and the ingestion of melatonin.

12.5.3.1 Behavioural therapy

In this, the sleep–wake cycle is accentuated. This entails promoting the patient's social or physical activities and discouraging sleeping or napping in the daytime; at night, it involves encouraging the individual to rest or sleep in bed (Okawa et al., 1993; Van Someren et al., 1997b). This has been shown to be effective in elderly subjects, both those who are otherwise healthy and those suffering from forms of dementia (Figure 12.6). Success has been assessed by better nocturnal sleep and less activity at night, by increased and more consistent activity in the daytime, and by improvements in the patient's symptoms. It has not been demonstrated that there is an increased amplitude or phase stability in other rhythms and, even if this were the case, such effects could be attributed to the endogenous timing system only if they continued during constant-routine conditions (see also Section 12.5.3).

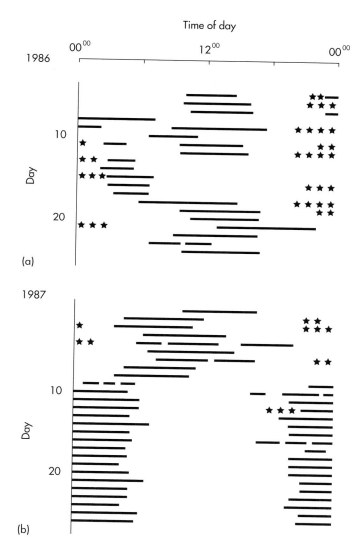

Figure 12.6 Times of sleep (bars) on successive days in an elderly patient suffering from Alzheimer's disease. (a) Before treatment; (b) during behavioural therapy the following year. Asterisks are assessments of the amount of general confusion suffered by the patient. Reproduced from Okawa et al. (1991).

A special example of behaviour therapy applies to cases of delayed sleep phase syndrome (DSPS), a disorder in which the rhythms are stable and entrained to a 24 h day but are abnormally delayed, and phase-advancing them appears to be difficult (see Table 12.1). Treatment consists of requiring the patient to go to bed later each day until sleep is at an acceptable time (say, 23.00–07.00). At this point, sleep times must be maintained regularly at this time, the individual being warned against going to bed late or 'lying in' in the morning.

12.5.3.2 Light therapy

Light therapy increases the consolidation of nocturnal sleep both in patients with senile dementia (Figure 12.7) and in sighted patients with free-running circadian rhythms (Van Someren *et al.*, 1997a; Chesson *et al.*, 1999). The light is sometimes given in the morning, but increasing light intensity indoors throughout the daytime has also been successful. The improvement in the sleep–wake cycle is often associated with a decrease in the severity of any pre-existing symptoms of disorientation and general confusion, and it should be remembered that this must be one of the main aims of treatment.

12.5.3.3 Melatonin therapy

Melatonin, generally taken in capsule form about 3 h before sleep is desired, has been used in subjects suffering from insomnia due to a

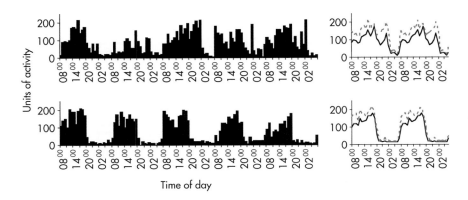

Figure 12.7 Raw (left) and double-plotted (right) average rest–activity patterns (solid line) and one standard deviation above this level (dashed line) of a patient with Alzheimer's disease assessed two times for five days: before (upper panels) and during (lower panels) light treatment. Adapted from Van Someren *et al.* (1997a).

variety of causes. In addition to insomnia associated with jet lag and shift work (see Sections 12.5.1.4 and 12.5.2.1), other examples of successful treatment include insomnia in otherwise healthy aged subjects; in patients with senile dementia; in blind subjects (Figure 12.8) and sighted subjects with 'free-running' rhythms; in patients with DSPS; and in young and old subjects with sleep problems (Garfinkel *et al.*, 1995; Lewy and Sack, 1997; Okawa *et al.*, 1998; Sack *et al.*, 2000). It has been successful as judged by subjective and objective estimates of sleep, and by objective measures of movement. Melatonin has an advantage over light therapy insofar as taking a capsule regularly is far less intrusive than having to be near bright light for a prescribed time each day (Hayakawa *et al.*, 1998).

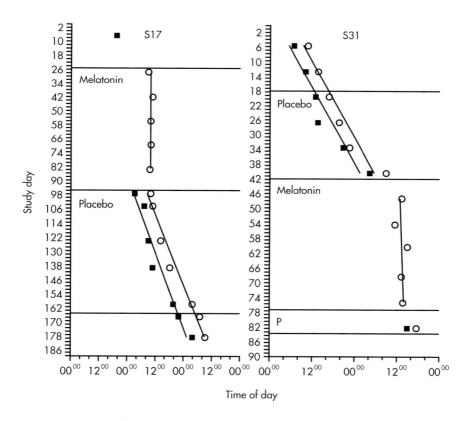

Figure 12.8 Acrophases on successive days of urinary rhythm of the metabolite of melatonin (solid squares) and urinary rhythm of cortisol (open circles) in two blind patients. Subject S17 was given melatonin therapy before placebo; subject S31 was treated in the reverse sequence. Best-fitting lines have been inserted. Reproduced from Lockley *et al.* (2000).

Most authors regard the success of this form of treatment as due to the natural hypnotic properties of melatonin. Melatonin might also be acting as an 'internal zeitgeber'. To investigate this, the response to treatment of the phase of the circadian rhythms of core temperature and other variables has been measured in a few studies, but the picture at the moment is unclear. In one study (Nakamura et al., 1997), the sleep–wake cycle in a sighted man with 'free-running' rhythms was stabilised by melatonin ingestion, associated with which were stabilisations of the circadian rhythms of core temperature and plasma cortisol. By contrast, in other studies (Folkard et al., 1990; Lushington et al., 1997; Nagtegaal et al., 1998), improvements in the sleep–wake cycle were not accompanied by similar improvements in other rhythms. More work is required to demonstrate the means by which melatonin exerts its effects and how widely it affects circadian rhythmicity.

12.6 Summary

Even though the body clock of normal subjects is entrained to a period of 24 h, the exact timing varies day-by-day in any individual and between individuals. This latter difference, when applied to the 'tails' of the distribution of the timing of a particular rhythm in the population as a whole, distinguishes between 'larks' and 'owls'. Whereas these differences are rarely a source of inconvenience, for some the differences are marked enough to cause difficulties in dealing with a conventional lifestyle and are considered as clinical abnormalities. These abnormalities include patients with delayed sleep phase syndrome and those, often blind, whose circadian rhythms appear to 'free-run'. In all cases, the main inconvenience is the difficulty in initiating and maintaining sleep when the circadian rhythms are phased inappropriately.

Another group of problems exists in healthy subjects who possess normal circadian rhythms but who then make substantial changes to their sleep–wake cycle, as after a time-zone transition (when 'jet lag' is suffered) or during night work (when the symptoms are called 'shift-workers' malaise'). In both cases there are difficulties with sleep, but there are also problems – some of which might have important implications for the general public – with staying awake and working effectively. There is some evidence for inter-individual differences in the severity of symptoms suffered in these two conditions.

The main treatments for all of these problems are, first, the use of natural or artificial hypnotics and, second, the regular ingestion of melatonin or exposure to appropriately-timed rhythms in the environment.

The first group of treatments is effective in dealing directly with some of the sleep problems faced by the individuals, but is unlikely to address the cause of the problems. The second group of treatments is aimed at strengthening exposure to zeitgebers and, therefore, stabilising the phasing of the body clock, but a more direct effect upon the body as a whole is an additional or alternative explanation of the effectiveness of this form of treatment. The ability of drugs to affect clock mechanisms is considered in the next chapter.

Acknowledgements

We thank T. Reilly for his constructive comments on an earlier draft of the manuscript, and E. Van Someren for advice with Figure 12.6.

References

Akerstedt T (1995). Work hours, sleepiness and the underlying mechanisms. *J Sleep Res* 4(Suppl 2): 15–22.
Akerstedt T, Ficca G (1997). Alertness-enhancing drugs as a counter-measure to fatigue in irregular hours. *Chronobiol Int* 14: 145–158.
Akerstedt,T, Folkard S (1995). Validation of the S and C components of the three-process model of alertness regulation. *Sleep* 18: 1–6.
Akerstedt T, Froberg J (1975). Work hours and 24 h temporal patterns in sympathetic-adrenal medullary activity and self-rated activation. In: Colquhoun P, Folkard S, Knauth P, eds. *Experimental Studies of Shiftwork*. Opladen: Westdeutscher Verlag, 78–93.
Baehr E, Fogg L, Eastman C (1999). Intermittent bright light and exercise to entrain human circadian rhythms to night work. *Am J Physiol* 46: R1598–R1604.
Boggild H, Knutsson A (1999). Shift work, risk factors and cardiovascular disease. *Scand J Work Environ Health* 25: 85–99.
Buxton O, Copinschi G, Van Onderbergen A, *et al.* (2000). A benzodiazepine hypnotic facilitates adaptation of circadian rhythms and sleep–wake homeostasis to an eight hour delay shift simulating westward jet lag. *Sleep* 23: 915–927.
Cagnacci A (1997). Influences of melatonin on human circadian rhythms. *Chronobiol Int* 14: 205–220.
Campbell S, Murphy P, Vanden Heuval C, *et al.* (1999). Etiology and treatment of intrinsic circadian rhythm sleep disorders. *Sleep Med Rev* 3: 179–200.
Chesson A, Littner M, Davila D, *et al.* (1999). Practice parameters for the use of light therapy in the treatment of sleep disorders. *Sleep* 22: 641–660.
Costa G, Ghirlanda G, Tarondi G, *et al.* (1994). Evaluation of a rapidly rotating shift system for tolerance of nurses to night work. *Int Arch Occup Environ Health* 65: 305–311.
Costa G, Kovacic M, Bertoldi A, *et al.* (1997). The use of a light visor during night work by nurses. *Biol Rhythm Res* 28: 16–25.

Czeisler C, Johnson M, Duffy J, *et al*. (1990). Exposure to bright light and darkness to treat physiologic maladaptation to night work. *N Engl J Med* 322: 1253–1259.

Dagan Y, Eisenstein M (1999). Circadian rhythm sleep disorders: toward a more precise definition and diagnosis. *Chronobiol Int* 16: 213–222.

Daurat A, Benoit O, Buguet A (2000). Effects of zopiclone on the rest/activity rhythm after a westward flight across five time zones. *Psychopharmacology* 149: 241–245.

Dawson D, Armstrong S (1996). Chronobiotics – drugs that shift rhythms. *Pharmacol Ther* 69: 15–36.

Dinges D (1995). An overview of sleepiness and accidents. *J Sleep Res* 4(Suppl 2): 4–14.

Eastman C, Martin S (1999). How to use light and dark to produce circadian adaptation to night shift work. *Ann Med* 31: 87–98.

Edwards B, Waterhouse, J, Atkinson G, Reilly T (2002). Exercise does not necessarily influence the phase of the circadian rhythm in temperature in humans. *J Sport Sci* 20: 725–732.

Elliott A, Mills J, Waterhouse J (1970). A man with too long a day. *J Physiol* 212: 30P–31P.

Folkard S (1990). Circadian performance rhythms: some practical and theoretical implications. *Philos Trans R Soc Lond B Biol Sci* 327: 543–553.

Folkard S, Condon R (1987). Night shift paralysis in air traffic control officers. *Ergonomics* 30: 1353–1363.

Folkard S, Arendt J, Aldhous M, Kennett H (1990). Melatonin stabilises sleep onset time in a blind man without entrainment of cortisol or temperature rhythms. *Neurosci Lett* 113: 193–198.

Folkard S, Arendt J, Clark M (1993). Can melatonin improve shift workers' tolerance of the night shift? Some preliminary findings. *Chronobiol Int* 10: 315–320.

Garfinkel D, Laudon M, Nof D, Zisapel N (1995). Improvement of sleep quality in elderly people by controlled-release melatonin. *Lancet* 346: 541–544.

Guardiola-Lemaître B (1997). Toxicology of melatonin. *J Biol Rhythms* 12: 697–706.

Haimov I, Arendt J (1999). The prevention and treatment of jet lag. *Sleep Med Rev* 3: 229–240.

Harma M (1995). Sleepiness and shiftwork: individual differences. *J Sleep Res* 4(Suppl 2): 57–61.

Harma M, Ilmarinen J, Knauth P, *et al*. (1988). Physical training intervention in female shift workers: II. The effects of intervention on the circadian rhythms of alertness, short-term memory, and body temperature. *Ergonomics* 31: 51–63.

Harma M, Laitinen J, Partinen M, Savanto S (1994). The effect of four-day round trip flights over 10 time zones on the circadian variation of salivary melatonin and cortisol in air-time flight attendants. *Ergonomics* 37: 1479–1489.

Hayakawa T, Kamei Y, Urata J, *et al*. (1998). Trials of bright light exposure and melatonin administration in a patient with non-24 hour sleep–wake syndrome. *Psychiatry Clin Neurosci* 52: 261–262.

Hofman M (2000). The human circadian clock and aging. *Chronobiol Int* 17: 245–259.

Honma S, Shirakawa T, Nakamura T, Honma K (2000). Synaptic communication of cellular oscillations in the rat suprachiasmatic neurons. *Neurosci Lett* 294: 113–116.
Kerkhof G (1985). Inter-individual differences in the human circadian system: a review. *Biol Psychol* 20: 83–112.
Kerkhof G, Van Dongen H (1996). Morning-type and evening-type individuals differ in the phase-position of their endogenous circadian oscillator. *Neurosci Lett* 218: 153–156.
Klein K, Wegmann H, Hunt B (1972). Desynchronization of body temperature and performance circadian rhythm as a result of outgoing and homegoing transmeridian flights. *Aerospace Med* 43: 119–132.
Leathwood P (1989). Circadian rhythms of plasma amino acids, brain neurotransmitters and behaviour. In: Arendt J, Minors D, Waterhouse J, eds. *Biological Rhythms in Clinical Practice*. London: Wright (Butterworth), 136–159.
Lennernas M, Akerstedt T, Hambreus L (1994). Nocturnal eating and serum cholesterol of three-shift workers. *Scand J Work Environ Health* 20: 401–406.
Lennernas M, Hambreus L, Akerstedt T (1995). Shift related dietary intake in day and shift workers. *Appetite* 25: 253–265.
Lewy A, Sack R (1997). Exogenous melatonin's phase-shifting effects on the endogenous melatonin profile in sighted humans: a brief review and critique of the literature. *J Biol Rhythms* 12: 588–594.
Lockley S, Skene D, Butler L, Arendt J (1999). Sleep and activity rhythms are related to circadian phase in the blind. *Sleep* 22: 616–623.
Lockley S, Skene D, James K, *et al*. (2000). Melatonin administration can entrain the free-running circadian system of blind subjects. *J Endocrinol* 164: R1–R6.
Lowden D, Akerstedt T (1998). Retaining home-base sleep hours to prevent jet lag in connection with a westward flight across nine time zones. *Chronobiol Int* 15: 365–376.
Lushington K, Dawson D, Morris M, Lack L (1989). The reliability of phase position of the circadian pacemaker when determined by constant routine. *Sleep Res* 17: 429.
Lushington K, Pollard K, Lack L, *et al*. (1997). Daytime melatonin administraton in elderly good and poor sleepers: effects on core body temperature and sleep latency. *Sleep* 20: 1135–1144.
McArthur A, Lewy A, Sack R (1996). Non-24 hour sleep–wake syndrome in a sighted man – circadian rhythm studies and efficacy of melatonin treatment. *Sleep* 19: 544–553.
Minors D, Waterhouse J (1981). *Circadian Rhythms and the Human*. Bristol: Wright PSG.
Minors D, Atkinson G, Bent N, *et al*. (1998). The effects of age upon some aspects of lifestyle and implications for studies on circadian rhythmicity. *Age Ageing* 27: 67–72.
Mirmiran M, Kok J (1991). Circadian rhythms in early human development. *Early Hum Dev* 26: 121–128.
Mishima K, Okawa M, Hozumi S, Hishikawa Y (2000). Supplementary administration of artificial bright light and melatonin as potent treatment for disorganized circadian rest-activity and dysfunctional autonomic and neuroendocrine systems in institutionalised demented elderly persons. *Chronobiol Int* 17: 419–432.

Monk T, Buysse D, Reynolds C, *et al.* (1995). Circadian temperature rhythms of older-people. *Exp Gerontol* 30: 455–474.

Mrosovsky N, Salmon P (1987). A behavioural method for accelerating re-entrainment of rhythms to new light–dark cycles. *Nature* 330: 372–373.

Nagtegaal J, Kerkho G, Smits M, Swart A (1998). Delayed sleep phase syndrome: a placebo-controlled cross-over study on the effects of melatonin administered five hours before the individual dim light melatonin onset. *J Sleep Res* 7: 135–143.

Nakagawa H, Sack R, Lewy A (1992). Sleep propensity free-runs with the temperature, melatonin and cortisol rhythms in a totally blind person. *Sleep* 15: 330–336.

Nakamura K, Hashimoto S, Honma S, Honma K (1997). Daily melatonin intake resets circadian rhythms of a sighted man with non-24 hour sleep–wake syndrome who lacks the nocturnal melatonin rise. *Psychiatry Clin Neurosci* 51: 121–127.

Okawa M, Mishima K, Hishikawa Y, *et al.* (1991). Circadian rhythm disorders in sleep–waking and body temperature in elderly patients with dementia and their treatment. *Sleep* 14: 478–485.

Okawa M, Mishima K, Hishikawa Y, *et al.* (1993). Sleep disorder in elderly patients with dementia and trials of new treatments – enforcement of social interaction and bright light therapy. In: Kumar V, Mallick H, Nayar U, eds. *Sleep–Wakefulness*. New Delhi: Wiley Eastern, 129–132.

Okawa M, Uchiyama M, Ozaki S, *et al.* (1998). Melatonin treatment for circadian sleep disorders. *Psychiatry Clin Neurosci* 52: 259–260.

Ozaki S, Uchiyama M, Shirakawa S, Okawa M (1996). Prolonged interval from body temperature nadir to sleep offset in patients with delayed sleep phase syndrome. *Sleep* 19: 36–40.

Redfern P (1989). 'Jet-lag': strategies for prevention and cure. *Hum Psychopharmacol* 4: 159–168.

Reilly T, Atkinson G, Waterhouse J (1997). *Biological Rhythms and Exercise*. Oxford: Oxford University Press.

Rittig S, Knudsen U, Norgaard J, *et al.* (1989). Abnormal diurnal rhythm of plasma vasopressin and urinary output in patients with enuresis. *Am J Physiol* 256: F664–F667.

Robertson G, Rittig S, Kovacs L, *et al.* (1999). Pathophysiology and treatment of enuresis in adults. *Scand J Urol Nephrol* 33: 36–39.

Rosekind M, Smith R, Miller D, *et al.* (1995). Alertness management: strategic naps in operational settings. *J Sleep Res* 4(Suppl 2): 62–66.

Sack R, Brandes R, Kendall A, Lewy A (2000). Entrainment of free-running circadian rhythms in blind people. *N Engl J Med* 343: 1070–1077.

Shibui K, Uchiyama M, Okawa M (1999). Melatonin rhythms in delayed sleep phase syndrome. *J Biol Rhythms* 14: 72–76.

Spitzer R, Terman M, Williams J, *et al.* (1999). Jet lag: clinical features, validation of a new syndrome-specific scale, and lack of response to melatonin in a randomised, double-blind trial. *Am J Psychiatry* 156: 1392–1396.

Takahashi T, Sasaki M, Itoh H, *et al.* (2000). Effect of 3 mg melatonin on jet lag syndrome in an 8-h eastward flight. *Psychiatry Clin Neurosci* 54: 377–378.

Van Cauter E, Plat L, Leproult R, Copinschi G (1998). Alterations of circadian rhythmicity and sleep in aging: endocrine consequences. *Horm Res* 49: 147–152.

Van Reeth O (1998). Sleep and circadian disturbances in shift work: strategies for their management. *Horm Res* 49: 158–162.

Van Someren E, Kessler A, Mirmiran M, Swaab D (1997a). Indirect bright light improves circadian rest–activity rhythm disturbances in demented patients. *Biol Psychiatry* 41: 955–963.

Van Someren E, Lijzenga C, Mirmiran M, Swaab D (1997b). Long-term fitness training improves the circadian rest–activity rhythm in healthy elderly males. *J Biol Rhythms* 12: 146–156.

Walsh J, Muehlbach M, Schweitzer P (1995). Hypnotics and caffeine as countermeasures for shiftwork-related sleepiness and sleep disturbance. *J Sleep Res* 4(Suppl): 80–83.

Waterhouse J, Reilly T, Atkinson G (1997). Jet lag. *Lancet* 350: 1611–1616.

Waterhouse J, Weinert D, Minors D, *et al.* (2000). Estimates of the daily phase and amplitude of the endogenous component of the circadian rhythm of core temperature in sedentary humans living nychthemerally. *Biol Rhythm Res* 31: 88–107.

Waterhouse J, Edwards B, Nevill A, *et al.* (2002). Identifying some determinants of 'jet lag' and its symptoms; a study of athletes and other travellers. *Br J Sports Med* 36: 54–60.

Wedderburn A (1987). Sleeping on the job: the use of anecdotes for recording rare but serious events. *Ergonomics* 30: 1229–1233.

Weinert D, Sitka U, Minors D, *et al.* (1997). Twenty-four-hour and ultradian rhythmicities in healthy full-term neonates: exogenous and endogenous influences. *Biol Rhythm Res* 28: 441–452.

Zhou J, Swaab D (1999). Activation and degeneration during aging: a morphometric study of the human hypothalamus. *Microsc Res Tech* 44: 36–48.

13

Drug effects on suprachiasmatic nuclei function

David J Kennaway

13.1 Introduction

There is a growing appreciation of the importance of circadian rhythms in health and disease and this is clearly illustrated by preceding chapters in this book. The most visible rhythm as far as the general public is concerned is that of the sleep–wake cycle. Other less obvious daily cycles of temperature and hormone levels clearly underpin the sleep–wake cycle. And behind all these rhythms is the master clock, the suprachiasmatic nuclei (SCN), which have their intrinsic, near 24 h rhythmicity entrained by the retinal perception of environmental light.

Why would it be useful to be able to control the clock with drugs? There is a significant proportion of the population who have disrupted sleep–wake rhythms. The causes are many and diverse and it is outside the scope of this review to address them all. There are, however, people who have sleep–wake problems that are clearly related to dysfunctional circadian rhythm control such as delayed sleep phase syndrome (DSPS), advanced sleep phase syndrome (ASPS) and non-24 h sleep syndrome. The incidence of these disorders may be underestimated. For example, in the case of ASPS, while the quality and even duration of sleep may be acceptable, the fact that the timing of the sleep outside 'normal' societal expectations is distressing, although it may not be seen as a reason for attending a clinic. Western society's demands for greater industrial efficiency have meant that another large section of the community is being forced to work and therefore to sleep and spend their recreation time outside the traditional 17.00 to 9.00. The toll that such shift work has on sleep and general health is well documented. Finally, the huge volume of international travel means that large numbers of people are attempting to sleep at their destinations at times when their internal clock is set for wakefulness.

There are few practical strategies that can be used to reset rhythms, apart from appropriately timed bright light exposure and possibly the use of melatonin. An understanding of the neural mechanisms underpinning the entrainment of rhythms by light may lead to the development of pharmacological approaches for altering rhythms.

13.1.1 What are the challenges?

There are a number of important defining characteristics of rhythms, including the period (time from peak to peak), amplitude and phase position. In terms of treatment of circadian-based sleep disorders, jet lag or shift work, it may prove beneficial to address one or more of these rhythm characteristics. The question is, of course, 'Are these rhythm parameters likely to be susceptible to drugs?' Before addressing this question, a brief outline of the neural pathways projecting to the SCN and our current understanding of cellular rhythmicity is required.

13.1.2 Innervation of the SCN

The neurochemistry of the SCN have been discussed in detail in Chapter 2; to summarise briefly, the primary innervation of the mammalian SCN arises from retinal ganglion cells that project to the SCN via the optic nerve, forming the retinohypothalamic tract (RHT) terminating in the ventrolateral subdivision of the SCN (Moore and Lenn, 1972). It is generally believed that the transmitter in these neurons is an excitatory amino acid, although whether it is glutamate, aspartate, both, or N-acetylaspartylglutamate has not been clarified. An indirect multisynaptic connection involves bifurcating neurons of the optic tract projects to the intergeniculate leaflet (IGL). Neurons from the IGL project to the ventrolateral SCN where they release γ-aminobutyric acid (GABA) and neuropeptide-Y. Another major projection to the SCN is from the raphe nuclei (dorsal and medial) providing serotonergic input to the nuclei as well as the IGL. It remains to be shown whether the neurons projecting from the retina to the dorsal raphe nucleus that have now been identified in many species (Shen and Semba, 1994; Kawano et al., 1996; Fite et al., 1999) make synaptic contact with nerves projecting to the SCN to constitute a retino–raphe–SCN pathway. Finally, there is some evidence for another multisynaptic retinal projection to the SCN via forebrain structures, in this case utilising acetylcholine. These four pathways and their transmitters and receptors are the most likely targets for drugs that might be expected to affect the input to the SCN.

As will be discussed later, other transmitter and intracellular systems may also prove to be effective targets for drugs.

13.1.3 Generation of the endogenous period

The molecular basis of time keeping has been discussed in Chapter 3 and so I will only briefly discuss the cellular events as they may relate to changing aspects of rhythmicity, e.g. period and phase. Our contemporary understanding of cellular rhythmicity centres around the induction of the period genes, *per1* and *per2*. This induction is gated by the CLOCK and BMAL1 proteins, which form heterodimers and act as activators at E-boxes in the promoter region of the *period* genes. The *period* genes influence their own transcription via a feedback action of PER(1/2)/CRY(1/2) protein heterodimer competition with CLOCK/BMAL1 at the E-box.

How intracellular cycling of mRNA and protein is translated across to the circadian rhythm characteristics mentioned above is not clear, but some information is emerging. The period of the SCN cycle is known to be species-specific and is maintained within a fairly narrow range when the endogenous period is expressed in constant darkness. The period is around 24.2 h in laboratory rats, while it is less than 24 h in mice. In hamsters there appears to be a particularly wide range in period. The endogenous period is heritable and has been shown to co-select with certain other traits not normally associated with rhythms (Ferguson and Kennaway, 1999). Endogenous periodicity is different in constant darkness compared to continuous light, which suggests that there are internal, perhaps neuronal, constraints upon the full expression of the period. At the genetic level, it is clear that mutations of the *Clock* gene that result in defective transcriptional activation manifest as a longer period in animals carrying one mutant allele and very long periods (27–28 h and eventual arhythmicity) in homozygous mutants (Vitaterna *et al.*, 1994). Recent elegant studies on the TAU mutant hamster showed that the endogenous period may be susceptible to the rate of phosphorylation of the PER proteins by casein kinase IΣ (*ckIΣ*) (Lowrey *et al.*, 2000). The TAU mutant hamster was shown to have a defective *ckIΣ* protein, which decreased the phosphorylation of PER, thus permitting PER/CRY heterodimers to enter the nucleus, accumulate and inhibit period gene transcription early. The phenotypic consequence of alterations in the phosphorylation of PER1 is seen in the very short endogenous period of the mutants (~20 h). Familial ASPS, characterised by early morning awakening, was recently shown to be caused by a mutation in the *per2* gene that resulted in hypophosphorylation, highlighting the critical role of posttranslational processing in rhythmicity (Toh *et al.*, 2001).

To date there are no drugs that are known to alter endogenous period; in some studies there is evidence of altered free-running period following a single dose of a drug (Edgar *et al.*, 1993), but it is difficult to interpret these studies of animals kept for very long periods in constant darkness, especially since there is some evidence of chronic morphological changes in the SCN under such conditions (Guldner *et al.*, 1997).

13.1.4 Rhythm amplitude

Alterations in amplitude of rhythms might be expected to arise from the 'strength' of an output signal, which is then manifested in the physiological end point being studied. In animal studies, the output measures are the number of rotations of a running wheel, the magnitude of the temperature change during the day or the amount of hormone (e.g. melatonin) produced. In humans, in addition to the hormonal rhythms, one might also consider some aspects of sleep (more arousals, less REM, etc.). Clearly the condition of the output system is a confounding issue in studying changes in amplitude, especially in the case of age-related changes. The ageing process may impinge on the ability of an animal to run long distances every night, just as deterioration of the sympathetic system may be responsible for lower melatonin production. At the cellular and gene level, 'strength' of the output signal will depend upon the number and health of SCN cells.

13.1.5 Phase response

This characteristic is essentially cryptic since it is not observed until animals are challenged by unexpected light exposure at night or by certain drugs. It refers to the observation of qualitative and quantitative differences in the physiological response to a challenge that is dependent upon the time of the challenge. A simple example is the observation that light presented unexpectedly early in the night causes rhythms to shift to a later phase position the next night, whereas light presented just prior to dawn will tend to shift rhythms to an earlier time the next night. It is intriguing that this photic responsiveness to light is apparently universal and independent of the endogenous period (although quantitative differences may exist). The physiological significance of this phenomenon is not clear, since animals are rarely exposed to high intensity light flashes (except perhaps for lightning) during the night. It may, however, have some role in seasonal reproduction, when animals must detect changes in the length of the night.

At the cellular level, unexpected light exposure at night results in the induction of a number of genes as a result of the release of neurotransmitters within the SCN. Depending upon which transmitters are involved (see below), a second messenger cascade is triggered, followed by induction of the immediate early genes c-*fos* and *jun*-B. After their dimerisation and translocation to the nucleus, the protein products of these genes trigger the induction of other genes possessing AP1 binding sites in the promoter region. One such gene may be *per1*, since *per1* mRNA has been shown to increase within a few minutes of a light pulse (Albrecht *et al.*, 1997). Exactly how this induction is permitted is not known; at the time of the nocturnal pulses, CLOCK/BMAL1 is presumed to be complexed by the PER/CRY heterodimers, with the consequential low rate of *per* transcription. Perhaps PER phosphorylation is increased, targeting the phospho-PER for destruction and diminishing the nuclear entry of PER. Alternatively, the stimulus may result in activation of *per* transcription at sites other than at E-boxes by other transcription factors.

It is interesting that the mammalian period genes *per1* and *per2* but not *per3* are induced by light. Even more interesting is the observation that while *per1* is induced by light both early and late in the night, *per2* is only induced by light in the early part of the night in the mouse (Albrecht *et al.*, 1997) and rat (Miyake *et al.*, 2000), while in hamsters both genes are induced at CT (circadian time) 14 and CT20 (Horikawa *et al.*, 2000), This differential inducibility of the *per1* and *per2* genes has recently been described other systems outside the SCN (Miyake *et al.*, 2000).

With this background, it is worthwhile to reconsider the multiple input pathways to the SCN and their various transmitters and receptors. In the following sections I will discuss the evidence that drugs interacting with neurotransmitters affect circadian rhythmicity. It will be clear from the previous sections that the main focus has been on 'shifting' rhythms into a new phase position rather than altering period or amplitude.

13.2 Methodology

13.2.1 *In vitro* methods

The most widely used procedure for detecting the phase-shifting effects of drugs *in vitro* involves the sampling of the neuronal firing rate of individual cells across time, following administration of vehicle or drugs or the stimulation of the optic nerves. Cells sampled in the mid-night period (Zeitgeber Time 6, ZT6; where ZT12 is defined as the onset of activity in nocturnal animals or onset of darkness) have the highest firing rate and the rhythm is maintained in culture for up to 3–4 days while the slices are perfused with glucose-supplemented salts, although it is rare to see the results of long-term experiments reported. It is worth mentioning some important aspects of this technique that may help in the interpretation of the results of these studies. First, with few exceptions, the slices are prepared during the light period and treatments in the form of bath changes or microdrops are administered within a few hours of establishing the culture. The sampling of neurons then commences the next subjective day. When sampling commences and how long it continues varies with the experiment. Second, the procedure for determining the time of peak firing rate involves averaging the firing rate data for individual cells in 2 h intervals and the use of a moving average with 1 h lags. The data are then graphed and the average ± SEM is plotted at 1 h intervals for single experiments.

A related *in vitro* method that shows some promise involves multiunit recording of SCN neurons (Tcheng and Gillette, 1996). One advantage of the procedure is that it samples more than one neuron at a time and lends itself to automation. Another advantage is the observation that the *in vitro* recordings may be generalisable to the *in vivo* situation (Meijer et al., 1997). Surprisingly, it has not been widely used for the investigation of the effects of drugs on rhythms. Prosser and colleagues were unable to replicate phase shifts obtained in single-unit rhythms with serotonergic drugs (e.g. 8-OH-DPAT, 8-hydroxy-2-(di-*n*-propylamino)tetralin) when they attempted multiunit recording (Prosser, 1998). Whether this reflects a fundamental methodological problem with one of the techniques has not been determined. It is interesting in view of growing evidence of species differences that there are very few studies on drug effects on hamster SCN slices. Indeed. studies often report *in vivo* responses to drugs in hamsters and *in vitro* effects on the SCN in rats in the same paper as though they are identical.

13.2.2 *In vivo* methods

The rhythm of wheel-running activity has tended to be the method of choice for studying the circadian system of rodents. Generally the animals are allowed to free run in constant darkness or dim red light for about 14 days, and light, drugs or some other intervention is applied at a specific time after the spontaneous onset of wheel running. Phase shifts in the onset of wheel running are then calculated in relation to the pre-treatment onsets. Wheel running itself is a poorly understood phenomenon; no-one really knows why hamsters, mice and to a lesser extent rats engage in an activity that uses so much energy for no apparent reward (Sherwin, 1998). It may be misleading to consider the rhythm equivalent to the sleep–wake cycle. In using the rhythm for drug studies, it is also important to be aware that induced wheel-running activity alone at certain times of the day can shift the running rhythm in hamsters (Mrosovsky and Salmon, 1987) but not in rats (Mistlberger, 1991). Similarly, in many cases the activating or immobilising effects of a drug may indirectly induce shifts in rhythms, leading to conclusions that may not be transferable across species. For example, benzodiazepines administered during the subjective day activate hamsters and as a consequence phase-advance the wheel-running rhythm (Turek and Losee-Olson, 1986; van Reeth and Turek, 1989). By contrast, benzodiazepines are widely known to be potent hypnotics in rats and humans and are not known to shift rhythms under similar conditions. Behavioural effects of drugs may have other hidden consequences in rhythm research, for example in cases where a drug is being investigated for its ability to block the effects of a light pulse on rhythms. The 5-HT$_{1A/7}$ agonist 8-OH-DPAT causes flat body posture in rats (and probably in hamsters) along with other behavioural effects (the serotonin syndrome). It is not always clear that care has been taken to ensure that the eyes of the animals were actually exposed to the light pulse and that, for example, the animal was not immobilised under its food container! We have observed a similar flat posture in rats treated with the NMDA (*N*-methyl-D-aspartate) antagonist dizocilpine (MK-801) (Rowe and Kennaway, 1996).

The pineal gland produces melatonin rhythmically in response to signals from the SCN and it is clear that, apart from the previously discussed effects of light, the rhythm remains unaffected by other external or internal stimuli (i.e. it is not masked). Although melatonin is easily measured in the blood or saliva of large animals such as humans, sheep, pigs, etc., it has not proved practical to study plasma melatonin rhythms in laboratory species. We and others have instead studied melatonin production in rodents by monitoring the urinary excretion rate rhythm of a melatonin metabolite, 6-sulfatoxymelatonin (Kennaway, 1993). A major advantage of the technique is that it is semiautomated and allows complex studies to be performed on small numbers of animals. *In vivo* pineal microdialysis has recently been developed, which directly measures melatonin continuously for several days (Drijfhout *et al.*,

1997). A disadvantage, however, is the labour-intensive nature of the procedure. Nevertheless, both procedures allow an investigator to determine the time of onset/offset of melatonin production, which is a rhythm that has direct applicability in human studies.

13.2.3 SCN c-*fos* induction

Upon exposure to light at night, the immediate early gene c-*fos* is induced in SCN cells and the induced c-FOS protein can be detected in the nucleus by immunohistochemistry. Intensity–response studies of light indicate that, with an increase in stimulus strength, the number of c-FOS positive cells increases. This observation has been very valuable in drug studies of the circadian system. Typically, a drug expected to block the effects of light on rhythms will be expected also to block the light induction of c-*fos* in the SCN. Although there are some examples of c-*fos* induction in the absence of phase shifts (Rea *et al.*, 1993b), intracerebroventricular administration of antisense c-*fos* and *jun*-B prevented the light-induced shifts of the wheel-running rhythm in rats, confirming the central role of these immediate early genes in the control of rhythmicity (Wollnik *et al.*, 1995).

13.3 Neurotransmitter and drug effects on rhythms

13.3.1 Excitatory amino acids

As the primary neural pathway to the SCN, the RHT and its excitatory amino acid neurotransmitters are believed to be pre-eminent in the entrainment of rhythms to a new light–dark cycle. The localisation of a range of metabotropic and ionotropic excitatory amino acid (EAA) receptors in the SCN is consistent with it being a major target for EAA input.

Phase dependency of the effects of EAA is evident in the induction of the clock genes, *per* and *BMAL1*. Glutamate, like light, induces *per1* in the rat SCN, 2 h and 8 h after lights off in a 12L:12D photoperiod. Interestingly there are apparently different intracellular pathways leading to this induction, possibly through a 'cAMP/PKA gate' (Tischkau *et al.*, 2000). In studies in the hamster, light pulses 1.5 and 8 h after lights off in 12L:12D induced *per1* and *per2*, but not *per3* in the SCN (Moriya *et al.*, 2000). Prior administration of the NMDA receptor antagonist dizocilpine (MK-801) (5 mg/kg) blocked the light induction of *per1* and *per2* at both times. When impact of light and glutamate on BMAL1 immunoreactivity in the rat SCN was investigated (Tamaru *et al.*, 2000), there was a rapid decrease in BMAL1 protein following 30 min exposure to 1000 lux, 3 h and 9 h after lights off, and following local administration of glutamate or NMDA 2 h after subjective dark onset. This observation of light-induced BMAL1 reduction, presumably by degradation, is an intriguing new control point for affecting rhythms, although the mechanisms are unknown.

Again at the intracellular level, it has been found that either systemic or local administration of dizocilpine (MK-801) blocked the photic induction of c-*fos* in the hamster SCN at times when light provoked phase advances and delays (Rea *et al.*, 1993a). By contrast, administration of MK-801 (3 mg/kg) to rats failed to block the light induction of c-*fos* in the SCN during the early subjective night, at a time when light exposure causes rhythm delays (Rowe and Kennaway, 1996). During the second half of the night, however, MK-801 became progressively more potent, such that by the time of expected lights on MK-801 almost completely blocked c-*fos* induction.

Elegant studies by the Gillette group (Ding *et al.*, 1998) have provided compelling evidence for the fundamental role of glutamate in the mediation of nocturnal phase shifts using the rat SCN slice preparation. Administration of glutamate between the time of expected lights off/activity onset (CT12) and CT16 provoked phase advances in the neuronal firing rate rhythm, whereas the same drug caused substantial delays in the rhythm between CT17 and CT23 (Ding *et al.*, 1994). Subsequent studies by the group have provided further insight into the intracellular second messenger systems, in particular nitric oxide, underlying these phase shifts.

In vivo experiments testing the effects of EAA agonists on rhythmicity require that the drugs be administered directly into the SCN or the ventricles because they fail to cross the blood–brain barrier and/or have undesirable effects on other parts of the brain. Early experiments suggested that glutamate administration during the subjective day (~CT10), provoked phase advances in hamster running activity (Meijer *et al.*, 1988), which is counterintuitive if glutamate mediates light responses. These experiments were conducted before it was recognised that induced activity could cause phase shifts. It is possible that the injections, handling or other small experimental manipulations stimulated activity in the hamsters. More recently it has been reported that *in vivo* administration of NMDA into the SCN of hamsters in constant darkness caused phase delays in the wheel-running activity rhythm when administered early in the night and advances in the late night (Mintz *et al.*, 1999).

Most experiments in this area have utilised the noncompetitive NMDA antagonist dizocilpine (MK-801) because it crosses the blood–brain barrier. This drug, however, has complex behavioural effects that may impinge upon rhythm studies. When administered to rats at a dose of 0.05 mg/kg it has no effect on locomotor activity; between 0.1 and 0.2 mg/kg it stimulates locomotion 8-fold; while at doses above 0.5 mg/kg it evokes sniffing and ataxia (Maj *et al.*, 1991). Nevertheless, MK-801 administration to hamsters was shown to attenuate light-induced phase advances and delays at doses above 1.6 mg/kg, while complete blockade was achieved at a dose of 5 mg/kg (Colwell *et al.*, 1990). Central infusion of γ-D-glutamylglycine, a competitive antagonist of NMDA- and non-NMDA-type glutamate receptors, markedly inhibited light-induced phase advances and delays in wheel running (Vindlacheruvu *et al.*, 1992). These experiments proved the role of glutamate transmission in the control of rhythmicity in the absence of possible confounding behavioural side effects.

Despite the apparently fundamental role played by EAA in the control of circadian rhythms, no human studies have been reported to date on administration of drugs active at EAA receptors. This is clearly due to the profound and dangerous side effects of both agonists and antagonists. Nevertheless, there is an EAA antagonist, memantine (1-amino-3,5-dimethyladamantane) (Parsons *et al.*, 1999), that could be worth studying in relation to circadian rhythms.

13.3.2 Serotonin

The SCN have a rich innervation of serotonergic fibres originating in the medial and dorsal raphe nucleus and it is also indirectly innervated via a raphe projection to the IGL. An interesting feature of the serotonergic innervation is that it terminates in what is generally regarded as the retino-recipient area of the SCN (the ventrolateral area). The function of the 5-HT innervation is controversial because there appear to be considerable species differences. In hamsters it is thought that the raphe input *modulates* circadian rhythms and their responses to light (Morin, 1994). In rats, we and others have provided evidence that the serotonin pathways provide a complementary pathway to the RHT *mediating* the effects of light on rhythms (Kennaway *et al.*, 2001).

The serotonergic input to the SCN may provide a promising target for drugs that alter circadian rhythms. There are at least 14 serotonin receptors that could be potential targets for drugs because of their different specificities and regional localisation in the brain (Barnes and Sharp, 1999). Some of these receptors are expressed in the SCN. The following section discusses the effects of serotonergic drugs on circadian rhythms in relation to the receptor subtypes that have been identified in the SCN to date. Unfortunately, research in this area has been hampered by the lack of drugs specific for each receptor subtype.

13.3.2.1 5-HT$_{1A}$ receptors

In 1990, Prosser and colleagues reported that the application of the serotonin agonist quipazine to rat SCN slices advanced or delayed the rhythm of neuronal firing depending upon the time of administration (Prosser *et al.*, 1990). Treatment 3–6 h before expected lights off caused a 3–4 h advance in the time of peak firing rate, whereas quipazine exposure 3–6 h after expected darkness resulted in 3–4 h delays in the rhythm. The ED$_{50}$ for the advance shift was approximately 5×10^{-7} mol/L. Based upon what was known about receptors at the time, this suggested that 5-HT$_1$ or 5-HT$_2$ receptors may be involved.

It was subsequently reported that application of serotonin to slices caused phase advances (3–7 h) only when administered 3–9 h before the expected time of lights out and caused no delays during the subjective night (Medanic and Gillette, 1992). Two drugs, 8-OH-DPAT and 5-CT (5-carboxamidotryptamine), mimicked the serotonin responses, with advances but no delays. These two drugs were used in concentrations estimated by the authors to be around 10^{-6} mol/L, thus potentially activating a range of receptors. A subsequent dose–response study by another group indicated an ED$_{50}$ of around 5×10^{-7} mol/L for 8-OH-DPAT (Shibata *et al.*, 1992). Further pharmacological characterisation of the slice technique led to the conclusion that 5-HT$_{1A}$ receptor agonists phase-advanced rhythms when administered during subjective day (Prosser *et al.*, 1993). The authors also concluded that the delays during the night were via non-5-HT$_{1A}$ receptor activation.

These early results on 5-HT$_{1A}$ drugs have been confirmed *in vivo* in rats following intraventricular injection of 8-OH-DPAT (Edgar *et al.*, 1993) and following systemic injection in hamsters (Cutrera *et al.*, 1996). We are aware of no other studies in rats that have shown 8-OH-DPAT-induced phase advances *in vivo*. Quipazine administered at CT6 failed to advance the melatonin rhythm (Kennaway *et al.*, 1996) and wheel-running rhythm (Kohler *et al.*, 1999) even when administered intraventricularly (Kalkowski and Wollnik, 1999). We also found that 8-OH-DPAT

(5 mg/kg) administered at CT6 failed to alter the melatonin rhythm (Kennaway, 1997). More recently it has been reported that while systemic injection of 8-OH-DPAT into hamsters during the late subjective day advanced the wheel-running rhythm, intra-SCN injection of the drug did not (Mintz et al., 1997). Interestingly, injection of 8-OH-DPAT into the dorsal or medial raphe induced phase shifts. The implication of these and previous results is that activity feedback from the raphe nuclei caused the phase advances (but see Antle et al., 1998). This interpretation is consistent with the profound effects of behaviour on rhythms in hamsters but not in rats (Mistlberger, 1991), especially since 8-OH-DPAT induces profound immobility.

13.3.2.2 5-HT_{1A} receptors and light

Some early studies suggested that serotonin may affect light-induced changes in SCN function. When the serotonin agonist 8-OH-DPAT was administered to hamsters prior to exposure to light pulses at CT14 or CT19, it completely blocked the light-induced delays and advances (Rea et al., 1994). Similarly, at CT19, 8-OH-DPAT (5 mg/kg) reduced by 50% the c-*fos* induction in the SCN (Rea et al., 1994). Administration at CT14 was not investigated. While this effect of 8-OH-DPAT has been replicated in several different laboratories, a recent report has provided some confusion with the demonstration that a 'selective' 5-HT_{1A} receptor agonist, MKC-242 (5-{-[((2S)-1,4-benzodioxan-2-ylmethyl) amino]-propyl}-1,3-benzodioxole) potentiated the phase-advancing effects of light pulses in hamsters (Moriya et al., 1998). This raises the possibility that the effects of 8-OH-DPAT on rhythms are mediated not by 5-HT_{1A} receptors but perhaps by 5-HT_7 receptors (see below).

13.3.2.3 5-HT_{1B} receptors

Although 5-HT_{1B}-receptor mRNA is not highly expressed in the SCN (Roca et al., 1993), binding of a novel 5-HT_{1B} radioligand within the rat SCN was increased following chemical lesion with 5,7-dihydroxytryptamine (Manrique et al., 1994). This is consistent with the presynaptic localisation of these receptors, possibly on EAA neurons. Interestingly, there have been no studies on the rhythm responses to 5-HT_{1B} receptor activation in the rat. In our laboratory, administration of CGS 12066A (7-trifluoromethyl-4-(4-methyl-1-piperazinyl)pyrrole[1,2-a]quinoxaline) failed to affect the melatonin rhythm in rats when the drug was administered at CT18 (D. J. Kennaway et al., unpublished results). By contrast, CGS 12066A and the less specific agonist TFMPP (N-(3-trifluoromethylphenyl)piperazine) blocked light-induced phase delays (CT14) and advances (CT19) in hamsters and blocked light-induced c-*fos* induction in the SCN at CT19 (Pickard et al., 1996) and light suppression of pineal melatonin content (Rea and Pickard, 2000). Pharmacological characterisation with antagonists suggested that the responses were in fact due to 5-HT$_{1B}$ activation (Pickard et al., 1996). Similar effects were reported in mice for the light-induced phase delay of morning activity and c-*fos* induction in the SCN, but extremely high doses of TFMPP (25 mg/kg) were required (Pickard and Rea, 1997). These effects of TFMPP were not observed in 5-HT$_{1B}$ receptor knockout mice (Pickard et al., 1999).

13.3.2.4 5-HT$_2$ receptors

This class of serotonin receptor has three members, 5-HT$_{2A}$, 5-HT$_{2B}$ and 5-HT$_{2C}$, and they are uniquely characterised by the fact that they are coupled positively to phospholipase C and increase the accumulation of inositol phosphates, rather than acting on adenylyl cyclase or ion channels. Note that the former 5-HT$_{1C}$ receptor is now known as the 5-HT$_{2C}$ receptor. Contrary to the recurrent assertion in papers and reviews, the SCN, at least in the rat, are a very rich site of 5-HT$_{2C}$-receptor mRNA (Roca et al., 1993) and immunoreactive protein (Moyer and Kennaway, 1999). By contrast, there is little evidence of significant amounts of 5-HT$_{2A}$ or 5-HT$_{2B}$ receptor mRNA or protein in the SCN.

The first study addressing whether 5-HT$_2$-receptor agonists shifted rhythms used quipazine and DOB (1-(2,5-dimethoxy-4-bromophenyl)-2-aminopropane) as test drugs (Prosser et al., 1993). As indicated earlier, quipazine (a nonspecific serotonin agonist having highest affinity for 5-HT$_{2C}$ receptors) was shown to provoke both phase advances and phase delays of the neuronal firing rate rhythms in the rat SCN slice. It alone caused delays among the drugs tested. DOB administered at CT6 (n = 1), CT12 (n = 1) or CT15 (n = 2) had no effect; on this basis (together with reported low binding of the nonspecific 5-HT$_2$ receptor ligand ^{125}I-LSD), it was concluded that 5-HT$_{2C}$ receptors had no role in rhythm control (Prosser et al., 1993). Starting first with quipazine and later with the 5-HT$_{2A/2C}$ agonist DOI (1-(2,5-dimethoxy-4-iodophenyl)-2-aminopropane), Kennaway and colleagues showed that 5-HT$_{2C}$ agonists administered to rats during darkness mimicked the effects of light pulses by acutely suppressing melatonin production (Kennaway et al., 1996; Kennaway and Moyer, 1998) and shifting melatonin rhythms

(Kennaway et al., 1996; Kennaway and Moyer, 1998), activity rhythms (Kennaway and Moyer, 1998) and temperature rhythms (Kennaway and Moyer, 1998) and induced c-*fos* in the ventrolateral SCN (Moyer et al., 1997). The 5-HT$_{2A/2C}$ agonist DOI is the most potent drug reported to shift rhythms *in vivo*, shifting the onset of melatonin production at doses as low as 0.1 mg/kg (Kennaway and Moyer, 1998). The phase-shifting effects of serotonin agonist activation during the night have been confirmed in rats (Kohler et al., 1999; Kalkowski and Wollnik, 1999).

Administration of DOI (5 mg/kg) to hamsters 30 min before a light pulse at CT19 had no effect on the light-induced phase advances, and DOI alone 'produced small and variable phase shifts' (Pickard et al., 1996). We are not aware of reports of DOI administration at an earlier time point in species other than rats. We have recently conducted a study of the effects of DOI administration on c-*fos* induction in the hamster and found that a dose of 2 mg/kg failed to induce c-*fos* at ZT14 or ZT18 or to modify the effects of a light pulse on c-FOS in the SCN (R. W. Moyer, P. Pevet and D. J. Kennaway, unpublished results).

13.3.2.5 5-HT$_{5A}$ receptors

The localisation of 5-HT$_{5A}$ receptor immunoreactivity in the hamster SCN has been reported (Duncan et al., 2000). There are currently no specific 5-HT$_{5A}$ agonists or antagonists available and so there are no reports on the functional significance (if any) of these receptors in SCN rhythmicity.

13.3.2.6 5-HT$_7$ receptors

The cloning and characterisation of the 5-HT$_7$ receptor (Lovenberg et al., 1993) has had an enormous impact on circadian rhythm research because it was discovered that 8-OH-DPAT bound with high affinity to 5-HT$_7$ receptors. Furthermore, serotonin agonists that provoked phase advances in the neuronal firing rate rhythm of rat SCN slices when administered at CT6 had pharmacological profiles consistent with the profile of binding to 5-HT$_7$ receptors (Lovenberg et al., 1993). It is, however, interesting to note that in the original report (Lovenberg et al., 1993) it was stated that 'We have been unable to establish definitively whether 5-HT$_7$ mRNA is expressed in the SCN. Expression of 5-HT$_7$ mRNA was detected by PCR amplification from histologically identified SCN tissue punches (data not shown). However we have not consistently detected 5-HT$_7$ (receptor) in SCN by *in situ* hybridisation, perhaps because expression is near the threshold of detection by this method.' Subsequently, two other groups have reported that 5-HT$_7$ mRNA was absent from the rat SCN as determined by *in situ* hybridisation (Gustafson *et al.*, 1996; Kinsey *et al.*, 2001). There is, however, a report of 5-HT$_7$ receptor mRNA in the rat SCN identified by *in situ* hybridisation (Heidmann *et al.*, 1998). There are also recent reports of 5-HT$_7$ receptor immunoreactivity in the mouse (Belenky and Pickard, 2001) and rat (Neumaier *et al.*, 2001), although in our own studies we were unable to detect the receptor by immunohistochemistry (Moyer and Kennaway, 1999).

In a thorough review of the evidence for the localisation and function of 5-HT$_7$ receptors in the rodent SCN, Gannon (2001) wrote 'The function of 5-HT$_7$ receptors in the SCN is not evident from any of the studies conducted to date ... the biggest challenge in this area is to determine what, if anything, 5-HT$_7$ receptors within the rodent SCN have to do with circadian rhythms.'

13.3.3 GABA

The inhibitory neurotransmitter γ-aminobutyric acid (GABA) is present in large amounts in the SCN with all, or nearly all, of the neurons in the IGL and SCN being GABA producing. According to Moore and Speh (1993), GABA should be considered the principal neurotransmitter of the circadian system. The SCN are also the site of both GABA$_A$ and GABA$_B$ receptor subunits.

The role that GABA plays in circadian rhythmicity and entrainment is complex. Intra-SCN injection of the GABA$_A$ agonist muscimol and the GABA$_B$ agonist baclofen reduced the phase-delaying and phase-advancing effects of light and induction of *c-fos* by light at both CT13.5 and CT19 (Gillespie *et al.*, 1999). Injection of the GABA$_A$ antagonist bicuculline and the GABA$_B$ antagonist CGP-35348 (3-aminopropyl (diethoxymethyl) phosphinic acid) enhanced the phase-delaying, but not the phase-advancing effects of light. Bicuculline increased the number of c-FOS-positive cells at both phases, whereas CGP-35348 was without effect. I am aware of no studies of systemic administration of GABA$_B$ antagonists on circadian rhythms, but bicuculline was shown to block light-induced delays but not advances (Ralph and Menaker, 1985), which is the opposite of the effects described above for intra-SCN administration.

When muscimol was added to SCN cultures between CT4 and CT8, it caused phase advances and had no effect at any other time

(Tominaga et al., 1994). Baclofen similarly caused phase advances during subjective day but in addition caused delays during the night (Biggs and Prosser, 1998).

The GABA receptor complex has binding sites for benzodiazepines; when occupied, they facilitate GABA-stimulated transfer of chloride ion into cells. Thus it might be expected that benzodiazepines would have effects on circadian rhythms through the GABA receptor complex. Triazolam administration to hamsters during late subjective day was spectacularly effective in phase-advancing wheel-running rhythms (Turek and Losee-Olson, 1986); however, it was shown subsequently that the effect was due to an idiosyncratic stimulation of running by the drug. Diazepam, by contrast, caused slight phase delays during the night (Ralph and Menaker, 1986).

13.3.4 Melatonin

The SCN control melatonin production by the pineal gland and it is also a likely site of action of the hormone since it expresses melatonin receptor mRNA and binds melatonin radioligands (Vanecek et al., 1987). This raises the question whether melatonin may affect SCN rhythmicity. Daily melatonin injections into rats free-running in constant darkness did indeed entrain the wheel-running activity when the time of injection coincided with the onset of activity (Redman et al., 1983). Siberian hamsters, but not Syrian hamsters, entrain to melatonin injections (Hastings et al., 1998). There are two times of day when melatonin administration has entraining effects, a few hours before lights off and an hour before lights on, with the administration causing advances at both times.

Melatonin also has entraining effects on the neuronal firing rate rhythm of the SCN *in vitro* (McArthur et al., 1991) when applied around ZT10 and for a very short time around ZT23. Curiously, it was found that while the MT_1 receptor subtype accounts for all of the detectable ^{125}I-2-iodomelatonin binding in the SCN, mice deficient in this receptor still responded to melatonin by phase-advancing the neuronal firing rate rhythm (Liu et al., 1997b). It has been suggested that the MT_2 receptor may be responsible for the phase shifts (Dubocovich et al., 1998). The mechanism of action of melatonin in the SCN is not clear. There is no question that administration of melatonin can alter the timing of rhythms *in vivo* and that it has practical use in alleviating jet lag (Herxheimer and Petrie, 2001) and possibly delayed sleep phase syndrome. Whether endogenous melatonin has a role in the maintenance of normal entrainment is not clear. Certainly it is difficult to explain the remarkably narrow window of effectiveness around ZT10 since it is approximately 6 h prior to the rise of endogenous melatonin in the rat and as much as 10 h earlier than the rise in mice. The window of responsiveness around ZT23 corresponds to the time when melatonin production has ceased and is being cleared from the circulation.

13.3.5 Acetylcholine

Acetylcholine has been identified in the SCN, as having both muscarinic and nicotinic receptors (Swanson *et al.*, 1987; van den Pol and Tsujimoto, 1985; Van der Zee *et al.*, 1991), together with its synthetic enzyme, choline acetyltransferase (Brownstein *et al.*, 1975). Acetylcholine levels were reported to rise 3-fold in the rat SCN after a light pulse presented at night (Murakami *et al.*, 1984), implicating a role for acetylcholine in light-mediated effects on SCN rhythmicity. This transmitter is not present in the major retinal connections to the SCN, but presumably is part of projections from the brainstem and basal forebrain (Bina *et al.*, 1993).

The extent to which either muscarinic or nicotinic cholinergic receptors, or both, are involved in cholinergic modulation of circadian rhythms is controversial. Muscarinic receptor activation (possibly M_1 subtype) caused large phase advances in the *in vitro* SCN neuronal firing rate rhythm of rats (Liu and Gillette, 1996). Interestingly, the advances were observed across a very large period from ZT13 to ZT22 (Liu *et al.*, 1997a). Clearly this phase response does not fit with the *in vivo* observations where intracerebroventricular cholinergic agonist administration in hamsters produced light-like phase–response curves for wheel-running activity (Wee *et al.*, 1992).

Application of nicotine to SCN slices resulted in phase advances of the neuronal firing rate rhythm at most times of the subjective day and night (Trachsel *et al.*, 1995). By contrast, administration of nicotine to rats *in vivo* in the early night (CT16) induced *c-fos* in the SCN and delayed the onset of the melatonin rhythm on the subsequent night, which is similar to the effects observed with light pulses (Ferguson *et al.*, 1999).

13.3.6 Purinergic

Adenosine is an inhibitory neuromodulator in the mammalian brain and acts through two receptor subtypes (A_1 and A_2; Fredholm, 1995). Caffeine and other methylxanthine drugs interact with these receptors and may be expected to affect circadian rhythmicity if the receptors are located within the input pathways. The adenosine A_1 receptor mRNA was reported to be absent from the rat SCN, although there was abundant expression in retinal ganglion cells (Reppert et al., 1991). Nevertheless, administration of the A_1 receptor agonist N^6-cyclohexyladenosine (N-CHA) inhibited light-induced phase shifts in hamsters, while the A_2 receptor agonist N^6-[2-(3,5-dimethoxyphenyl)-2-(2-methylphenyl)ethyl]adenosine (DPMA) was without effect (Watanabe et al., 1996). N-CHA but not DPMA also inhibited light-induced c-*fos* induction in the SCN. Bath application of N-CHA but not DPMA inhibited optic nerve stimulation-evoked field potentials in rat SCN slices. In studies in our laboratory, systemic administration of N-CHA, DPMA and caffeine in the early subjective night had no effect on melatonin rhythmicity, and N-CHA and DMPA also failed to block the phase-delaying effects of light pulses (K. McFadden-Lewis and D. J. Kennaway, unpublished results). It is worth stressing, however, that systemic administration of N-CHA (0.5 mg/kg) appeared to completely inhibit wheel-running activity of hamsters for 24 h (Watanabe et al., 1996) and immobilised rats in our studies for a similar period. It is not known whether this immobility in the running wheels affected the exposure of the animals to the light pulse or alone affected the response.

There have been no studies of the effects of A_1 or A_2 agonists on human rhythms, but there has been a study on the effects of the adenosine-receptor antagonist caffeine on melatonin and temperature rhythms. Administration of 200 mg caffeine at 20.00 and again at 02.00 prevented the nocturnal rise of saliva melatonin and altered core body temperature (Wright et al., 1997). The study did not address the rhythm on subsequent nights (to test for possible phase delays). The authors discussed their results in terms of the interference of melatonin production by the pineal gland. Another possibility is that the lowered melatonin levels in saliva were the result of changes in salivary gland function.

13.3.7 Histamine

The rat SCN receive dense histaminergic innervation from the posterior hypothalamus and histamine release in the hypothalamus is highest at night (Panula et al., 1989). By contrast, hamsters have undetectable amounts of histamine in the SCN (Scott et al., 1998). Nevertheless, histamine treatment of SCN slices caused phase shifts in the neuronal firing rate rhythm (Cote and Harrington, 1993) with a phase response similar to that for light pulses *in vivo*. Histamine synthesis inhibition (Eaton *et al.*, 1995) but not histamine antagonists (Eaton *et al.*, 1996) blocked light-induced phase shifts of wheel running in hamsters. Further *in vitro* studies suggested that histamine-induced phase shifts occurred via NMDA-dependent mechanisms (Meyer *et al.*, 1998). A totally different conclusion was reached by others. On the basis of the failure to detect histamine in the hamster SCN and small phase shifts in wheel-running activity in response to SCN microinjections, Scott *et al.* (1998) concluded that histamine does not play a prominent role in circadian rhythm regulation in this species. This issue is yet to be fully resolved.

13.3.8 Dopamine

There is considerable evidence that dopaminergic systems in the SCN are functional during embryonic and early neonatal life in rats and hamsters. For example, administration of the dopamine D_1 receptor agonist SKF-38393 to pregnant rats on embryonic day 20 results in c-*fos* induction in the fetal SCN (Weaver *et al.*, 1992). The c-*fos* induction by SKF-38393 was no longer present by postnatal day 4 (Weaver and Reppert, 1995). Adult rats failed to induce c-*fos* following SKF-38393 administration in the early subjective night (S. A. Ferguson and D. J. Kennaway, unpublished studies), despite the expression of D_1-receptor mRNA in the rat SCN (Weaver *et al.*, 1992). In primates, both D_1 and D_5 receptors are expressed in the SCN and injection of SKF-38393 increased glucose uptake in the SCN but failed to induce c-*fos* (Rivkees and Lachowicz, 1997).

In follow-up studies on the embryonic induction of SCN c-*fos* by cocaine (a dopamine agonist), pregnant hamsters injected with cocaine from embryonic day 15 to postnatal day 5 delivered pups that had altered responsiveness to light pulses when they reached adulthood (Strother *et al.*, 1998). In similar experiments, administration of a single dose of SKF-38393 to pregnant rats on embryonic day 20 delayed the development of light induction of c-*fos* in the SCN during the first few days of life (Ferguson *et al.*, 2000). In subsequent studies, chronic administration of SKF-38393 from embryonic day 15 to day 20 resulted in a blunted c-*fos* response to light when the offspring were adults (Ferguson and Kennaway, 2000).

13.4 Conclusion

The suprachiasmatic nuclei are clearly potential targets for pharmacological manipulation. Activation or antagonism of receptor systems mediating the effects of the neural input to the SCN is a very promising approach. Unfortunately, drugs affecting many of the systems, when administered systemically, have unpleasant or unwanted side effects due to the wide spread tissue distribution of the receptors. Identification of receptor subtypes mediating changes in rhythms may allow specific drugs to be used that have benign side effects. Although there are many excitatory amino acid receptor subtypes within the SCN, compounds have not yet been reported that cross the blood–brain barrier, activate/inhibit the SCN and are free from side effects on other systems. The serotonergic system is far more promising, since even general agonists or re-uptake inhibitors and antagonists have a relatively benign spectrum of side effects. If the effects on rhythms obtained with the 5-HT_{2C} and 5-HT_7 receptor agonists are confirmed in humans, specific short-acting agonists for these receptors may prove useful for shifting rhythms in cases where this is desirable, for example in sleep disorders. Until then, melatonin is the only drug that can be used to alter rhythms in humans.

Another approach, for which there are currently no data, could involve the development of compounds that interact at critical levels of the intracellular pathway generating the endogenous rhythms. One example might be a drug that alters the activity of casein kinase IΣ, which would then affect *period* gene stability and perhaps rhythm period or phase. The clock genes are active in many other parts of the brain as well as the periphery and, in the absence of knowledge of their function in extra-SCN tissue, the whole body consequences of such an approach cannot be predicted. Nevertheless, there are major health and community benefits to be gained from further research on the control of the biological clock by drugs.

Acknowledgements

Original research reported in this review was supported by a grant from the National Health and Medical Research Council of Australia.

References

Albrecht U, Sun Z S, Eichele G, et al. (1997). A differential response of two putative mammalian circadian regulators, mper1 and mper2, to light. Cell 91: 1055–1064.

Antle M C, Marchant E G, Niel L, et al. (1998). Serotonin antagonists do not attenuate activity-induced phase shifts of circadian rhythms in the Syrian hamster. Brain Res 813: 139–149.

Barnes N M, Sharp T (1999). A review of central 5-HT receptors and their function. Neuropharmacology 38: 1083–1152.

Belenky M A, Pickard G E (2001). Subcellular distribution of 5-HT$_{1B}$ and 5-HT$_7$ receptors in the mouse suprachiasmatic nucleus. *J Comp Neurol* 432: 371–388.

Biggs K R, Prosser R A (1998). GABA$_B$ receptor stimulation phase-shifts the mammalian circadian clock *in vitro*. *Brain Res* 807: 250–254.

Bina K G, Rusak B, Semba K (1993). Localization of cholinergic neurons in the forebrain and brainstem that project to the suprachiasmatic nucleus of the hypothalamus in rat. *J Comp Neurol* 335: 295–307.

Brownstein M, Kobayashi R, Palkovits M, *et al.* (1975). Choline acetyltransferase levels in diencephalic nuclei of the rat. *J Neurochem* 24: 35–38.

Colwell C S, Ralph M R, Menaker M (1990). Do NMDA receptors mediate the effects of light on circadian behavior? *Brain Res* 523: 117–120.

Cote N K, Harrington M E (1993). Histamine phase shifts the circadian clock in a manner similar to light. *Brain Res* 613: 149–151.

Cutrera R A, Saboureau M, Pevet P (1996). Phase-shifting effect of 8-OH-DPAT, a 5-HT$_{1A}$/5-HT$_7$ receptor agonist, on locomotor activity in golden hamster in constant darkness. *Neurosci Lett* 210: 1–4.

Ding J M, Chen D, Weber E T, *et al.* (1994). Resetting the biological clock: mediation of nocturnal circadian shifts by glutamate and NO. *Science* 266: 1713–1717.

Ding J M, Buchanan G F, Tischkau S A, *et al.* (1998). A neuronal ryanodine receptor mediates light-induced phase delays of the circadian clock. *Nature* 394: 381–384.

Drijfhout W J, Brons H F, Oakley N, *et al.* (1997). A microdialysis study on pineal melatonin rhythms in rats after an 8-h phase advance: new characteristics of the underlying pacemaker. *Neuroscience* 80: 233–239.

Dubocovich M L, Yun K, Al G W, *et al.* (1998). Selective MT2 melatonin receptor antagonists block melatonin-mediated phase advances of circadian rhythms. *FASEB J* 12: 1211–1220.

Duncan M J, Jennes L, Jefferson J B, *et al.* (2000). Localization of serotonin (5A) receptors in discrete regions of the circadian timing system in the Syrian hamster. *Brain Res* 869: 178–185.

Eaton S J, Cote N K, Harrington M E (1995). Histamine synthesis inhibition reduces light-induced phase shifts of circadian rhythms. *Brain Res* 695: 227–230.

Eaton S J, Eoh S, Meyer J, *et al.* (1996). Circadian rhythm photic phase shifts are not altered by histamine receptor antagonists. *Brain Res Bull* 41: 227–229.

Edgar D M, Miller J D, Prosser R A, *et al.* (1993). Serotonin and the mammalian circadian system 2. Phase- shifting rat behavioral rhythms with serotonergic agonists. *J Biol Rhythms* 8: 17–31.

Ferguson S A, Kennaway D J (1999). Emergence of altered circadian timing in a cholinergically supersensitive rat line. *Am J Physiol* 277: R1171–R1178.

Ferguson S A, Kennaway D J (2000). Prenatal exposure to SKF-38393 alters the response to light of adult rats. *Neuroreport* 11: 1539–1541.

Ferguson S A, Kennaway D J, Moyer R W (1999). Nicotine phase shifts the 6-sulphatoxymelatonin rhythm and induces c-Fos in the SCN of rats. *Brain Res Bull* 48: 527–538.

Ferguson S A, Rowe S A, Krupa M, *et al.* (2000). Prenatal exposure to the dopamine agonist SKF-38393 disrupts the timing of the initial response of the suprachiasmatic nucleus to light. *Brain Res* 858: 284–289.

Fite K V, Janusonis S, Foote W, et al. (1999). Retinal afferents to the dorsal raphe nucleus in rats and Mongolian gerbils. *J Comp Neurol* 414: 469–484.

Fredholm B B (1995). Adenosine, adenosine receptors and the actions of caffeine. *Pharmacol Toxicol* 76: 93–101.

Gannon R L (2001). 5HT$_7$ receptors in the rodent suprachiasmatic nucleus. *J Biol Rhythms* 16: 19–24.

Gillespie C F, Van-Der B E, Mintz E M, et al. (1999). GABAergic regulation of light-induced c-Fos immunoreactivity within the suprachiasmatic nucleus. *J Comp Neurol* 411: 683–692.

Guldner F H, Bahar E, Young C A, et al. (1997). Structural plasticity of optic synapses in the rat suprachiasmatic nucleus: adaptation to long-term influence of light and darkness. *Cell Tissue Res* 287: 43–60.

Gustafson E L, Durkin M M, Bard J A, et al. (1996). A receptor autoradiographic and *in situ* hybridization analysis of the distribution of the 5-HT$_7$ receptor in rat brain. *Br J Pharmacol* 117: 657–666.

Hastings M H, Duffield G E, Ebling F J, et al. (1998). Non-photic signalling in the suprachiasmatic nucleus. *Biol Cell* 89: 495–503.

Heidmann D E A, Szot P, Kohen R, et al. (1998). Function and distribution of three rat 5-hydroxytryptamine 7 (5-HT$_7$) receptor isoforms produced by alternative splicing. *Neuropharmacology* 37: 1621–1632.

Herxheimer A, Petrie K J (2001). Melatonin for preventing and treating jet lag (Cochrane Review). In: *The Cochrane Library*, Issue 1. Oxford: Update Software.

Horikawa K, Yokota S, Fuji K, et al. (2000). Nonphotic entrainment by 5-HT$_{1A/7}$ receptor agonists accompanied by reduced *per1* and *per2* mRNA levels in the suprachiasmatic nuclei. *J Neurosci* 20: 5867–5873.

Kalkowski A, Wollnik F (1999). Local effects of the serotonin agonist quipazine on the suprachiasmatic nucleus of rats. *Neuroreport* 10: 3241–3246.

Kawano H, Decker K, Reuss S (1996). Is there a direct retina-raphe-suprachiasmatic nucleus pathway in the rat? *Neurosci Lett* 212: 143–146.

Kennaway D J (1993). Urinary 6-sulphatoxymelatonin excretory rhythms in laboratory rats – effects of photoperiod and light. *Brain Res* 603: 338–342.

Kennaway D J (1997). Light, neurotransmitters and the suprachiasmatic nucleus control of pineal melatonin production in the rat. *Biol Signals* 6: 247–254.

Kennaway D J, Moyer R W (1998). Serotonin 5-HT$_{2C}$ agonists mimic the effect of light pulses on circadian rhythms. *Brain Res* 806: 257–270.

Kennaway D J, Rowe S A, Ferguson S A (1996). Serotonin agonists mimic the phase shifting effects of light on the melatonin rhythm in rats. *Brain Res* 737: 301–307.

Kennaway D J, Moyer R W, Voultsios A, et al. (2001). Serotonin, excitatory amino acids and the photic control of melatonin rhythms and SCN c-FOS in the rat. *Brain Res* 897: 36–43.

Kinsey A M, Wainwright A, Heavens R, et al. (2001). Distribution of 5-HT$_{5A}$, 5-HT$_{5B}$, 5-HT$_6$ and 5-HT$_7$ receptor mRNAs in the rat brain. *Mol Brain Res* 88: 194–198.

Kohler M, Kalkowski A, Wollnik F (1999). Serotonin agonist quipazine induces photic-like phase shifts of the circadian activity rhythm and c-*fos* expression in the rat suprachiasmatic nucleus. *J Biol Rhythms* 14: 131–140.

Liu C, Gillette M U (1996). Cholinergic regulation of the suprachiasmatic nucleus circadian rhythm via a muscarinic mechanism at night. *J Neurosci* 16: 744–751.

Liu C, Ding J M, Faiman L E, *et al.* (1997a). Coupling of muscarinic cholinergic receptors and cGMP in nocturnal regulation of the suprachiasmatic circadian clock. *J Neurosci* 17: 659–666.

Liu C, Weaver D R, Jin X, *et al.* (1997b). Molecular dissection of two distinct actions of melatonin on the suprachiasmatic circadian clock. *Neuron* 19: 91–102.

Lovenberg T W, Baron B M, de-Lecea L, *et al.* (1993). A novel adenylyl cyclase-activating serotonin receptor ($5-HT_7$) implicated in the regulation of mammalian circadian rhythms. *Neuron* 11: 449–458.

Lowrey P L, Shimomura K, Antoch M P, *et al.* (2000). Positional syntenic cloning and functional characterization of the mammalian circadian mutation tau. *Science* 288: 483–492.

Maj J, Rogoz Z, Skuza G (1991). Antidepressant drugs increase the locomotor hyperactivity induced by MK-801 in rats. *J Neural Transm Gen Sect* 85: 169–179.

Manrique C, Francois Bellan A M, Segu L, *et al.* (1994). Impairment of serotoninergic transmission is followed by adaptive changes in $5HT_{1B}$ binding sites in the rat suprachiasmatic nucleus. *Brain Res* 663: 93–100.

McArthur A J, Gillette M U, Prosser R A (1991). Melatonin directly resets the rat suprachiasmatic circadian clock *in vitro*. *Brain Res* 565: 158–161.

Medanic M, Gillette M (1992). Serotonin regulates the phase of the rat suprachiasmatic circadian pacemaker *in vitro* only during the subjective day. *J Physiol* 450: 629–642.

Meijer J H, van-der-Zee E A, Dietz M (1988). Glutamate phase shifts circadian activity rhythms in hamsters. *Neurosci Lett* 86: 177–183.

Meijer J H, Schaap J, Watanabe K, *et al.* (1997). Multiunit activity recordings in the suprachiasmatic nuclei: *in vivo* versus *in vitro* models. *Brain Res* 753: 322–327.

Meyer J L, Hall A C, Harrington M E (1998). Histamine phase shifts the hamster circadian pacemaker via an NMDA dependent mechanism. *J Biol Rhythms* 13: 288–295.

Mintz E M, Gillespie C F, Marvel C L, *et al.* (1997). Serotonergic regulation of circadian rhythms in Syrian hamsters. *Neuroscience* 79: 563–569.

Mintz E M, Marvel C L, Gillespie C F, *et al.* (1999). Activation of NMDA receptors in the suprachiasmatic nucleus produces light-like phase shifts of the circadian clock *in vivo*. *J Neurosci* 19: 5124–5130.

Mistlberger R E (1991). Effects of daily schedules of forced activity on free-running rhythms in the rat. *J Biol Rhythms* 6: 71–80.

Miyake S, Sumi Y, Yan L, *et al.* (2000). Phase-dependent responses of *per1* and *per2* genes to a light-stimulus in the suprachiasmatic nucleus of the rat. *Neurosci Lett* 294: 41–44.

Moore R Y, Lenn N J (1972). A retinohypothalamic projection in the rat. *J Comp Neurol* 146: 1–14.

Moore R Y, Speh J C (1993) GABA is the principle neurotransmitter of the circadian system. *Neurosci. Lett.* 150: 112–116.

Morin L P (1994). The circadian visual system. *Brain Res Rev* 67: 102–127.

Moriya T, Yoshinobu Y, Ikeda M, *et al.* (1998). Potentiating action of MKC-242, a selective 5-HT$_{1A}$ receptor agonist, on the photic entrainment of the circadian activity rhythm in hamsters. *Br J Pharmacol* 125: 1281–1287.

Moriya T, Horikawa K, Akiyama M, *et al.* (2000). Correlative association between N-methyl-D-aspartate receptor-mediated expression of period genes in the suprachiasmatic nucleus and phase shifts in behavior with photic entrainment of clock in hamsters. *Mol Pharmacol* 58: 1554–1562.

Moyer R W, Kennaway D J (1999). Immunohistochemical localisation of serotonin receptors in the rat suprachiasmatic nucleus. *Neurosci Lett* 271: 147–150.

Moyer R W, Kennaway D J, Ferguson S A, *et al.* (1997). Quipazine and light have similar effects on c-*fos* induction in the rat suprachiasmatic nucleus. *Brain Res* 765: 337–342.

Mrosovsky N, Salmon P A (1987). A behavioural method for accelerating re-entrainment of rhythms to new light–dark cycles. *Nature* 330: 372–373.

Murakami N, Takahashi K, Kawashima K (1984). Effect of light on the acetylcholine concentrations of the suprachiasmatic nucleus in the rat. *Brain Res* 311: 358–360.

Neumaier J F, Sexton T J, Yracheta Y, *et al.* (2001). Localization of 5-HT$_7$ receptors in rat brain by immunocytochemistry, in situ hybridization, and agonist stimulated c-Fos expression. *J Chem Neuroanat* 21: 63–73.

Panula P, Pirvola U, Auvinen S, *et al.* (1989). Histamine-immunoreactive nerve fibers in the rat brain. *Neuroscience* 28: 585–610.

Parsons C G, Danysz W, Quack G (1999). Memantine is a clinically well tolerated N-methyl-D-aspartate (NMDA) receptor antagonist – a review of preclinical data. *Neuropharmacology* 38: 735–767.

Pickard G E, Rea M A (1997). TFMPP, a 5HT$_{1B}$ receptor agonist, inhibits light-induced phase shifts of the circadian activity rhythm and c-*fos* expression in the mouse suprachiasmatic nucleus. *Neurosci Lett* 231: 95–98.

Pickard G E, Weber E T, Scott P A, *et al.* (1996). 5HT$_{1B}$ receptor agonists inhibit light-induced phase shifts of behavioral circadian rhythms and expression of the immediate-early gene c-*fos* in the suprachiasmatic nucleus. *J Neurosci* 16: 8208–8220.

Pickard G E, Smith B N, Belenky M, *et al.* (1999). 5-HT$_{1B}$ receptor-mediated presynaptic inhibition of retinal input to the suprachiasmatic nucleus. *J Neurosci* 19: 4034–4045.

Prosser R A (1998). In vitro circadian rhythms of the mammalian suprachiasmatic nuclei: comparison of multi-unit and single-unit neuronal activity recordings. *J Biol Rhythms* 13: 30–38.

Prosser R A, Miller J D, Heller H C (1990). A serotonin agonist phase-shifts the circadian clock in the suprachiasmatic nuclei *in vitro*. *Brain Res* 534: 336–339.

Prosser R A, Dean R R, Edgar D M, *et al.* (1993). Serotonin and the mammalian circadian system. 1. *In vitro* phase shifts by serotonergic agonists and antagonists. *J Biol Rhythms* 8: 1–16.

Ralph M R, Menaker M (1985). Bicuculline blocks circadian phase delays but not advances. *Brain Res* 325: 362–365.

Ralph M R, Menaker M (1986). Effects of diazepam on circadian phase advances and delays. *Brain Res* 372: 405–408.

Rea M A, Pickard G E (2000). A 5-HT$_{1B}$ receptor agonist inhibits light-induced suppression of pineal melatonin production. *Brain Res* 858: 424–428.

Rea M A, Buckley B, Lutton L M (1993a). Local administration of EAA antagonists blocks light-induced phase shifts and c-*fos* expression in hamster SCN. *Am J Physiol* 265: R1191–R1198.

Rea M A, Michel A M, Lutton L M (1993b). Is Fos expression necessary and sufficient to mediate light-induced phase advances of the suprachiasmatic circadian oscillator? *J Biol Rhythms* 8(Suppl): S59–S64.

Rea M A, Glass J D, Colwell C S (1994). Serotonin modulates photic responses in the hamster suprachiasmatic nuclei. *J Neurosci* 14: 3635–3642.

Redman J, Armstrong S, Ng K T (1983). Free-running activity rhythms in the rat: entrainment by melatonin. *Science* 219: 1089–1091.

Reppert S M, Weaver D R, Stehle J H, *et al.* (1991). Molecular cloning and characterization of a rat A1-adenosine receptor that is widely expressed in brain and spinal cord. *Mol Endocrinol* 5: 1037–1048.

Rivkees S A, Lachowicz J E (1997). Functional D1 and D5 dopamine receptors are expressed in the suprachiasmatic, supraoptic, and paraventricular nuclei of primates. *Synapse* 26: 1–10.

Roca A L, Weaver D R, Reppert S M (1993). Serotonin receptor gene expression in the rat suprachiasmatic nuclei. *Brain Res* 608: 159–165.

Rowe S A, Kennaway D J (1996). Effect of NMDA receptor blockade on melatonin and activity rhythm responses to a light pulse in rats. *Brain Res Bull* 41: 351–358.

Scott G, Piggins H D, Semba K, *et al.* (1998). Actions of histamine in the suprachiasmatic nucleus of the Syrian hamster. *Brain Res* 783: 1–9.

Shen H, Semba K (1994). A direct retinal projection to the dorsal raphe nucleus in the rat. *Brain Res* 635: 159–168.

Sherwin C M (1998). Voluntary wheel running: a review and novel interpretation. *Anim Behav* 56: 11–27.

Shibata S, Tsuneyoshi A, Hamada T, *et al.* (1992). Phase-resetting effect of 8-OH-DPAT, a serotonin 1A receptor agonist, on the circadian rhythm of firing rate in the rat suprachiasmatic nuclei *in vitro*. *Brain Res* 582: 353–356.

Strother W N, Vorhees C V, Lehman M N (1998). Long-term effects of early cocaine exposure on the light responsiveness of the adult circadian timing system. *Neurotoxicol Teratol* 20: 555–564.

Swanson L W, Simmons D M, Whiting P J, *et al.* (1987). Immunohistochemical localization of neuronal nicotinic receptors in the rodent central nervous system. *J Neurosci* 7: 3334–3342.

Tamaru T, Isojima Y, Yamada T, *et al.* (2000). Light and glutamate-induced degradation of the circadian oscillating protein BMAL1 during the mammalian clock resetting. *J Neurosci* 20: 7525–7530.

Tcheng T K, Gillette M U (1996). A novel carbon fiber bundle microelectrode and modified brain slice chamber for recording long-term multiunit activity from brain slices. *J Neurosci Methods* 69: 163–169.

Tischkau S A, Gallman E A, Buchanan G F, *et al.* (2000). Differential cAMP gating of glutamatergic signaling regulates long-term state changes in the suprachiasmatic circadian clock. *J Neurosci* 20: 7830–7837.

Toh K L, Jones C R, He Y, *et al.* (2001). An h*per2* phosphorylation site mutation in familial advanced sleep-phase syndrome. *Science* 291: 1040–1043.

Tominaga K, Shibata S, Hamada T, *et al.* (1994). GABA$_A$ receptor agonist muscimol can reset the phase of neural activity rhythm in the rat suprachiasmatic nucleus *in vitro*. *Neurosci Lett* 166: 81–84.

Trachsel L, Heller H C, Miller J D (1995). Nicotine phase-advances the circadian neuronal activity rhythm in rat suprachiasmatic nuclei explants. *Neuroscience* 65: 797–803.

Turek F W, Losee-Olson S (1986). A benzodiazepine used in the treatment of insomnia phase-shifts the mammalian circadian clock. *Nature* 321: 167–168.

van den Pol A N, Tsujimoto K L (1985). Neurotransmitters of the hypothalamic suprachiasmatic nucleus: immunocytochemical analysis of 25 neuronal antigens. *Neuroscience* 15: 1049–1086.

Van der Zee E A, Streefland C, Strosberg A D, *et al.* (1991). Colocalization of muscarinic and nicotinic receptors in cholinoceptive neurons of the suprachiasmatic region in young and aged rats. *Brain Res* 542: 348–352.

Vanecek J, Pavlik A, Illnerova H (1987). Hypothalamic melatonin receptor sites revealed by autoradiography. *Brain Res* 435: 359–362.

van-Reeth O, Turek F W (1989). Stimulated activity mediates phase shifts in the hamster circadian clock induced by dark pulses or benzodiazepines. *Nature* 339: 49–51.

Vindlacheruvu R R, Ebling F JP, Maywood E S, *et al.* (1992). Blockade of glutaminergic neurotransmission in the suprachiasmatic nucleus prevents cellular and behavioural response of the circadian system to light. *Eur J Neurosci* 4: 673–679.

Vitaterna M H, King D P, Chang A M, *et al.* (1994). Mutagenesis and mapping of a mouse gene, *Clock*, essential for circadian behavior. *Science* 264: 719–725.

Watanabe A, Moriya T, Nisikawa Y, *et al.* (1996). Adenosine A1-receptor agonist attenuates the light-induced phase shifts and *fos* expression *in vivo* and optic nerve stimulation-evoked field potentials in the suprachiasmatic nucleus *in vitro*. *Brain Res* 740: 329–336.

Weaver D R, Reppert S M (1995). Definition of the developmental transition from dopaminergic to photic regulation of c-*fos* gene expression in the rat suprachiasmatic nucleus. *Mol Brain Res* 33: 136–148.

Weaver D R, Rivkees S A, Reppert S M (1992). D1-dopamine receptors activate c-*fos* expression in the fetal suprachiasmatic nuclei. *Proc Natl Acad Sci USA* 89: 9201–9204.

Wee B E, Anderson K D, Kouchis N S, *et al.* (1992). Administration of carbachol into the lateral ventricle and suprachiasmatic nucleus (SCN) produces dose-dependent phase shifts in the circadian rhythm of locomotor activity. *Neurosci Lett* 137: 211–215.

Wollnik F, Brysch W, Uhlmann E, *et al.* (1995). Block of c-*fos* and *jun*-B expression by antisense oligonucleotides inhibits light-induced phase shifts of the mammalian circadian clock. *Eur J Neurosci* 7: 388–393.

Wright K P, Badia P, Myers B L, *et al.* (1997). Caffeine and light effects on night-time melatonin and temperature levels in sleep-deprived humans. *Brain Res* 747: 78–84.

14

The effect of ageing on melatonin secretion: clinical aspects

Yvan Touitou

14.1 Melatonin secretion and control

14.1.1 Synthesis

Melatonin (N-acetyl-5-methoxytryptamine) is the main hormone secreted by the pineal gland. The name is derived from its effect on melanin pigmentation because it lightens the skin colour of amphibians. Melatonin is synthesised in the pineal gland. The precursor tryptophan is hydroxylated to 5-hydroxytryptophan and then decarboxylated to 5-hydroxytryptamine (5-HT, serotonin). The 5-HT is N-acetylated by the enzyme N-acetyltransferase (NAT) to N-acetylserotonin and finally converted to melatonin by the enzyme hydroxyindole O-methyltransferase (HIOMT) (Klein, 1979) (See Figure 14.1).

The pineal gland, a forebrain structure (~5–10 mm long and weighing 100–150 mg in humans) situated close to the upper midbrain, receives afferents from postganglionic sympathetic fibres that arise from the paired superior cervical ganglia (SGG) in the neck. The gland is highly vascularised, with a flow rate of around 4 mL/min/g. In humans, as in most mammals, the pineal gland is outside the blood–brain barrier and is thereby susceptible to the influence of peripherally active drugs (Vollrath, 1981).

14.1.2 Noradrenergic control

The main neurotransmitter regulating the synthesis of melatonin in the pineal gland is noradrenaline (norepinephrine; NA) released into the synaptic clefts between the sympathetic nerve endings and the pinealocyte. NA is released at night in response to stimulatory signals originating in the suprachiasmatic nuclei (SCN), and activates adenylate cyclase, which induces cAMP production (Klein, 1985). This in turn activates

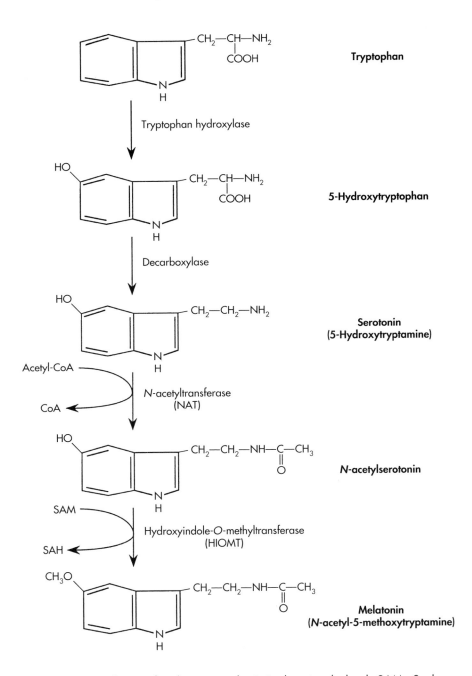

Figure 14.1 Pathway of melatonin synthesis in the pineal gland. SAM, S-adenosylmethionine; SAH, S-adenosylhomocysteine.

N-acetyltransferase (NAT), the key enzyme in melatonin synthesis. Whereas HIOMT displays negligible rhythmicity, pineal NAT exhibits a dramatic rhythm with a nocturnal activity 50 to 100 times higher than the diurnal activity.

β-Adrenoreceptor antagonists such as propranolol and atenolol can reduce melatonin secretion in humans (Vaughan *et al.*, 1976; Cowen *et al.*, 1983; Arendt *et al.*, 1985). By contrast, β-adrenoreceptor agonists, at least at the doses used, do not stimulate secretion (Lewy, 1984; Berlin *et al.*, 1995) when given in the morning, when melatonin secretion is physiologically low.

14.1.3 Light control

The light–dark cycle is the main synchroniser of hormone control: it entrains the SCN and thus a number of circadian rhythms, including that of melatonin. Under normal environmental conditions, light transmitted through the eyes is the key zeitgeber for the entrainment of the circadian clock. In the absence of light as a synchroniser, the melatonin circadian rhythm free-runs: it is no longer synchronised with the environment, and becomes out of phase with the external light–dark cycle (Reinberg and Touitou, 1996). This rhythm desynchronisation occurs in, among other people, the blind, shift workers, night workers, and travellers on or immediately after transmeridian flights (see Chapter 12). Besides its effect as a synchroniser, bright light can acutely suppress melatonin secretion in humans (Lewy *et al.*, 1980).

Although extraocular photoreception can entrain the circadian rhythm of invertebrates and nonmammalian vertebrates, the mammalian circadian system seems to be affected only by light to the eyes. It has been reported that light directed to the back of the knee shifted melatonin and core body temperature circadian rhythms in humans (Campbell and Murphy, 1988). However, other studies on extraocular light failed to show a phase shift of melatonin, cortisol and thyroid-stimulating hormone (TSH) circadian rhythms in humans (Lindblom *et al.*, 2000a; Eastman *et al.*, 2000), or a reduction of melatonin secretion (Lockley *et al.*, 1998; Hébert *et al.*, 1999; Rogers *et al.*, 1999; Lindblom *et al.*, 2000b).

14.1.4 Circadian rhythm

The high-amplitude circadian rhythm of melatonin observed in the pineal gland is reflected in plasma levels, with low levels during the day and high levels at night. This circadian rhythmicity is comparable in humans and other primates as well as in rodents (either diurnal or nocturnal), with plasma concentrations up to ten times (or more) higher at night than during the day. For individuals of any age, the circadian rhythm of melatonin is highly consistent from day to day. Melatonin circadian rhythmicity is maintained in the elderly (Touitou *et al.*, 1981, 1985; Iguchi *et al.*, 1982) (see Figure 14.2) regardless of the season (Touitou *et al.*, 1984) (see Figure 14.3).

14.1.5 Metabolism

Circulating melatonin is metabolised primarily by the liver P450 enzymes, which hydroxylate melatonin at the carbon-6 position, followed by sulfate (mainly) and glucuronic acid conjugation. The main urinary metabolite is 6-sulfatoxymelatonin, and the first morning urine collection provides a good index of nocturnal secretion. Melatonin clearance from the peripheral circulation is biphasic, with half-lives of about 3 min and 45 min.

14.2 Pharmacokinetics and toxic effects

Melatonin is usually administered orally at doses of 1–5 mg, which results in pharmacological levels in plasma. It has also been found that an oral administration of around 0.3 mg given 2–4 h before habitual bedtime results in plasma melatonin levels similar to normal night-time levels (Zhdanova *et al.*, 1996). Thirty minutes to 180 min after oral administration (80 mg), plasma melatonin concentration increases with an absorption half-life of 0.4 h and an elimination half-life of 0.8 h (Waldhauser *et al.*, 1984; Aldhous *et al.*, 1985). Large inter-individual variations that can reach a 25-fold difference between subjects have been observed for melatonin absorption (Waldhauser *et al.*, 1984). Plasma melatonin concentrations and melatonin half-lives depend on the dose, the time of administration and the type of oral preparation used (Guardiola-Lemaître, 1997). The time of administration is critical because melatonin receptor sensitivity has a time window of 17.00 to 20.00, in both rats and humans (Masson-Pévet *et al.*, 1993). Melatonin's high-amplitude circadian rhythm, with its high nocturnal and low daytime concentrations, must therefore be preserved when exogenous melatonin is administered for therapeutic purposes. It remains to be determined whether the characteristics of melatonin pharmacokinetics in the elderly are the same as those found in young adults.

Melatonin administered orally (10 mg/day for 4 weeks) to healthy

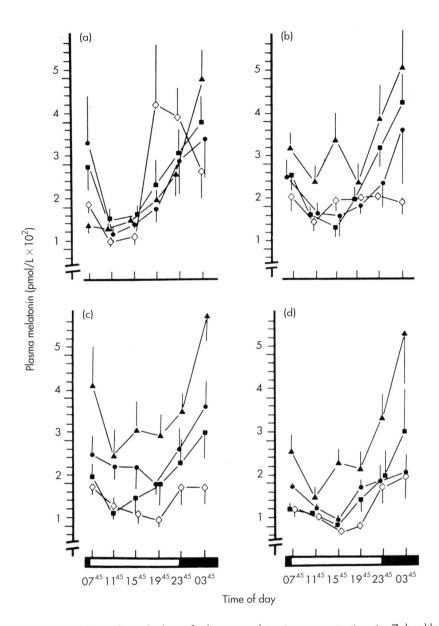

Figure 14.2 Circadian rhythm of plasma melatonin concentration in 7 healthy young men 24 years old (solid triangles), 6 elderly men (solid squares), 6 elderly women (solid circles) and 6 elderly patients of both sexes (open diamonds) in their eighties with senile dementia of the Alzheimer's type. The experiment was performed four times a year, in (a) January, (b) March, (c) June and (d) October. Reproduced from Touitou et al. (1981).

volunteers did not produce any alterations in a large set of biochemical variables (Seabra *et al.*, 2000). The most relevant side effect mentioned in various papers is somnolence and headache (Seabra *et al.*, 2000; Avery *et al.*, 1998).

14.3 Melatonin and ageing: physiological aspects

14.3.1 Fetal period and infancy

Studies in sheep (Yellon and Longo, 1988; McMillen and Nowak, 1989) and rats (Blazquez *et al.*, 1989) failed to show any production of melatonin by the fetus. In humans, no data are available on the fetal period.

During the fetal period and in the weeks after birth, human fetuses and infants rely on the circadian rhythm in their mother's melatonin through the placenta or milk. Since melatonin has sleep-inducing properties in humans, it would be interesting to know whether breast-fed babies are more synchronised to their mothers' sleep–wake cycle than are bottle-fed babies.

The circadian rhythm of melatonin develops between the second and third months of life; night-time melatonin is low or undetectable up to 2–3 months of age, then increases steadily (Waldhauser *et al.*, 1988; Kennaway *et al.*, 1992). The highest night-time serum melatonin levels have been observed in children aged 1–3 years. During puberty, blood melatonin concentration declines progressively by about 80% from its childhood levels (Waldhauser *et al.*, 1988).

14.3.2 Melatonin in the elderly

Although the circadian rhythm in melatonin is maintained in the elderly, some differences have been reported in the secretory pattern of the hormone. The main difference is that plasma melatonin concentration decreases with age in humans (see Figures 14.2 and 14.4) whether healthy (Touitou and Haus, 2000) or not (Touitou *et al.*, 1981; Luboshitzky *et al.*, 2001). Data dealing with the effects of age on melatonin secretion arise mainly from cross-sectional studies comparing two or more groups of human subjects differing in age from 20 to 90 years old (see Tables 14.1 and 14.2). The rate of the nocturnal decrease of melatonin in the elderly is, on average, 40–50%, with an overall range from 20% to 80% (see Table 14.1).

Zeitzer *et al.* (1999), using a constant-routine protocol, failed to

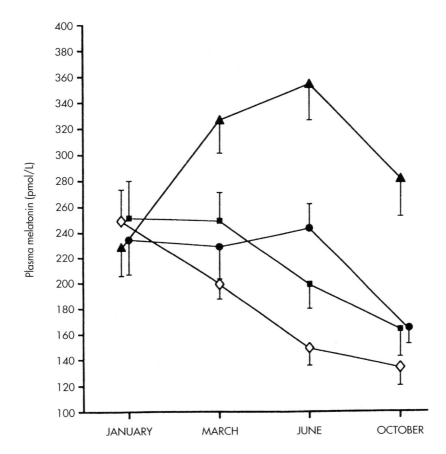

Figure 14.3 Seasonal variation of plasma melatonin concentration in 7 healthy young men 24 years old (solid triangles), 6 elderly men (solid squares), 6 elderly women (solid circles) and 6 elderly patients of both sexes (open diamonds) in their eighties with senile dementia of the Alzheimer's type. The experiment was performed in January, March, June and October. Reproduced from Touitou et al. (1984).

show a nocturnal decline of plasma melatonin in subjects in their seventies when compared to young men in their twenties. Nonetheless, they did find a small group of elderly subjects whose plasma melatonin cycle had a significantly lower amplitude. It is likely that the main reason for this discrepancy is to be found in the study conditions. Throughout the constant-routine protocol, which lasted at least 30 h, subjects remained awake, in bed in a semirecumbent position under constant dim ambient illumination of less than 15 lux and received equicaloric snacks and fluids hourly. All other studies were performed in subjects living in their normal

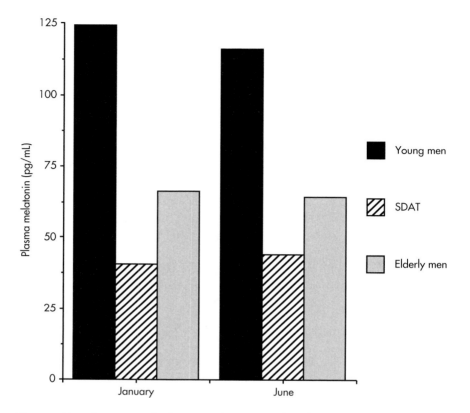

Figure 14.4 Decrease of plasma melatonin peak values in the elderly at two opposite seasons. SDAT, senile dementia of the Alzheimer type. Reproduced from Touitou et al. (1984).

environment and most of them were controlled for the use of melatonin-suppressing drugs (β-adrenoreceptor antagonists, NSAIDs), for the lighting regimen and for medical conditions. Further research is therefore needed first to confirm then to explain this difference and to assess the effect of constant-routine conditions on elderly subjects.

In another study, Kennaway et al. (1999) reported an effect of age on the amount of 6-sulfatoxymelatonin excreted in subjects aged between 21 and 82 years, with a decrease of 40% in the oldest subjects. They suggest from their data and their re-evaluation in the same paper of data in the literature (ANOVA analysis) that after the age of 30 the decrease of melatonin concentration is no longer significant (Kennaway et al., 1999).

The mechanism by which melatonin levels appear to fall with age

Table 14.1 Comparison of night-time plasma melatonin concentrations (or 24 h urinary 6-sulfatoxymelatonin concentrations) in young and elderly subjects

Authors	Mean age (years) and no. of subjects				Sex	Per cent decrease in the elderly	Comments and additional findings in the elderly compared to young subjects
	Young		Elderly				
	Age	No.	Age	No.			
Touitou et al. (1981)	24	7	72	18	M, F	45	Rhythm conservation in SDAT patients with low amplitude (plasma)
Iguchi et al. (1982)	26	5	84	5	M	88	Plasma
Claustrat et al. (1984)	27	8	62	8	M	–	Plasma
Bartsch et al. (1985)	33	10	70	13	M	30	Plasma
Cowen et al. (1985)	30	15	56	14	M, F	–	Plasma
Thomas and Miles (1989)	26	10	73	9	M, F	62	Plasma
Van Coevorden et al. (1991)	23	8	75	8	M	47	Phase advance (plasma)
Ferrari et al. (1995)	25	14	78	16	F	35	Young women in early follicular phase (plasma)
Cagnacci et al. (1995)	27	8	58	6	F	18	Phase advance in the elderly. Young women in early follicular phase (plasma)
Uchida et al. (1996)	21	12	66	13		26	–
Hajak et al. (1996)	26	5	49	5	M	31	After flunitrazepam administration (plasma)
Magri et al. (1997)	27	22	80	52	F	45	Low-amplitude circadian rhythm in SDAT patients (plasma)
Mazzaccoli et al. (1997)	45	7	69	7		–	Plasma
Ohashi et al. (1997)	21	10	66	13	M, F	–	Offset time advances in the elderly (plasma)
Zeitzer et al. (1999)	23	98	68	34	M, F	–	Constant-routine protocol (plasma)
Luboshitzky et al. (2001)							
AD	33	12	73	22	M	61	24 h collection of urine
Healthy	33	12	61	12	M	40	

SDAT, senile dementia of the Alzheimer type; AD, patients with Alzheimer's disease.

Table 14.2 Transverse studies on plasma melatonin (Mel) and 24 h urinary 6-sulfatoxymelatonin (6SMel) concentrations in different age groups of human subjects

Author	Fluid	Characteristics	Age/age range (years) Young → Middle-aged → Old					Conclusions
			19–25	45–50	51–55	56–65	66–75	
Nair et al. (1986)	Plasma Mel	Age (years)						Decrease with age correlated for the 24 h secretion and the peak concentration from 19 to 65 years old. Phase delay. No correlation with height, weight, obesity.
		No. of subjects	12	5	10	9	10	
		Men/women[a]	5/7	–/5	6/4	5/4	5/5	
		Mean age (years)	24/22	–/46	51/52	61/63	69/68	
		Value (pg/mL)	112/125				62/58	
			19–25		42–65		66–89	
Sharma et al. (1989)	Plasma Mel	Age (years)						40% decrease in plasma peak value and 24 h mean secretion. Phase delay.
		No. of subjects	19		27		25	
		Men/women[a]	13/6		15/12		16/9	
		Mean age (years)	22		57		70	
		Value (pg/mL)	118.9		71.5		58.8	
			20–39	40–59		60–79	80	
Sack et al. (1986)	Urine 6SMel	Age (years)						Decline by about 50% in the two oldest groups compared to the youngest one. Sex not taken into account.
		No. of subjects	20	15		15	10	
		Men/women[a]	10/10	7/8		8/7	5/5	
		Value (nmol/24 h)	12.1/11.4	11.1/6.0		6.3/5.1	6.2/5.8	

Table 14.2 Continued

Author	Fluid	Characteristics	Age/age range (years) Young → Middle-aged → Old	Conclusions
Kennaway et al. (1999)	Urine 6SMel	Age (years)	20–35 / 35–50 / 50–65 / 65–80	Decrease with age of the three oldest groups of 16%, 30% and 37%, respectively. Sex not taken into account (105 men, 146 women).
		No. of subjects (M+W)	64 / 48 / 97 / 42	
		Value (nmol/24 h)	49 ± 29 / 41 ± 29 / 34 ± 26 / 31 ± 22	
Bojkowski and Arendt (1990)	Urine 6SMel	Age (years)	20–24.9 / 25–29.9 / 30–39.9 / 40–49.9 / 50–59.9 / 60–80	No sex difference. No difference according to pineal calcification.
		No. of subjects (M+W)	31 / 15 / 16 / 9 / 8 / 11	
		Value (pg/mL)	9.5 ± 4.5 / 10.2 ± 6.1 / 10.8 ± 5.0 / 8.0 ± 4.0 / 7.8 ± 4.1 / 4.0 ± 2.7	
Skene et al. (1990)	Urine 6SMel	Age (years)	40–44 / 45–49 / 50–54 / 55–59 / 60–64 / 65–69	Decline in the three oldest groups when compared to the youngest.
		No. of subjects (W)	37 / 29 / 26 / 43 / 10 / 6	
		Value (pg/mL)	6.5 ± 4.2 / 5.1 ± 2.7 / 5.3 ± 3.0 / 4.2 ± 2.6 / 4.2 ± 2.6 / 3.3 ± 1.4	
Liu et al. (1999)	CSF Mel	Age (years)	41–80 / >80	
		No. of subjects	53 / 29	
		Value (pg/mL)	330 / 176	

[a]Values represent data for men/women.

is not yet fully understood. In aged rodents, Greenberg and Weiss (1978) showed that β-adrenoreceptors in the pinealocyte membranes were fewer and/or were less responsive to noradrenaline. Progressive alterations in the concentrations of α_1-adrenoreceptors occur in the hypothalamus in the ageing rat. These alterations involve both a decrease in receptor concentration and changes in the diurnal rhythm; they appear to be limited to those brain regions associated with rhythmic functions (Weiland and Wise, 1990). Increased clearance could also account for the decline of plasma melatonin with age, but the finding by Sack et al. (1986) of reduced urinary excretion of 6-hydroxymelatonin in aged human subjects makes this explanation less likely. In addition, the positive correlation between plasma melatonin and urinary 6-hydroxymelatonin excretion found by Markey et al. (1985) makes it clear that, among subjects with normal liver function, low plasma melatonin levels do not result primarily from more rapid metabolism and clearance but rather from a lower pineal production. Calcareous deposits become prominent in the pineal gland from adolescence onwards. These concretions consist mainly of calcium and phosphorus in the form of hydroxyapatite, and variable amounts of magnesium, trace elements and organic substances. Calcification of the pineal gland is common with ageing and concretions increase in number and size with age (Vollrath, 1981), but the degree of calcification varies between individuals and can start very early in life: 3% of 1-year-olds and 7% of 10-year-olds examined by computer tomography showed pineal concretions (Winkler and Helmke, 1987). We do not yet know whether this pineal calcification has any physiological significance in humans. Rodent species such as Syrian hamsters, which characteristically develop heavy pineal calcification, also have substantially lower pineal melatonin levels (Reiter et al., 1980). Studies in humans, however, suggest that calcification of the pineal body alters neither the histology of the pinealocyte (Tapp and Huxley, 1972) nor the pineal activity of HIOMT (Wurtman et al., 1964). It has been shown in old rats that the activity of NAT, the key enzyme responsible for the synthesis of melatonin, exhibits a dramatic decrease in the pineal (see Figure 14.5), which parallels the decline of rat serum melatonin (see Figure 14.6) irrespective of circadian stage (Selmaoui and Touitou, 1999). This decline in pineal NAT activity may explain the decrease of plasma melatonin observed with ageing.

In conclusion, although the mechanisms by which melatonin levels are reduced with age are not yet clearly understood, the explanation appears to lie in modifications of pineal melatonin production.

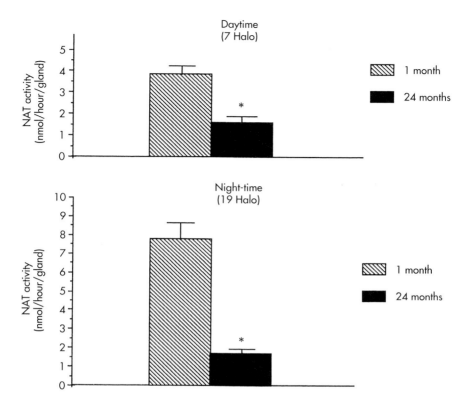

Figure 14.5 Decrease of N-acetyltransferase (NAT) activity in old (24 months) rats compared to young (1 month) rats. This decrease is observed during both daytime at 7 HALO (hours after light onset) and night-time at 19 HALO. (7 HALO = middle of light span; 19 HALO = middle of dark span.) Data are means ± SEM ($n = 8$). *Old rats had 2 to 3.6 times lower pineal NAT activity than young rats ($p < 0.05$). Reproduced from Selmaoui and Touitou (1999).

14.4 Effects of melatonin on sleep

One of the hallmarks of ageing is change in the sleep–wake pattern. Forty per cent to 70% of the elderly population present chronic sleep disturbances, with nocturnal awakening and difficulties in falling asleep (Cutler *et al.*, 1997; Maggi *et al.*, 1998; Uchimura *et al.*, 1998). There is no strong evidence that physiological secretion of melatonin regulates the sleep cycle, although it has been suggested that impaired melatonin secretion may contribute to the increased frequency of sleep disorders in the elderly. In one study, urinary 6-sulfatoxymelatonin excretion was

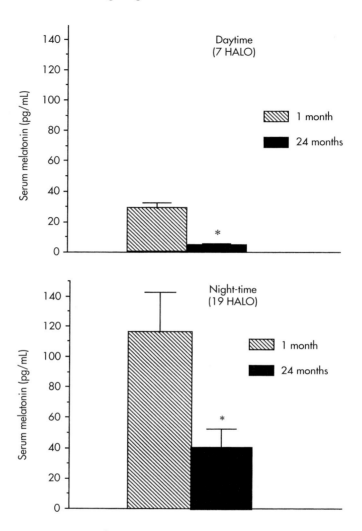

Figure 14.6 Decrease of serum melatonin in old (24 months) rats compared to young (1 month) rats. This decrease is observed during both daytime at 7 HALO (hours after light onset) and night-time at 19 HALO. (7 HALO = middle of light span; 19 HALO = middle of dark span.) Data are means ± SEM ($n = 8$). *Old rats had 3 to 6 times lower serum levels than young rats ($p < 0.05$). Reproduced from Selmaoui and Touitou (1999).

significantly lower, with a delayed peak time, in insomniac elderly subjects than in age-matched noninsomniac individuals (Haimov et al., 1994). In another study, electrophysiological recordings showed that the steepest increase in nocturnal sleepiness correlated with the rise in urinary 6-sulfatoxymelatonin excretion (Tzischinsky *et al.*, 1992). It has been reported that disrupted sleep and decreased excretion of 6-sulfatoxymelatonin evaluated at a single time point were the same in healthy elderly individuals and in patients with Alzheimer's disease (Luboshitzky *et al.*, 2001).

The literature indicates that supraphysiological or pharmacological doses of melatonin may treat sleep disturbances effectively in elderly people (review in Dawson and Encel, 1993), although these effects are not as pronounced as those obtained with classical hypnotic drugs such as benzodiazepines. In elderly insomniacs, melatonin replacement therapy may be beneficial in the initiation and maintenance of sleep. However, sleep quality deteriorates after cessation of treatment (Haimov *et al.*, 1995). It may nonetheless be the circadian phase rather than the amount of melatonin produced that is related to sleep quality. More studies on a large number of elderly insomniac patients are needed to assess the effect of melatonin in these patients.

Administration of melatonin decreases the time to sleep onset and to stage 2 sleep, without altering sleep architecture (Reid *et al.*, 1996; Gilbert *et al.*, 1999). Actual sleep time and sleep efficiency are increased (Tzischinsky and Lavie, 1994; Nave *et al.*, 1995, 1996; Attenburrow *et al.*, 1996). The duration of REM sleep does not seem to be affected by melatonin (Dijk *et al.*, 1995). Its effect on non-REM sleep is more controversial: one study reported no effect (Dijk *et al.*, 1995), while another reported an increase (Attenburrow *et al.*, 1996). Because melatonin decreases body temperature, it may also regulate the sleep–wake cycle through thermoregulatory mechanisms (Dawson and Encel, 1993; Dawson *et al.*, 1996; Hughes and Badia, 1997). It is unclear whether a similar link exists between the hypothermic effect of melatonin and its sedative action.

Melatonin has also been used to treat sleep disorders in subjects with total blindness (Tzischinsky *et al.*, 1992; Sack *et al.*, 2000) or jet lag (Suhner *et al.*, 1998; Arendt *et al.*, 1997; see also Chapter 12) and in those with neurological diseases such as Alzheimer's disease (Brusco *et al.*, 1999) and tuberous sclerosis (O'Callaghan *et al.*, 1999).

14.5 Melatonin as a resynchronising agent

As has been fully described in Chapter 1, the organism is normally synchronised with, and by, environmental factors (zeitgebers) that include the light–dark cycle, the sleep–wake cycle, meal schedules and seasonal factors (modifications of photoperiod and outside temperature).

Desynchronisation occurs when the biological clock is no longer in step with its surroundings. This situation is found not only with jet lag, shift work and night work and in totally blind people (Touitou and Haus, 1994) but also in some elderly people and in some diseases including depression and cancer. In those situations, both light and melatonin, given at the proper time, are signals that can resynchronise the biological clock. Depending on the time of day melatonin is given, the hormone delays or advances the phase of its own rhythm according to a phase–response curve (see Chapter 1; and also Lewy *et al.*, 1992, 1996, 1998; Zaidan *et al.*, 1994). Exogenous melatonin administration in the evening can advance the phase of the circadian system by about 45–60 min (Lewy *et al.*, 1992, 1996), whereas morning administration causes a phase delay (Mallo *et al.*, 1988; Lewy *et al.*, 1992). The phase advance is more effective than the phase delay (Zaidan *et al.*, 1994). To delay the circadian rhythm, melatonin should be administered when endogenous melatonin levels drop (Sack *et al.*, 1996). To phase-advance the circadian rhythm, i.e. to obtain an earlier sleep period, melatonin seems most effective when administered about 5 h before the time that endogenous melatonin starts to increase (Lewy *et al.*, 1992). Data reported by Dubocovich *et al.* (1998) showed that melatonin phase-advances circadian rhythms by activation of a membrane-bound receptor, the MT_2 receptor subtype.

14.6 Melatonin as a free-radical scavenger

Abnormal production of free radicals is observed in several human diseases as different as atherosclerosis, diabetes, neurodegenerative disorders, Parkinson's and Alzheimer's diseases, rheumatoid arthritis, senile cataract and cancers (Harman, 1984). These aggressive oxygen species (hydroxyl radicals, superoxide radicals, hydrogen peroxide and singlet oxygen, $^1O_2$) can damage all of the biochemical components of the organism – DNA, RNA, proteins, carbohydrates, unsaturated lipids, etc.

Free radicals are produced continuously in cells as by-products of oxidative phosphorylation in mitochondria and during fatty acid oxidation in peroxisomes. Cells are protected to a certain degree from the detrimental effects of free radicals by antioxidants, which include enzymes (superoxide dismutase, catalase, glutathione peroxidase), vitamins (vitamin E, vitamin C, vitamin A and its precursor, carotene) and various other molecules such as glutathione and uric acid (Harman, 1995). The extremely short half-life of free radicals (the estimated half-life within cells of the hydroxyl radical •OH is around 1×10^{-9} s) prevents direct investigation of their activity. Indirect information is available from measuring antioxidant levels (enzymes, vitamins, glutathione).

The free radical theory of ageing states that age-related dysfunctions are related in part to the effects of free radicals on organs, cells and subcellular constituents (Harman, 1995). Melatonin has recently been shown to possess antioxidative properties (Reiter *et al.*, 2000). In one study, melatonin appeared to be more effective as an antioxidant than other molecules including vitamin E, glutathione and mannitol (Reiter *et al.*, 1993). It is an efficient direct scavenger of the highly toxic hydroxyl radical, •OH (Tan *et al.*, 1993; Susa *et al.*, 1997). Melatonin detoxifies this radical by electron donation and itself becomes a radical (melatonyl cation radical), albeit less reactive and less toxic (Hardeland *et al.*, 1995). *In vitro* it also reduces oxidative breakdown of various molecules in the cell by activating enzymes related to the antioxidative defence system, including superoxide dismutase, catalase and glutathione peroxidase (Antolin *et al.*, 1996; Kotler *et al.*, 1998). Melatonin has been shown to inhibit the activity of the pro-oxidative enzyme nitric oxide synthase (NOS), which reduces NO formation (Pozo *et al.*, 1994). Melatonin present in cellular membranes is localised in lipid bilayers (Ceraulo *et al.*, 1999) and can act locally as a free-radical scavenger. It has also been shown that melatonin stabilises cell membrane fluidity, which is essential for cell function (Garcia *et al.*, 1997). The antioxidant effects of melatonin occur at high concentrations, thousands of times higher than physiological levels. It has been shown that melatonin can be both antioxidant and pro-oxidant *in vitro* in live human liver cell line HepG$_2$, depending on the concentration and incubation time (Osseni *et al.*, 2000). Whether physiological concentrations of melatonin can affect the human antioxidative defence system *in vivo* remains to be determined. Because melatonin is a free-radical scavenger, it has been proposed as of potential benefit in treating cancer and in modulating ageing.

14.7 Conclusion

Melatonin is undoubtedly useful in humans as a drug for manipulation of the circadian clock to resynchronise subjects desynchronised by jet lag, shift work, night work, blindness or old age (Touitou, 2001). Timed melatonin may also be beneficial to patients suffering from delayed sleep phase syndrome. Because melatonin levels decline with age in humans, as does the antioxidant status of the organism, it has been assumed that the hormone may play a part in some ageing processes. Other suggested applications include its use as a prophylactic antioxidant treatment for ischaemia, for example. Further assessment on an appropriate number of patients is required for this indication, which is of interest from a clinical point of view. Melatonin has also been proposed as a drug to extend lifespan in rodents (Pierpaoli and Regelson, 1994), but a number of shortcomings of this study have been reported (Reppert and Weaver, 1995). In any case, ageing in humans is far from thoroughly understood yet and, in the absence of experimental data, no conclusion can be drawn about the effect of melatonin on human lifespan. Further large-scale human research is needed to understand the specific beneficial effects of melatonin in the aged as well as the optimal conditions for its administration (treatment duration, dosage, possible chronic toxicity, side effects and interactions with other drugs).

References

Aldhous M, Franey C, Wright J, *et al.* (1985). Plasma concentrations of melatonin in man following oral absorption of different preparations. *Br J Clin Pharmacol* 19: 517–521.

Antolin I, Rodriguez C, Sainz R M, *et al.* (1996). Neurohormone melatonin prevents cell damage effect on gene expression for antioxidant enzymes. *FASEB J* 10: 882–890.

Arendt J, Bojkowski C, Franey C, *et al.* (1985). Immunoassay of 6-hydroxymelatonin sulphate in human plasma and urine: abolition of the urinary 24-hour rhythm with atenolol. *J Clin Endocrinol Metab* 60: 1166–1173.

Arendt J, Skene D J, Middleton B, *et al.* (1997). Efficacy of melatonin treatment in jet lag, shift work, and blindness. *J Biol Rhythms* 12: 604–617.

Attenburrow M E, Cowen P J, Sharpley A L (1996). Low dose melatonin improves sleep in healthy middle-aged subjects. *Psychopharmacology* 126: 179–181.

Avery D, Lenz M, Landis C (1998). Guidelines for prescribing melatonin. *Ann Med* 30: 122–130.

Bartsch C, Bartsch H, Fluchter S H, *et al.* (1985). Evidence for modulation of melatonin secretion in men with benign and malignant tumors of the prostate: relationship with pituitary hormones. *J Pineal Res* 2: 121–132.

Berlin I, Touitou Y, Guillemant S, *et al.* (1995). Beta-adrenoceptor agonists do not stimulate daytime melatonin secretion in healthy subjects. A double blind placebo controlled study. *Life Sci* 156: PL325–PL331.

Blazquez E, Lopez Gil A, Alvarez E, *et al.* (1989). Biochemical and ultrastructural approaches to the onset of the pineal melatonin rhythm in the rat. *Neuroendocrinology* 5: 500–505.

Bojkowski C J, Arendt J (1990). Factors influencing urinary 6-sulfatoxymelatonin, a major melatonin metabolite, in normal human subjects. *Clin Endocrinol* 33: 435–444.

Brusco L I, Fainstein I, Marquez M, *et al.* (1999). Effect of melatonin in selected populations of sleep-disturbed patients. *Biol Signals Recept* 8: 126–131.

Cagnacci A, Soldani R, Yen SS (1995). Hypothermic effect of melatonin and nocturnal core body temperature decline are reduced in aged women. *J Appl Physiol* 78: 314–317.
Campbell S S, Murphy P J (1998). Extraocular circadian phototransduction in humans. *Science* 279: 396–399.
Ceraulo L, Ferrugia M, Tesoriere L, et al. (1999). Interactions of melatonin with membrane models: portioning of melatonin in AOT and lecithin reversed unicells. *J Pineal Res* 26: 108–112.
Claustrat B, Chazot G, Brun J, et al. (1984). A chronobiological study of melatonin and cortisol secretion in depressed subjects: plasma melatonin, a biochemical marker in major depression. *Biol Psychiatry* 19: 1215–1228.
Cowen P J, Fraser S, Sammons R, et al. (1983). Atenolol reduces plasma melatonin concentration in man. *Br J Clin Pharmacol* 15: 579–581.
Cowen P J, Bevan J S, Gosden B, et al. (1985). Treatment with beta-adrenoceptor blockers reduces plasma melatonin concentration. *Br J Clin Pharmacol* 19: 258–260.
Cutler N L, Hughes R J, Singer C M, et al. (1997). Insomnia in older individuals: is early morning awakening the main problem? *Sleep Res* 26: 348.
Dawson D, Encel N (1993). Melatonin and sleep in humans. *J Pineal Res* 15: 1–12.
Dawson D, Gibbon S, Singh P (1996). The hypothermic effect of melatonin on core body temperature: is more better? *J Pineal Res* 20: 192–197.
Dijk D J, Roth C, Landolt H P, et al. (1995). Melatonin effect on daytime sleep in men: suppression of EEG low frequency activity and enhancement of spindle frequency activity. *Neurosci Lett* 201: 13–16.
Dubocovich M L, Yun K, Al-Ghoul W M, et al. (1998). Selective MT2 melatonin receptor antagonists block melatonin-mediated phase advances of circadian rhythms. *FASEB J* 12: 1211–1220.
Eastman C I, Martin S K, Hebert M (2000). Failure of extraocular light to facilitate circadian rhythm reentrainment in humans. *Chronobiol Int* 17: 807–826.
Ferrari E, Magri F, Dori D, et al. (1995). Neuroendocrine correlates of the ageing brain in humans. *Neuroendocrinology* 61: 464–470.
Garcia J J, Reiter R J, Guerrero J M, et al. (1997). Melatonin presents changes in microsomal membrane fluidity during induced lipid peroxidation. *FEBS Lett* 408: 297–308.
Gilbert S S, van den Heuvel C J, Dawson D (1999). Daytime melatonin and temazepam in young adult humans: equivalent effects on sleep latency and body temperatures. *J Physiol Lond* 514: 905–914.
Greenberg L H, Weiss B (1978). β-Adrenergic receptors in aged rat brain: reduced number and capacity of pineal to develop supersensitivity. *Science* 201: 61–63.
Guardioloa-Lemaître B (1997). Toxicology of melatonin. *J Biol Rhythms* 12: 697–706.
Haimov I, Landon M, Zisapel N, et al. (1994). Impaired 6-sulfatoxymelatonin rhythms in the elderly: coincidence with sleep disorders. *Br Med J* 309: 167.
Haimov I, Lavie P, Landon M, et al. (1995). Melatonin replacement therapy of elderly insomniacs. *Sleep* 18: 598–603.
Hajak G, Rodenbeck A, Bandelow B, et al. (1996). Nocturnal plasma melatonin levels after flunitrazepam administration in healthy subjects. *Eur Neuropsychopharmacol* 6: 149–153.

Hardeland R, Balzer I, Poeggeler B, et al. (1995). On the primary functions of melatonin in evolution: mediation of photoperiodic signals in a unicel photo-oxidation and scavenging of free radicals. *J Pineal Res* 18: 104–111.

Harman D (1984). Free radical theory of ageing: the 'free radical' diseases. *Age* 7: 111–131.

Harman D (1995). Role of antioxidant nutrients in ageing: overview. *Age* 18: 51–62.

Hébert M, Martin S K, Eastman C I (1999). Nocturnal melatonin secretion is not suppressed by light exposure behind the knee in humans. *Neurosci Lett* 274: 127–130.

Hughes R J, Badia P (1997). Sleep-promoting and hypothermic effects of daytime melatonin administration in humans. *Sleep* 20: 124–131.

Iguchi H, Kato K I, Ibayashi H (1982). Age-dependent reduction in serum melatonin concentrations in healthy human subjects. *J Clin Endocrinol Metab* 55: 27–29.

Kennaway D J, Stamp G E, Goble F C (1992). Development of melatonin production in infants and the impact of prematurity. *J Clin Endocrinol Metab* 75: 367–369.

Kennaway D J, Lushington K, Dawson D, et al. (1999). Urinary 6-sulfatoxymelatonin excretion and ageing: new results and a critical review of the literature. *J Pineal Res* 27: 210–220.

Klein D C (1979). Circadian rhythms in the pineal gland. In: Krieger D T, ed. *Endocrine Rhythms*. New York: Raven Press, 203–223.

Klein D C (1985). Photoneural regulation of the mammalian pineal gland. In: Evered D, Clark S, eds. *Photoperiodism, Melatonin and Pineal*. London: Pitman, 38–56.

Kotler M, Rodriguez C, Sainz R M, et al. (1998). Melatonin increases gene expression for antioxidant enzymes in rat brain cortex. *J Pineal Res* 24: 83–89.

Lewy A J (1984). Human melatonin secretion (I): a marker for adrenergic function. In: Post R M, Ballenger J C, eds. *Frontiers of Clinical Neuroscience*, vol. 1, *Neurobiology of Mood Disorders*. Baltimore: Williams and Wilkins, 207–214.

Lewy A J, Wehr T A, Goodwin F K, et al. (1980). Light suppresses melatonin secretion in humans. *Science* 210: 1267–1269.

Lewy A J, Ahmed S, Jackson J M, et al. (1992). Melatonin shifts human circadian rhythms according to a phase–response curve. *Chronobiol Int* 9: 380–392.

Lewy A J, Ahmed S, Sack R L (1996). Phase shifting the human circadian clock using melatonin. *Behav Brain Res* 73: 131–134.

Lewy A J, Bauer V K, Ahmed S, et al. (1998). The human phase response curve (PRC) to melatonin is about 12 hours out of phase with the PRC to light. *Chronobiol Int* 15: 71–83.

Lindblom N, Heiskala H, Hätonen T, et al. (2000a). No evidence for extraocular light induced phase shifting of human melatonin, cortisol and thyrotropin rhythms. *Neuro Report* 11: 713–717.

Lindblom N, Heiskala H, Hätonen T, et al. (2000b). Bright light exposure of a large skin area does not affect melatonin or bilirubin levels in humans. *Biol Psychiatry* 48: 1098–1104.

Liu R-Y, Zhou J-N, van Heerikhuize J, et al. (1999). Decreased melatonin levels in postmortem cerebrospinal fluid in relation to aging, Alzheimer's disease, and apolipoprotein E-ε4/4 genotype. *J Clin Endocrinol Metab* 83: 323–327.

Lockley S W, Skene D J, Thapan K, et al. (1998). Extraocular light exposure does not suppress melatonin in humans *J Clin Endocrinol Metab* 83: 3369–3372.

Luboshitzky R, Shen-Orr Z, Tzischinsky O, *et al.* (2001). Actigraphic sleep–wake patterns and urinary 6-sulfatoxymelatonin excretion in patients with Alzheimer's disease. *Chronobiol Int* 18: 513–524.

Maggi S, Langlois J A, Minicuci N, *et al.* (1998). Sleep complaints in community-dwelling older persons: prevalence, associated factors, and reported causes. *J Am Geriatr Soc* 46: 161–168.

Magri F, Locatelli M, Balza G, *et al.* (1997). Changes in endocrine circadian rhythms as markers of physiological and pathological brain aging. *Chronobiol Int* 14: 385–396.

Mallo C, Zaidan R, Faure A, *et al.* (1988). Effect of four-day nocturnal melatonin treatment on the 24 h plasma melatonin, cortisol and prolactin profiles in humans. *Acta Endocrinol* 119: 474–480.

Markey S P, Higa S, Shih S, *et al.* (1985). The correlation between plasma melatonin levels and urinary 6-hydroxymelatonin excretion. *Clin Chim Acta* 150: 221–225.

Masson-Pévet M, Gauer F, Pévet P (1993). Melatonin regulation of melatonin receptor density in the rat pars tuberalis and suprachiasmatic nuclei. In: Touitou Y, Arendt J, Pévet P, eds. *Melatonin and the Pineal Gland: From Basic Science to Clinical Application*. Amsterdam: Elsevier, 99–104.

Mazzoccoli G, Correra M, Bianco G, *et al.* (1997). Age-related changes of neuro-endocrine-immune interactions in healthy humans. *J Biol Regul Homeost Agents* 11: 143–147.

McMillen I C, Nowak R (1989). Maternal pinealectomy abolishes the diurnal rhythm in plasma melatonin concentrations in the fetal sheep and pregnant ewe during late gestation. *J Endocrinol* 120: 459–464.

Nair N P V, Hariharasubrahamian N, Pilapil C, *et al.* (1986). Plasma melatonin – an index of brain aging in humans? *Biol Psychiatry* 21: 141–150.

Nave R, Peled R, Lavie P (1995). Melatonin improves evening napping. *Eur J Pharmacol* 275: 213–216.

Nave R, Herer P, Haimov I, *et al.* (1996). Hypnotic and hypothermic effects of melatonin on daytime sleep in humans: lack of antagonism by flumazenil. *Neurosci Lett* 214: 123–126.

O'Callaghan F J, Clarke A A, Hancock E, *et al.* (1999). Use of melatonin to treat sleep disorders in tuberous sclerosis. *Dev Med Child Neurol* 41: 123–126.

Ohashi Y, Okamoto N, Uchida K, *et al.* (1997). Differential pattern of the circadian rhythm of serum melatonin in young and elderly healthy subjects. *Biol Signals* 6: 301–306.

Osseni R A, Rat P, Bogdan A, *et al.* (2000). Evidence of prooxidant and antioxidant action of melatonin on human liver cell line HepG2. *Life Sci* 68: 387–399.

Pierpaoli W, Regelson W (1994). Pineal control of aging: effect of melatonin and pineal grafting on aging mice. *Proc Natl Acad Sci USA* 91: 787–791.

Pozo D, Reiter R J, Calvo J R, *et al.* (1994). Physiological concentrations of melatonin inhibit nitric oxide synthase in rat cerebellum. *Life Sci* 55: PL455–PL460.

Reid K, van den Heuvel C, Dawson D (1996). Day-time melatonin administration: effects on core temperature and sleep onset latency. *J Sleep Res* 5: 150–154.

Reinberg A E, Touitou Y (1996). Synchronisation et dyschronisme des rhythmes circadiens humains. *Pathol Biol (Paris)* 44: 487–495.

Reiter R J, Richardson B A, Johnson L, et al. (1980). Pineal melatonin rhythm: reduction in aging Syrian hamsters. *Science* 210: 1372–1373.

Reiter R J, Poeggeler B, Tan D-X (1993). Antioxidant capacity of melatonin: a novel action not requiring a receptor. *Neuroendocrinol Lett* 15: 103–116.

Reiter R J, Tan D-X, Osuna C, et al. (2000). Actions of melatonin in the reduction of oxidative stress. *J Biomed Sci* 7: 444–458.

Reppert S M, Weaver D R (1995). Melatonin madness. *Cell* 83: 1059–1062.

Rogers N L, Kennaway D J, Dawson D (1999). The effect of extra-ocular light exposure on nocturnal salivary melatonin levels. *Sleep Res Online* 2(Suppl 1): 622.

Sack R L, Lewy A J, Erb D E, et al. (1986). Human melatonin production decreases with age. *J Pineal Res* 3: 379–388.

Sack R L, Lewy A J, Hughes R J, et al. (1996). Melatonin as a chronobiotic drug. *Drug News Perspect* 9: 325–332.

Sack R L, Brandes R W, Kendall A R, et al. (2000). Entrainment of free-running circadian rhythms by melatonin in blind people. *N Engl J Med* 343: 1070–1077.

Seabra M L V, Bignotto M, Pinto L R Jr, et al. (2000). Randomized, double-blind clinical trial, controlled with placebo, of the toxicology of chronic melatonin treatment. *J Pineal Res* 29: 193–200.

Selmaoui B, Touitou Y (1999). Age-related differences in serum melatonin and pineal NAT activity and in the response of rat pineal to a 50-Hz magnetic field. *Life Sci* 64: 2291–2297.

Sharma M, Palacios-Bois J, Schwartz G, et al. (1989). Circadian rhythms of melatonin and cortisol in aging. *Biol Psychiatry* 25: 305–319.

Skene D J, Bojkowski C J, Currie J E, et al. (1990). 6-Sulphatoxymelatonin production in breast cancer patients. *J Pineal Res* 8: 269–279.

Suhner A, Schlagenhauf P, Johnson R, et al. (1988). Comparative study to determine the optimal melatonin dosage form for the alleviation of jet lag. *Chronobiol Int* 15: 655–666.

Susa N, Ueno S, Furukawa Y, et al. (1997). Potent protective effects of melatonin on chromium(VI)-induced DNA strand breaks, cytotoxicity and lipid peroxidation in primary cultures of rat hepatocytes. *Toxicol Appl Pharmacol* 144: 377–384.

Tan D X, Chen L D, Poeggeler B, et al. (1993). Melatonin: a potent endogenous hydroxyl radical scavenger. *Endocrine J* 1: 57–60.

Tapp E, Huxley M (1972). The histological appearance of the human pineal gland from puberty to old age. *J Pathol* 108: 137–144.

Thomas D R, Miles A (1989). Melatonin secretion and age. *Biol Psychiatry* 25: 365–367.

Touitou Y (2001). Human aging and melatonin. Clinical relevance. *Exp Gerontol* 36: 1083–1100.

Touitou Y, Haus E (1994). *Biologic Rhythms in Clinical and Laboratory Medicine*. Heidelberg: Springer-Verlag.

Touitou Y, Haus E (2000). Alterations with aging of the endocrine and neuroendocrine circadian system in humans. *Chronobiol Int* 17: 369–390.

Touitou Y, Fèvre M, Lagoguey M, et al. (1981). Age- and mental health-related circadian rhythms of plasma levels of melatonin, prolactin, luteinizing hormone and follicle stimulating hormone in man. *J Endocrinol* 91: 467–475.

Touitou Y, Fèvre M, Bogdan A (1984). Patterns of plasma melatonin with ageing and mental condition: stability of nyctohemeral rhythms and differences in seasonal variations. *Acta Endocrinol* 106: 145–151.

Touitou Y, Fèvre-Motange M, Proust J, *et al.* (1985). Age- and sex associated modification of plasma melatonin concentration in man. Relationship to pathology, malignant or not, and autopsy findings. *Acta Endocrinol* 108: 135–144.

Tzischinksy O, Lavie P (1994). Melatonin possesses time-dependent hypnotic effects. *Sleep* 17: 638–645.

Tzischinsky O, Pal I, Epstein R, *et al.* (1992). The importance of timing in melatonin administration in a blind man. *J Pineal Res* 12: 105–108.

Uchida K, Okamoto N, Ohara K, *et al.* (1996). Daily rhythm of serum melatonin in patients with dementia of the degenerate type. *Brain Res* 717: 154–159.

Uchimara N, Hirano T, Mukai M, *et al.* (1998). Insomnia in the aged. In: Meier-Ewert K, Okawa M, eds. *Sleep–Wake Disorders*. New York: Plenum Press, 79–90.

Van Coevorden A, Mockel J, Laurent E, *et al.* (1991). Neuroendocrine rhythms and sleep in aging men. *Am J Physiol* 260: E651–E661.

Vaughan G M, Pelham M F, Pang S, *et al.* (1976). Nocturnal elevation of plasma melatonin and urinary 5-hydroxyindoleacetic acid in young men: attempts at modification by brief changes in environmental lighting and sleep by autonomic drugs. *J Clin Endocrinol Metab* 42: 579–581.

Vollrath I (1981). *The Pineal Organ*. Berlin: Springer-Verlag.

Waldhauser F, Waldhauser M, Lieberman H R, *et al.* (1984). Bioavailability of oral melatonin in humans. *Neuroendocrinology* 39: 307–313.

Waldhauser F, Weiszenbacher G, Tatzer E, *et al.* (1988). Alterations in nocturnal serum melatonin levels in humans with growth and aging. *J Clin Endocrinol Metab* 66: 648–652.

Weiland N G, Wise P M (1990). Aging progressively decreases the densities and later the diurnal rhythms of α_1-adrenergic receptors in selected hypothalamic regions. *Endocrinology* 126: 2392–2397.

Winkler P, Helmke K (1987). Age-related incidence of pineal gland calcification in children: a roentgenological study of 1,044 skull films and a review of the literature. *J Pineal Res* 4: 247–252.

Wurtman R, Axelrod J, Barchas J D (1964). Age and enzyme activity in the human pineal. *J Clin Endocrinol Metab* 24: 299–301.

Yellon S M, Longo L D (1988). Effect of maternal pinealectomy and reverse photoperiod on the circadian melatonin rhythm in the sheep and fetus during the last trimester of pregnancy. *Biol Reprod* 39: 1093–1099.

Zaidan R, Geoffriau M, Brun J, *et al.* (1994). Melatonin is able to influence its secretion in humans: description of a phase–response curve. *Neuroendocrinology* 60: 105–112.

Zeitzer J M, Daniels J E, Duffy J F, *et al.* (1999). Do plasma melatonin concentrations decline with age? *Am J Med* 107: 432–436.

Zhdanova I V, Wurtman R J, Morabito C, *et al.* (1996). Effects of low oral doses of melatonin, given 2–4 hours before habitual bedtime, on sleep in normal young humans. *Sleep* 19: 423–431.

15

Chronobiological mechanisms in seasonal affective disorder

Alexander Neumeister

15.1 Introduction

Seasonal changes in behaviour and physiology have been recognised since ancient times and their relevance to psychiatry was described at least as early as in the middle of the last century. The degree to which seasonal changes affect criteria such as mood, energy, sleep length, appetite, food preference, or the wish to socialise with other people can be called seasonality. Two characteristic syndromes of seasonal mood changes have been reported in the literature: recurrent depressions in autumn and winter, termed seasonal affective disorder (SAD) by Rosenthal *et al.* (1984); and the opposite pattern, recurrent depressions in the summer, described by Wehr *et al.* (1987) and Boyce and Parker (1988). Kasper *et al.* (1989) reported winter difficulties in individuals who neither met criteria for major affective disorder nor were seeking treatment for their difficulties but who nevertheless experienced mild dysfunction and vegetative changes similar to those found in SAD. This group has been termed sub-syndromal SAD (S-SAD). More recently, there has been described a subgroup of depressed patients with brief recurrent depressive episodes during winter, termed recurrent brief SAD (Kasper *et al.*, 1992). Whereas the winter depressive symptoms of patients with SAD and S-SAD respond well to light therapy (Neumeister *et al.*, 1999), a successful nonpharmacological treatment for summer depression has not yet been developed.

15.2 Diagnosis of SAD and light therapy

The diagnostic concept of SAD was first included in the *Diagnostic and Statistical Manual of Mental Disorders*, Revised Third Edition (DSM-III-R) and has been updated in the DSM-IV (APA, 1994) as 'seasonal pattern', an adjectival modifier of any form of seasonally recurrent mood disorder. The combination of depressed mood and a characteristic cluster of vegetative symptoms, together with the occurrence of depressive episodes during autumn and winter with full remission or hypomanic episodes during spring and summer, distinguishes SAD and its sub-syndromal form from other mood disorders. The demographic and clinical characteristics of patients with SAD are given in Table 15.1.

The characteristic neurovegetative symptoms in patients with SAD include abnormalities in eating behaviour and food preference (Rosenthal et al., 1984). Hyperphagia and carbohydrate craving are typical symptoms of SAD and have also been described in patients suffering from atypical depression. These patients differ from anorectic depressive patients by being more often female and more mildly depressed.

Table 15.1 Demographic and clinical characteristics of female and male patients with seasonal affective disorder in Bonn and Vienna

n	610
Age (years)	41.1 ± 12.9
Diagnosis (%)	
UP	77.0
BP-II	21.7
BP-I	1.3
GSS	15.4 ± 3.5
Age at onset (years)	29.8 ± 13.1
Psychiatric comorbidity (%)	37.4
Family history (%)	
Depression	40.0
Alcohol dependency	6.9
Schizophrenia	2.1
Affect (%)	
Depressed	93.0
Irritable	75.1
Anxious	65.6
Loss of Energy (%)	98.4
Appetite (%)	
Increased	64.6
Reduced	18.4
Carbohydrate craving	66.5
Sleep (%)	
Hypersomnia	72.2
Daytime drowsiness	93.7
Loss of libido (%)	74.3
Difficulties at workplace	69.2

UP, unipolar depression; BP-II, bipolar-II affective disorder; BP-I, bipolar-I affective disorder; GSS, global seasonality score.

Moreover, these atypical depressed patients show a more pronounced reduction in sexual interest. These clinical features are also observed in patients with SAD. Interestingly, in SAD hyperphagia often correlates with increasing severity of depression. SAD patients have also been shown to demonstrate a significant increase in their well-being after intake of carbohydrates. It has been postulated (Wurtman, 1981) that carbohydrate craving may reflect a functional serotonin (5-HT) deficiency and that carbohydrate craving in SAD patients during autumn and winter may represent a behavioural–biochemical feedback loop for raising the availability of brain 5-HT (Fernstrom and Hirsch, 1977).

Another characteristic symptom of SAD is hypersomnia. It has been speculated (Kupfer et al., 1972) that hypersomnic and hyposomnic depressed patients constitute two biologically distinct groups. 5-HT has been implicated in regulation of sleep (Jouvet, 1969). Several investigators have studied the relationship between diet and sleep and have shown that changes in diet may induce changes in total sleep time, delta sleep and REM sleep. It can be hypothesised that some of the changes in sleep observed in SAD patients during winter episodes of depressed mood may be related to changes in diet and weight, and that serotonergic mechanisms may be involved.

There is agreement in the literature that light therapy is the first-line treatment for patients with SAD. This is based on numerous studies showing efficacy and safety, including randomised, controlled trials (Eastman et al., 1998; Terman et al., 1998) and meta-analyses (Lee and Chan, 1999; Terman et al., 1989). Such studies have shown that administration of light therapy in the morning is more effective than at any other time of the day (see Figure 15.1).

It was noted in the first large controlled trial of the effects of light therapy in SAD (Rosenthal et al., 1984) that light therapy is not only an effective treatment for the condition but may also serve as a research tool for exploring further the pathophysiology of the disorder. Since then, the study of the pathophysiology of SAD has been intimately linked to investigations into the mechanisms of action of light therapy. In 1980, Lewy et al. demonstrated that melatonin secretion in humans can be suppressed with artificial bright light (Lewy et al., 1980). Based on this finding, artificial light was used to treat a patient with winter depression (Lewy et al., 1982), subsequently stimulating the first systematic studies on the effects of light therapy in SAD (Rosenthal et al., 1985). Light therapy has also been found useful in nonseasonal depression, when combined with other treatment modalities, e.g. sleep deprivation (Neumeister et al., 1996) or pharmacotherapy (Kripke, 1998). This

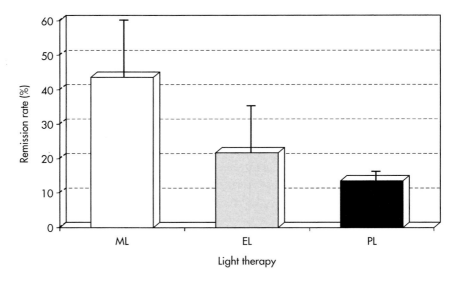

Figure 15.1 Comparison of morning and evening light therapy in the treatment of patients with SAD. The data from three controlled, randomised trials provide evidence for the superiority of morning light therapy (ML) over evening light therapy (EL) and placebo (PL). Data are derived from Terman *et al.* (1998), Eastman *et al.* (1998) and Lewy *et al.* (1998b).

chapter reviews the major biological hypotheses for SAD and light therapy, specifically focusing on circadian rhythms and neurotransmitter function.

15.3 Biological rhythms in SAD

Neurotransmitters have been postulated as playing a key role in the pathogenesis of SAD and also in the mechanisms of action of light therapy. Two neurotransmitter systems have been the main focus of interest during the past decade: serotonin (5-HT) and catecholamines, particularly noradrenaline (norepinephrine). Evidence is published in the literature for the role of each of these; however, it seems likely that the two transmitter systems interact and play important roles, in either different areas or different neurobiological systems of the brain.

15.3.1 Seasonality and brain serotonin function

There is considerable evidence in the literature that there exists a seasonal variation in several phenomena, such as mood and feeding behaviour, and that these variations may be related to changes in central and peripheral 5-HT function (Maes *et al.*, 1995). Changes in 5-HT function have been postulated also to be involved in the pathogenesis of SAD. Thus, it is pertinent to ask whether possible seasonal fluctuations in 5-HT function exist only in SAD patients or whether these fluctuations are also found in healthy controls.

Several studies have described seasonal variations in central and peripheral 5-HT function in healthy subjects and nonpsychiatric patients. Studies in humans can be divided in terms of whether measures are static (e.g. biochemical levels in body fluids or blood elements) or dynamic (e.g. neuroendocrine responses to pharmacological challenges). Several lines of evidence based on static measures support the hypothesis of seasonal fluctuations of 5-HT function in humans.

1. Hypothalamic 5-HT concentrations in human postmortem brain specimens are decreased in winter after values peak in autumn (Carlsson *et al.*, 1980).
2. Levels of plasma tryptophan, the precursor of 5-HT, show a bimodal seasonal pattern (Maes *et al.*, 1995).
3. Platelet 5-HT uptake and [^{3}H]imipramine binding show a seasonal pattern, albeit with some differences in seasonal peaks and troughs (Arora and Meltzer, 1988; DeMet *et al.*, 1989; Tang and Morris, 1985; Whitaker *et al.*, 1984).
4. Levels of 5-HT and its metabolites in cerebrospinal fluid show seasonal fluctuations, varying with latitude and population studied (Asberg *et al.*, 1980; Brewerton *et al.*, 1988).
5. Serum melatonin concentrations demonstrate summer and winter peaks in healthy males (Arendt *et al.*, 1977).
6. Neumeister *et al.* (2000) have reported *in vivo* a significantly reduced availability of hypothalamic 5-HT transporter sites in winter compared with summer in healthy female subjects (Figure 15.2).

There are only a few reports in the literature about seasonal variations in 5-HT function using dynamic measures. Joseph-Vanderpool *et al.* (1993) reported a seasonal variation in behavioural responses to the administration of *m*-chlorophenylpiperazine (m-CPP) in patients with SAD, with higher 'activation/euphoria' scores in SAD patients during winter compared with summer or after successful light therapy. More recently, Cappiello *et al.* (1996) demonstrated a seasonal variation in neuroendocrine (prolactin) response to intravenous tryptophan administration in unipolar, nonmelancholic depressed patients. Interestingly,

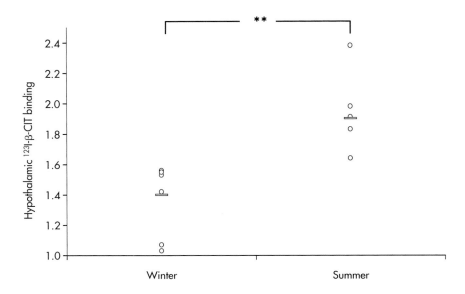

Figure 15.2 Hypothalamic serotonin transporter availability is significantly reduced in winter compared with summer in healthy female controls, as studied *in vivo* using ^{123}I-β-CIT (2β-carbomethoxy-3β-(4-iodophenyl)tropane) single-photon emission computed tomography (SPECT). **$p < 0.01$, Mann-Whitney U-test, 2-tailed. Reproduced from Neumeister *et al.* (2000).

seasonality was more pronounced in female than in male patients. No such seasonal variability was found in bipolar, melancholic or psychotic patients or in healthy controls.

It is clear from the above that there is substantial evidence arguing for a seasonal variation of central and peripheral 5-HT function in patients suffering from SAD and also in healthy controls. Thus, it can be hypothesised that seasonality of brain 5-HT function is physiological and may possibly represent a predisposing factor for nonseasonal depression and for seasonal depressions. Interestingly, seasonality was more pronounced in females than in males. However, it has to be acknowledged that this finding is preliminary and requires further study. Also, it has to be said that the variability in the specific seasonal peaks and nadirs reported by different researchers reflects the use of different study designs, methodologies, sample sizes and measures of 5-HT function. Consequently, further studies are needed to clarify the role of seasonal variations in central and peripheral 5-HT function in the regulation of human behaviour and in the pathogenesis of affective disorders, in particular of SAD.

15.3.2 Photoperiod and melatonin secretion

Early epidemiological studies suggested that the prevalence of SAD increases with increasing latitude, where the photoperiod is shorter in winter (Potkin *et al.*, 1986; Rosen *et al.*, 1990). On the basis of some of these earlier studies it was suggested that symptoms of SAD develop because of the shorter photoperiod (Rosenthal *et al.*, 1984). Treatment studies were designed to extend the daily photoperiod by administering light therapy for 3 h per day between 06.00 and 09.00 and between 16.00 to 19.00 (Rosenthal *et al.*, 1984). In support of this original hypothesis, the condition of most patients improved. However, subsequent studies have shown that photoperiod extension alone does not explain the therapeutic effects of light (Winton *et al.*, 1989). Moreover, it was shown that single daily pulses of light were as effective as the morning plus evening photoperiod extension. Also, more recent epidemiological studies suggest that the association between photoperiod and SAD is smaller than originally suspected (Blazer *et al.*, 1998; Mersch *et al.*, 1999). The photoperiod hypothesis of SAD has received renewed attention. It has been shown that the nocturnal duration of melatonin secretion reflects changes in the photoperiod of humans (Wehr, 1991). Healthy control subjects in normal living conditions did not show seasonal changes in melatonin profiles, suggesting that artificial indoor light may suppress the melatonin response to seasonal changes in photoperiod (Wehr *et al.*, 1995). In contrast, patients with SAD showed a significant seasonal variation in melatonin response to photoperiod, with a longer nocturnal melatonin duration in SAD than in controls. This agrees with findings from a study that used propranolol to block melatonin secretion, showing that truncation of the early-morning melatonin secretion normalises the melatonin profile in patients with SAD (Schlager, 1994). It has also been shown that photoperiod may be important in the onset of the atypical vegetative symptoms that reflect key symptoms in SAD (Young *et al.*, 1991, 1997).

The melatonin hypothesis has received substantial attention since it was shown that in many animals the photoperiod signal is mediated by the duration of nocturnal melatonin secretion, that light suppresses melatonin secretion, and that patients with bipolar mood disorders are supersensitive to light (Lewy *et al.*, 1985a). The sensitivity of melatonin to light suppression has been shown to depend on light intensity (Bojkowski *et al.*, 1987; McIntyre *et al.*, 1989b) and timing (McIntyre *et al.*, 1989a). Suppression was greater in bipolar disorder relative to normal controls (Lewy *et al.*, 1981, 1985a), but no difference was found between unipolar depressed subjects and healthy controls (Cummings *et al.*, 1989). However, no significant differences in 24 h melatonin rhythm were found between SAD patients and controls, and melatonin rhythm was not affected by light therapy (Checkley *et al.*, 1993; Partonen *et al.*, 1996). The therapeutic effects of light therapy are not explained by suppression of melatonin secretion (Wehr *et al.*, 1986). Drugs suppressing melatonin secretion, such as short-acting β-adrenoreceptor antagonists, have been found to be useful in the treatment of SAD (Schlager, 1994), possibly by truncating the early-morning melatonin secretion curve. However, atenolol, a long-acting β-adrenoreceptor antagonist, was not effective in SAD (Rosenthal *et al.*, 1988). Thus, it can be speculated that the suppression of melatonin secretion per se is not critical in achieving an antidepressant effects in SAD, but rather that the appropriate time point of melatonin suppression is crucial.

Administration of exogenous melatonin has also been tested for its potential antidepressant effect in SAD. One study, administering 5 mg melatonin, did not show beneficial effects in SAD (Wirtz-Justice *et al.*, 1990). Studies that used melatonin in more physiological doses at a specific time of day to induce a circadian phase advance in patients with SAD were found to be effective (Lewy *et al.*, 1998a), and it was also shown that the clinical response was correlated with the degree of phase advance.

15.3.3 Circadian phase shift

Several researchers have advanced hypotheses implicating abnormalities of circadian rhythms in the pathophysiology of SAD. Lewy *et al.* (1987b) suggested that patients with SAD have delayed circadian rhythms, whereas Czeisler and Kronauer (Czeisler *et al.*, 1987) suggested that the amplitude of the circadian rhythm is reduced in patients with SAD. According to these hypotheses, light therapy exerts its beneficial effects by restoring circadian rhythms to their normal phase position or amplitude.

In humans, the circadian pacemaker can be entrained and phase-shifted by light (Czeisler *et al.*, 1986; Minors *et al.*, 1991) and by administration of exogenous melatonin (Arendt and Broadway, 1987; Lewy *et al.*, 1992; Zaidan *et al.*, 1994) or melatonin receptor agonists (Krauchi *et al.*, 1997). The timing of light exposure relative to the circadian cycle dictates the direction and magnitude of circadian rhythm phase shifts.

On the basis of work done in patients with nonseasonal depressions (Kripke *et al.*, 1978; Wehr *et al.*, 1979), researchers have developed a phase-delay hypothesis for SAD (Lewy *et al.*, 1985b, 1988). The hypothesis is built on the assumption that SAD results from internal circadian rhythms that are phase-delayed relative to the external clock and to other rhythms, e.g. the sleep–wake cycle, and that light therapy mediates its antidepressant effects by correcting the abnormal phase delay. On this assumption, morning light therapy results in a corrective phase advance, while evening light exposure further phase-delays the circadian phase. Consequently, according to the phase-delay hypothesis, morning light therapy would be expected to be superior to mid-day or evening light therapy. Light therapy in the middle of the day generally has no effect on circadian rhythms, and thus would be presumed to exert no antidepressant effects. Most, but not all, studies support such an assumption, showing the superiority of morning light therapy over mid-day and evening light therapy (for an overview see Neumeister *et al.*, 1999). In one study directly comparing morning versus evening light therapy for SAD, the phase position of 6-sulfatoxymelatonin, the urinary metabolite of melatonin, was also assessed, and most patients exhibited a phase delay (Wirtz-Justice *et al.*, 1993). However, the phase position did not predict whether a patient would respond better to morning or evening light therapy. Similarly, no association was found between phase advance of nocturnal salivary melatonin secretion and treatment response to light therapy (Rice *et al.*, 1995).

One area of interest in circadian rhythms research in SAD involves the time at which melatonin begins to be secreted by the pineal gland during controlled, dim light conditions, called dim light melatonin onset (DLMO). Patients with SAD were found to have phase-delayed DLMO relative to healthy control subjects (Lewy *et al.*, 1987a, 1998b; Sack *et al.*, 1990). Light therapy induced a phase shift, with morning light exposure resulting in a phase advance and evening light exposure resulting in a phase delay. All studies consistently showed that morning light therapy was more effective than evening light therapy.

Further evidence for abnormal biological rhythms in SAD comes from studies showing delayed melatonin rhythms in SAD (Terman *et al.*, 1988). In this study the melatonin cycle phase-advanced with morning light therapy, and also during a combination of morning and evening light therapy. Another parameter that has been shown to be phase-delayed in SAD is the circadian activity–rest cycle (Glod *et al.*, 1997); light therapy phase-advanced cortisol, temperature and melatonin, although the sleep–wake cycle also advanced. To control for environmental factors, such as sleep, light exposure, activity or feeding, researchers used the constant-routine method, in which subjects are studied for 36 h in a controlled setting to unmask endogenous circadian rhythms. Such studies confirmed phase delays of the DLMO, core body temperature and cortisol rhythm (Avery et al., 1997; Dahl et al., 1993). Light therapy improved depression scores, but the magnitude of changes was not associated with the magnitude of phase advance.

The results of other studies do not agree with these findings and raise the question whether circadian phases are the biological basis of SAD. The 24 h circadian profiles of various hormones, including cortisol, prolactin and thyrotrophin in plasma (Oren et al., 1996) and the 24 h core body temperature profile (Rosenthal et al., 1990) did not differ between patients with SAD and healthy controls and were not affected by light therapy. Most notably, one study reported significant phase delays of temperature in SAD patients during summer compared with the winter, effects opposite to the phase advances found after light therapy in winter (Levendosky et al., 1991). Another study also showed no difference in core body temperature between patients with SAD and controls (Eastman et al., 1993). The authors report that morning light therapy advanced the phase of temperature rhythm more in SAD patients than in controls; however, the relation between phase changes and improvement in depression scores was opposite to that predicted by the phase-delay hypothesis.

The interpretation of the findings of these studies is often limited by small sample sizes and by sample selections. It is reasonable to assume that patients with hypersomnia differ from patients with hyposomnia, but both groups of subjects have been studied together in the past. Another factor that has to be considered is that group mean data do not necessarily represent individual circadian responses. The magnitude of light-induced phase shifts may vary considerably between subjects. Terman et al. (1988) found no relation between clinical response to light therapy and whether patients were initially phase-delayed or phase-advanced (as measured by DLMO). However, the magnitude of individual phase advances was significantly correlated with the degree of clinical improvement (Terman et al., 2001). It has also to be considered that any correlation between clinical response to light therapy and phase advance does not necessarily mean that they are causally related. Other factors associated with morning light therapy, such as better compliance with the treatment or greater sensitivity to light, may affect the outcome. However, these assumptions require further study.

15.4 Conclusion

In summary, studies using the most reliable measures of endogenous circadian phase, such as DLMO or constant routine, provide substantial evidence for circadian phase delays in SAD. Such an assumption is supported by the evidence for the superiority of morning light therapy over evening or mid-day light therapy. There is also evidence that the clinical response to light therapy and melatonin is related to the degree of corrective phase advances, although these findings do not necessarily reflect causality. However, it has also been shown that a subgroup of patients with SAD do not show abnormal phase-delayed circadian rhythms nor do they exhibit phase shifts of their biological rhythms in response to light but they still respond to light therapy with clinical improvement. This indicates that mechanisms other than circadian phase abnormalities play a role in the pathophysiology of SAD and the mechanisms of light therapy.

Acknowledgement

The author is supported by APART (Austrian Programme for Advanced Research and Technology).

References

Arendt J, Broadway J (1987). Light and melatonin as zeitgebers in man. *Chronobiol Int* 4: 273–282.

Arendt J, Wirz-Justice A, Bradtke J (1977). Circadian, diurnal and circannual rhythms of serum melatonin and platelet serotonin in man. *Chronobiologia* 4: 96–97.

Arora R C, Meltzer H Y (1988). Seasonal variation of imipramine binding in the blood platelets of normal controls and depressed patients. *Biol Psychiatry* 23: 217–226.

APA (1994). *Diagnostic and Statistical Manual of Mental Disorders*, 4th edn, DSM-IV. Washington, DC: American Psychiatric Association.

Asberg M, Bertilsson L, Rydin E, *et al.* (1980). Monoamine metabolites in cerebrospinal fluid in relation to depressive illness, suicidal behavior and personality. In: Angrist B, Burrows G D, Lader M, eds. *Recent Advances in Neuropharmacology*. Oxford: Oxford University Press, 257–271.

Avery D H, Dahl K, Savage M V, *et al.* (1997). Circadian temperature and cortisol rhythms during a constant routine are phase-delayed in hypersomnic winter depression. *Biol Psychiatry* 41: 1109–1123.

Blazer D G, Kessler R C, Swartz M S (1998). Epidemiology of recurrent major and minor depression with a seasonal pattern. The National Comorbidity Survey. *Br J Psychiatry* 172: 164–167.

Bojkowski C J, Aldhous M E, English J, *et al.* (1987). Suppression of nocturnal

plasma melatonin and 6-sulphatoxymelatonin by bright and dim light in man. *Horm Metab Res* 19: 437–440.
Boyce P, Parker G (1988). Seasonal affective disorder in the southern hemisphere. *Am J Psychiatry* 145: 96–99.
Brewerton T D, Berrettini W H, Nurnberger J I Jr, Linnoila M (1988). Analysis of seasonal fluctuations of CSF monoamine metabolites and neuropeptides in normal controls: findings with 5HIAA and HVA. *Psychiatry Res* 23: 257–265.
Cappiello A, Malison R T, McDougle C J, *et al.* (1996). Seasonal variation in neuroendocrine and mood responses to i.v. L-tryptophan in depressed patients and healthy subjects. *Neuropsychopharmacology* 15: 475–483.
Carlsson A, Svennerholm L, Winblad B (1980). Seasonal and circadian monoamine variations in human brains examined post mortem. *Acta Psychiatr Scand Suppl* 280: 75–85.
Checkley S A, Murphy D G, Abbas M, *et al.* (1993). Melatonin rhythms in seasonal affective disorder. *Br J Psychiatry* 163: 332–337.
Cummings M A, Berga S L, Cummings K L, *et al.* (1989). Light suppression of melatonin in unipolar depressed patients. *Psychiatry Res* 27: 351–355.
Czeisler C A, Allan J S, Strogatz S H, *et al.* (1986). Bright light resets the human circadian pacemaker independent of the timing of the sleep–wake cycle. *Science* 233: 667–671.
Czeisler C A, Kronauer R E, Mooney J J, *et al.* (1987). Biologic rhythm disorders, depression, and phototherapy. A new hypothesis. *Psychiatr Clin North Am* 10: 687–709.
Dahl K, Avery D H, Lewy A J, *et al.* (1993). Dim light melatonin onset and circadian temperature during a constant routine in hypersomnic winter depression. *Acta Psychiatr Scand* 88: 60–66.
DeMet E M, Chicz-DeMet A, Fleischmann J (1989). Seasonal rhythm of platelet ^{3}H-imipramine binding in normal controls. *Biol Psychiatry* 26: 489–495.
Eastman C I, Gallo L C, Lahmeyer H W, Fogg L F (1993). The circadian rhythm of temperature during light treatment for winter depression. *Biol Psychiatry* 34: 210–220.
Eastman CI, Young M A, Fogg L F, *et al.* (1998). Bright light treatment of winter depression: a placebo-controlled trial. *Arch Gen Psychiatry* 55: 883–889.
Fernstrom J D, Hirsch M J (1977). Brain serotonin synthesis: reduction in cornmalnourished rats. *J Neurochem* 28: 877–879.
Glod C A, Teicher M H, Polcari A, *et al.* (1997). Circadian rest–activity disturbances in children with seasonal affective disorder. *J Am Acad Child Adolesc Psychiatry* 36: 188–195.
Joseph-Vanderpool J R, Jacobsen F M, Murphy D L, *et al.* (1993). Seasonal variation in behavioral responses to m-CPP in patients with seasonal affective disorder and controls. *Biol Psychiatry* 33: 496–504.
Jouvet M (1969). Biogenic amines and the states of sleep. *Science* 163: 32–41.
Kasper S, Rogers S L, Yancey A, *et al.* (1989). Phototherapy in individuals with and without subsyndromal seasonal affective disorder. *Arch Gen Psychiatry* 46: 837–844.
Kasper S, Ruhrmann S, Haase T, Moller H J (1992). Recurrent brief depression and its relationship to seasonal affective disorder. *Eur Arch Psychiatry Clin Neurosci* 242: 20–26.

Krauchi K, Cajochen C, Mori D, *et al.* (1997). Early evening melatonin and S-20098 advance circadian phase and nocturnal regulation of core body temperature. *Am J Physiol* 272: R1178–R1188.

Kripke D F (1998). Light treatment for nonseasonal depression: speed, efficacy, and combined treatment. *J Affect Disord* 49: 109–117.

Kripke D F, Mullaney D J, Atkinson M, Wolf S (1978). Circadian rhythm disorders in manic-depressives. *Biol Psychiatry* 13: 335–351.

Kupfer D J, Himmelhoch J M, Swartzburg M, *et al.* (1972). Hypersomnia in manic-depressive disease (a preliminary report). *Dis Nerv Syst* 33: 720–724.

Lee T M, Chan C C (1999). Dose–response relationship of phototherapy for seasonal affective disorder: a meta-analysis. *Acta Psychiatr Scand* 99: 315–323.

Levendosky A A, Joseph-Vanderpool J R, Hardin T, *et al.* (1991). Core body temperature in patients with seasonal affective disorder and normal controls in summer and winter. *Biol Psychiatry* 29: 524–534.

Lewy A J, Wehr T A, Goodwin F K, *et al.* (1980). Light suppresses melatonin secretion in humans. *Science* 210: 1267–1269.

Lewy A J, Wehr T A, Goodwin F K, *et al.* (1981). Manic-depressive patients may be supersensitive to light. *Lancet* i: 383–384.

Lewy A J, Kern H A, Rosenthal N E, Wehr T A (1982). Bright artificial light treatment of a manic-depressive patient with a seasonal mood cycle. *Am J Psychiatry* 139: 1496–1498.

Lewy A J, Nurnberger J I Jr., Wehr T A, *et al.* (1985a). Supersensitivity to light: possible trait marker for manic-depressive illness. *Am J Psychiatry* 142: 725–727.

Lewy A J, Sack R L, Singer C M (1985b). Treating phase typed chronobiologic sleep and mood disorders using appropriately timed bright artificial light. *Psychopharmacol Bull* 21: 368–372.

Lewy A J, Sack R L, Miller L S, Hoban T M (1987a). Antidepressant and circadian phase-shifting effects of light. *Science* 235: 352–354.

Lewy A J, Sack R L, Singer C M, White D M (1987b). The phase shift hypothesis for bright light's therapeutic mechanism of action: theoretical considerations and experimental evidence. *Psychopharmacol Bull* 23: 349–353.

Lewy A J, Sack R L, Singer C M, *et al.* (1988). Winter depression and the phase-shift hypothesis for bright light's therapeutic effects: history, theory, and experimental evidence. *J Biol Rhythms* 3: 121–134.

Lewy A J, Ahmed S, Jackson J M, Sack R L (1992). Melatonin shifts human circadian rhythms according to a phase–response curve. *Chronobiol Int* 9: 380–392.

Lewy A J, Bauer V K, Cutler N L, Sack R L (1998a). Melatonin treatment of winter depression: a pilot study. *Psychiatry Res* 77: 57–61.

Lewy A J, Bauer V K, Cutler N L, *et al.* (1998b). Morning vs evening light treatment of patients with winter depression. *Arch Gen Psychiatry* 55: 890–896.

Maes M, Scharpe S, Verkerk R, *et al.* (1995). Seasonal variation in plasma L-tryptophan availability in healthy volunteers. Relationships to violent suicide occurrence. *Arch Gen Psychiatry* 52: 937–946.

McIntyre I M, Norman T R, Burrows G D, Armstrong S M (1989a). Human melatonin response to light at different times of the night. *Psychoneuroendocrinology* 14: 187–193.

McIntyre I M, Norman T R, Burrows G D, Armstrong S M (1989b). Quantal melatonin suppression by exposure to low intensity light in man. *Life Sci* 45: 327–332.

Mersch P P, Middendorp H M, Bouhuys A L, *et al.* (1999). Seasonal affective disorder and latitude: a review of the literature. *J Affect Disord* 53: 35–48.

Minors D S, Waterhouse J M, Wirz-Justice A (1991). A human phase–response curve to light. *Neurosci Lett* 133: 36–40.

Neumeister A, Goessler R, Lucht M, *et al.* (1996). Bright light therapy stabilizes the antidepressant effect of partial sleep deprivation. *Biol Psychiatry* 39: 16–21.

Neumeister A, Stastny J, Praschak-Rieder N, *et al.* (1999). Light treatment in depression (SAD, s-SAD and non-SAD). In: Holick M F, Jung E G, eds. *Biologic Effects of Light*. Basel: Kluwer Academic, 409–416.

Neumeister A, Pirker W, Willeit M, *et al.* (2000). Seasonal variation of availability of serotonin transporter binding sites in healthy female subjects as measured by [^{123}I]-2β-carbomethoxy-3β-(4-iodophenyl)tropane and single photon emission computed tomography. *Biol Psychiatry* 47: 158–160.

Oren D A, Levendosky A A, Kasper S, *et al.* (1996). Circadian profiles of cortisol, prolactin, and thyrotropin in seasonal affective disorder. *Biol Psychiatry* 39: 157–170.

Partonen T, Vakkuri O, Lamberg-Allardt C, Lonnqvist J (1996). Effects of bright light on sleepiness, melatonin, and 25-hydroxyvitamin D(3) in winter seasonal affective disorder. *Biol Psychiatry* 39: 865–872.

Potkin S G, Zetin M, Stamenkovic V, *et al.* (1986). Seasonal affective disorder: prevalence varies with latitude and climate. *Clin Neuropharmacol* 9: 181–183.

Rice J, Mayor J, Tucker H A, Bielski R J (1995). Effect of light therapy on salivary melatonin in seasonal affective disorder. *Psychiatry Res* 56: 221–228.

Rosen L N, Targum S D, Terman M, *et al.* (1990). Prevalence of seasonal affective disorder at four latitudes. *Psychiatry Res* 31: 131–144.

Rosenthal N E, Sack D A, Gillin J C, *et al.* (1984). Seasonal affective disorder. A description of the syndrome and preliminary findings with light therapy. *Arch Gen Psychiatry* 41: 72–80.

Rosenthal N E, Sack D A, Carpenter C J, *et al.* (1985). Antidepressant effects of light in seasonal affective disorder. *Am J Psychiatry* 142: 163–170.

Rosenthal N E, Jacobsen F M, Sack D A, *et al.* (1988). Atenolol in seasonal affective disorder: a test of the melatonin hypothesis. *Am J Psychiatry* 145: 52–56.

Rosenthal N E, Levendosky A A, Skwerer R G, *et al.* (1990). Effects of light treatment on core body temperature in seasonal affective disorder. *Biol Psychiatry* 27: 39–50.

Sack R L, Lewy A J, White D M, *et al.* (1990). Morning vs evening light treatment for winter depression. Evidence that the therapeutic effects of light are mediated by circadian phase shifts. *Arch Gen Psychiatry* 47: 343–351.

Schlager D S (1994). Early-morning administration of short-acting beta blockers for treatment of winter depression. *Am J Psychiatry* 151: 1383–1385.

Tang S W, Morris J M (1985). Variation in human platelet ^{3}H-imipramine binding. *Psychiatry Res* 16: 141–146.

Terman M, Terman J S, Quitkin F M, *et al.* (1988). Response of the melatonin cycle to phototherapy for seasonal affective disorder. Short note. *J Neural Transm* 72: 147–165.

Terman M, Terman J S, Quitkin F M, *et al.* (1989). Light therapy for seasonal affective disorder. A review of efficacy. *Neuropsychopharmacology* 2: 1–22.

Terman M, Terman J S, Ross D C (1998). A controlled trial of timed bright light and negative air ionisation for treatment of winter depression. *Arch Gen Psychiatry* 55: 875–882.

Terman J S, Terman M, Lo E S, Cooper T B (2001). Circadian time of morning light administration and therapeutic response in winter depression. *Arch Gen Psychiatry* 58: 69–75.

Wehr T A (1991). The durations of human melatonin secretion and sleep respond to changes in daylength (photoperiod). *J Clin Endocrinol Metab* 73: 1276–1280.

Wehr T A, Wirz-Justice A, Goodwin F K, *et al.* (1979). Phase advance of the circadian sleep–wake cycle as an antidepressant. *Science* 206: 710–713.

Wehr T A, Jacobsen F M, Sack D A, *et al.* (1986). Phototherapy of seasonal affective disorder. Time of day and suppression of melatonin are not critical for antidepressant effects. *Arch Gen Psychiatry* 43: 870–875.

Wehr T A, Sack D A, Rosenthal N E (1987). Seasonal affective disorder with summer depression and winter hypomania. *Am J Psychiatry* 144: 1602–1603.

Wehr T A, Giesen H A, Moul D E, *et al.* (1995). Suppression of men's responses to seasonal changes in day length by modern artificial lighting. *Am J Physiol* 269: R173–R178.

Whitaker P M, Warsh J J, Stancer H C, *et al.* (1984). Seasonal variation in platelet ^{3}H-imipramine binding: comparable values in control and depressed populations. *Psychiatry Res* 11: 127–131.

Winton F, Corn T, Huson L W, *et al.* (1989). Effects of light treatment upon mood and melatonin in patients with seasonal affective disorder. *Psychol Med* 19: 585–590.

Wirz-Justice A, Graw P, Krauchi K, *et al.* (1990). Morning or night-time melatonin is ineffective in seasonal affective disorder. *J Psychiatr Res* 24: 129–137.

Wirz-Justice A, Graw P, Krauchi K, *et al.* (1993). Light therapy in seasonal affective disorder is independent of time of day or circadian phase. *Arch Gen Psychiatry* 50: 929–937.

Wurtman R J (1981). The effects of nutritional factors on memory. *Acta Neurol Scand Suppl* 89: 145–154.

Young M A, Watel L G, Lahmeyer H W, Eastman C I (1991). The temporal onset of individual symptoms in winter depression: differentiating underlying mechanisms. *J Affect Disord* 22: 191–197.

Young M A, Meaden P M, Fogg L F, *et al.* (1997). Which environmental variables are related to the onset of seasonal affective disorder? *J Abnorm Psychol* 106: 554–562.

Zaidan R, Geoffriau M, Brun J, *et al.* (1994). Melatonin is able to influence its secretion in humans: description of a phase–response curve. *Neuroendocrinology* 60: 105–112.

Index

Page numbers in *italic* refer to figures and tables.

Page numbers in **bold** indicate major entries.

absorption rates, drugs 130–131
accidents, shift-workers' malaise 318
acetaminophen *see* paracetamol (acetaminophen)
acetylcholine **359**
   gastric acid stimulation 134
acetylcholine muscarinic-receptor (ACh mR) antagonists **184–185**
   see also atropine, nocturnal asthmatics
acetylsalicylic acid *see* aspirin
N-acetyltransferase activity, decrease with age 380–381, *381*
activities by categories, data purification method, field studies 21–22, 22–23, *24*
activity count, fractional desynchronisation protocol 19
activity–rest cycles *see* rest–activity cycles
acute lymphoblastic leukaemia, oral methotrexate and mercaptopurine, chronoefficacy 268
adenosine, effects on rhythms 359
adenylyl cyclase, β-adrenoreceptors in heart 114, *115*
administration, routes of
   and rhythms **86, 88**
adrenal cortisol production 177
adrenal gland suppression
   corticosteroid administration 177, 178
adrenaline (epinephrine)
   plasma concentrations
      circadian changes 159, 165

      mental performance and mood 12
   urinary excretion, night workers 322
adrenergic signal transduction pathway *112*–113
adrenocorticotrophic hormone (ACTH) 177
adriamycin *see* doxorubicin
advanced sleep phase syndrome (ASPS) 313, 343
aeroallergen exposure, late asthmatic response 166
afferent fibres, SCN 32, 33
ageing
   and circadian rhythms 310, 311–312
   free radical theory of 385
   and melatonin **369–386**
      physiological aspects **374–381**
   night work and 321
   NOx excretion, rodents, reduction in 119
   rhythm amplitude 346
   sleep–wake pattern, changes 381, 382–383
aircrew, time-zone transitions 317
airway cooling, asthma 167
airway hyperreactivity 153
airway inflammation, salmeterol therapy 173–174
airway resistance, asthma 156, *157*
albumin, drug binding 90, 92
albuterol sulfate *see* salbutamol sulfate
alertness, night work 319

allergic factors, nocturnal asthma
      166–167
α_1-glycoprotein, drug binding 90, 92
α-adrenoreceptor antagonists
   24-hour BP profile 200
   see also doxazosin
alveolar tissue, eosinophils, asthma
      160
Alzheimer's disease, rest-activity
      patterns 313
   behavioural therapy *333*
   light treatment *334*
ambulatory chronotherapy, cancer
      units 271–273
aminoglycosides, renal toxicity 101,
      102
amlodipine, 24-hour BP pattern 198,
      199
amphetamines, jet lag treatment 328
analgesics **218–228**
   chronotherapeutic use, guidelines
      228–229
angina
   β-adrenoreceptor antagonists 204
   oral nitrates 202–203
   pain, biological rhythms *215*
   variant angina 120, 194, 202–203
angiotensin-converting enzyme (ACE)
      117–118
angiotensin-converting enzyme (ACE)
      inhibitors **198, 200**
   24-hour BP pattern, effects *201*
   evening dosing 118
antacids 138
anti-inflammatory agents
   analgesia **221, 222–224**
   asthma 184
anticancer drugs
   chronoefficacy
      animal data **265–268**
      human data **268–271**
   chronotherapy centres, EORTC
      Chronotherapy group
      271–273
   chronotoxicity and chronopharma-
      cology
      animal data **245–251**
      human data **251–262**
   circadian rhythmicity, influence on
      264–265

combination treatment, chrono-
      pharmacology **250–251**
   single agents, chronopharmacology
      245–250
   tolerability and MTD 272–273
antidiuretic hormone levels,
      nocturnal enuresis 312, 331
antihypertensives
   duration of effect 202
   main target mechanisms **194**
   see also angiotensin-converting
      enzyme (ACE) inhibitors;
      calcium channel blockers
antral gastritis 146
apoptosis, circadian variation 236,
      241–242
arterial hyperreactivity 203
articaine, circadian rhythm of action
      220
artificial indoor light 399
aspirin
   enteric-coated 131, *132*
   gastric mucosal damage 146, *147*,
      224
asthma **153–185**
   allergic factors **166–167**
   chronobiology **154–166**
   histopathology 153
   treatment **167–185**
atenolol 85, *197*, 202, 203, 204, 205
   melatonin secretion reduction 371
atropine, nocturnal asthmatics 165,
      184
autoregulatory feedback loop,
      mammalian circadian clock 60

backache *215*, 217–218
baclofen, phase-shifting studies 358
bambuterol, comparison with
      salmeterol 176
basal acid secretion 134
   H_2 receptor antagonists, effects on
      138
BAX protein, circadian expression
      236, 241
Bcl-2 proteins, circadian expression
      236, 241
beat cycle 5
behavioural changes, pain intensity
      evaluation 211

behavioural therapy, circadian rhythm abnormalities **332–334**
benazepril 198, 200, *201*
benzodiazepines
　jet lag 328
　phase-shifting studies 358
　rats, humans and hamsters, effects on 349
β$_2$-adrenoreceptor agonists, asthma
　inhaled **172–176**
　oral **171–172**
　see also terbutaline; theophylline
β$_2$-adrenoreceptor function and gene regulation, nocturnal asthma **165–166**
β-adrenoreceptor antagonists
　coronary heart disease **203–205**
　heart rate 196, 204
　hypertension **195–196**
β-adrenoreceptors
　CNS **116**
　heart **111–116**
β-endorphins 212
bevantolol, coronary heart disease 204
bicuculline, phase-shifting studies 357
biliary colic, pain *215*, 217
biliary excretion, drugs *101*, *103*
bioavailability 130
biological clocks **31–45**
　identification of structures containing 53–54
　inputs **32–41**, *42*, *43*
　mutations affecting 56, 57–58
　outputs **41**, *43*, **44**
biphasic release drug formulation 288–289
bipolar mood disorders and light 399
blindness, circadian rhythm abnormalities 313, *314*, 331–332, 336
　melatonin therapy *335*, 383
blood pressure 193, *194*
　catecholamines 113
　congestive heart failure 115, 116
　hypertension *see* hypertension
　see also renin-angiotensin system
bmal1 gene/BMAL1 protein 58, 345, 350, 351

bmal2 58–59
body clock 3, **11–15**, 27
　adjustment, night work **321–323**
　and light exposure, phase shift *10*, 326, 330–331
　widespread effects **11–15**
　see also circadian clock; phase-shifting
body temperature *see* core temperature
bone marrow, mitotic activity 242
bowel motility, time-dependent variations 90, *91*
breast cancer
　light exposure at night 238–239
　metastatic
　　cytokine levels 244, 263–264
　　survival and conserved cortisol diurnal gradient 239
bright light, jet lag treatment 325–327
bromosulfophthalein, biliary concentration and excretion rate 103
bronchoconstriction, asthmatics 158–159
bronchodilation, endogenous, nocturnal asthmatics 165–166
bupivacaine, plasma clearance 220, *221*
bursting pellets, time-delay drug delivery **292–293**
bursting tablets, time-delay drug delivery **290–291**
busulfan, malignant solid tumours 262

c-fos 36, **350**, 354
caffeine
　disordered rhythms, treatment 328, 329
　rhythms, effects on 359–360
calcium channel blockers **205**
　24-hour BP pattern, effects on **196**, **198**, *199*
calcium channel, drugs acting on, pharmacodynamics 122
cAMP concentrations
　brain 116
　heart 114–115

cancer
    biological rhythms in host and/or tumours 240–244
        animal data 240–242
        human data 243–244
    chemotherapy 235–274
    pain 215, 218, 219, 227–228
        analgesics, effects of 225, 226–227
capsules, time-delay formulations 293–301
captopril 198
    time-delayed formulation 294
carbohydrate craving, SAD 394, 395
carboplatin 246–247, 266
carcinogenesis, tumour outcome and circadian coordination 238–240
cardiovascular disease 193–205
    cardiovascular events 194
    coronary heart disease
        angina 194, 202–203, 204, 215
        β-adrenoreceptor antagonists 203–205
        calcium channel blockers 205
        oral nitrates 202–203
    drugs 197
    hypertension 195–202
    night work 319–320
cardiovascular system, chronobiology 193–194
catecholamines
    blood pressure and heart rate 113
    SAD 396
    see also adrenaline (epinephrine); noradrenaline (norepinephrine)
categories, purification by, data purification methods 21–22, 22–23, 24
CD4+ lymphocytes, airway tissues, asthmatics 161
CD51, nocturnal asthmatics 162
celecoxib, solid tumours 250, 268
cell cycling 345
    and gene expression 236, 237, 243
    IFNAR expression, implanted tumour cells 241
central nervous system (CNS)
    5-HT-receptor subtypes 116–117
    β-adrenoreceptors in 116
central and peripheral oscillators, differences 66–67
cephalic-vagal component, meal response 11, 139, 141
chlorphenamine (chlorpheniramine) release, time-delay capsules 300
cholesterol, elevated serum, treatment 121
cholesterol synthesis 120–121, 320
chronic low back pain 217–218
chronic pain syndrome, endorphins in CSF 212–213
chronic rhinosinusitis, nocturnal asthma 167
chronopharmacokinetics 76–88
    human studies 76–84
    mechanisms of 88, 90–103
Chronset technology 296, 298
chronotherapy centres, EORTC Chronotherapy group 271–273
Chronotopic 288
circadian clock 43, 53–69
    mutations, effects of 61, 62
    peripheral cells 235–236
circadian coordination, carcinogenesis and tumour outcome 238–240
circadian oscillator 53
circadian rhythms 8–11
    abnormal variations 312–314, 324
        treatment 323, 324–337, 384
        see also jet lag; night work
    anticancer treatment, effects 264–265
    conserved, and tumour outcome 238–240
    melatonin levels 371–372, 373
    normal variations 309–312, 324
    time-free environments 8
    tumour development, influence of 262–264
    see also biological clocks; phase-shifting
circadian timing system 31–32
circannual variation
    CSF endorphin concentrations,

chronic pain syndrome 212–213
cisplatin **247**, 262
 chrononephrotoxicity 101
 plasma protein binding 93
clock 56, 57–58, 345
 expression, oral mucosa and skin 236, *237*, 243
 polymorphisms 68
clock function alteration as adverse drug reaction 265
clock inputs, SCN **32–41**
clock outputs, SCN **41**, 44
CLOCK proteins 345
CLOCK/BMAL1 heterodimer 60
colorectal cancer
 5-FU 252, 253, 255
 and oxaliplatin or carboplatin, chronoefficacy 269
 liver metastases
 floxuridine, hepatic artery infusion 257, 258, 260, 261
 metastatic, survival and circadian rhythmicity 239, 240
congestive heart failure
 ACE inhibitors 198
 blood pressure and heart rate rhythms 115, 116
conjugation reactions **97–99**
constancy and rhythmicity **1–2**
constant routine **3–4**, 16, 27, 310
 variants 7
control days protocol, Moog and Hildebrandt 16, 22
coping mechanisms, night work 321
core temperature **2**, 27, 44
 body clock studies
 constant routines **3–4**
 fractional desynchronisation 5, 6
 masking effects 21–22
 measurement methods 17–18
 performance, mental and physical 12–15
 purified data *24*
 and sleep 15
 spontaneous activity 25–27
 exogenous effects 4
coronary heart disease **202–205**
 nitrate treatment 120, **202–203**

corticosteroids, asthma
 dose timing
 and adrenal suppression 177
 and maximal efficacy 178, 180–182
 oral and inhaled **176–182**
 response to, circadian variations 162–163
corticotrophin-releasing factor (CRF) 177
cortisol 12
 circadian changes 159, 164
 diurnal slope and survival, metastatic breast cancer 239
covariance analysis, data purification by **25–27**
CPT-11 *see* irinotecan (CPT11)
cryptochrome 1 and 2 (cry1 and cry2) 59, 60, 65
CRYPTOCHROME (CRY 1 and CRY2) 56, 60
5-CT, phase studies, neuronal firing rhythms 353
cyclins, cell cycle control 236, 243
cyclooxygenase-2 (COX-2) inhibitor (celecoxib) 250, 268
cysteinyl leukotrienes, nocturnal asthmatics 163
cystemustine, chronotoxicity **248–249**
cytochrome P450 monooxygenase system **94–96**
cytokines
 circadian variations and tumour development 263–264
 peripheral activity 244

data purification methods, field studies **19**, **20–27**
delayed sleep phase syndrome (DSPS) 313, 334, 343, 358, 385
depression 117
 atypical patients 395
 light therapy 395, 396
 see also seasonal affective disorder (SAD)
dermal photoreceptors, possibility of 63–64
desynchronisation, biological clock and zeitgebers 384
 see also jet lag; night work

desynchronisation protocols, field studies 16
dexamethasone
  circadian gene expression 67
  parenteral drug injections, chronokinetics 78
diabetes, cardiovascular complications 118
diazepam, protein binding 92, 93
diclofenac, biphasic release 289
digoxin *197*
dihydropyrimidine dehydrogenase (DPD) enzyme 244
1,4-dihydropyridines 196, 198, *199*
  see also nifedipine
diltiazem 196, 202
dim light melatonin onset (DLMO) 7
  SAD 401
diuretics, hypertension 200
dizocilpine
  rhythm studies 350–351, 352
  SCN, drug effects on 349
DMLO (dim light melatonin onset) 7
DNA syntheses, tumour development studies 262–263
docetaxel **247–248**, 266–267
  and doxorubicin 251
DOI, phase shift studies 355–356
dopamine, effects on rhythms 361
dorsal raphe nuclei (DRN) 39, 344
dorsal-medial suprachiasmatic nuclei (DM-SCN) 32, *43*
  efferent projections 41, *43*, 44
  rodent, neurochemistry *33*
doubletime (dbt)/ DOUBLETIME (gene and protein) 55
doxazosin 200, 202
doxorubicin
  chronoefficacy 266–267
8-OH-DPAT, phase-shift studies 349, 353–354, 354, 356
Drosophila melanogaster 54, 55–56, 57, 59
drug absorption 88, 90, 130, 131
  see also gastrointestinal (GI) tract
drug action, principles of **111**
drug bioavailability, factors affecting 131, 133–134
drug delivery, chronopharmaceutical 283–303

drug distribution **90, 92–94**
drug excretion **100–103**
drug formulations and physico-chemical properties **85–86**
drug metabolism, time-dependent changes, factors modifying 99–100
drug-related adverse events
  acute side effects, plasma concentrations and 130–131
  theophylline and salmeterol, comparison 175
  timing to minimize 223, 229
duodenal ulcer patients
  gastric acid and food ingestion *140, 141*
  gastric acid secretion 134, 135, *136*
  mean gastric pH *139*
  perforation rates, seasonal 145

early morning hours and cardiac events 194, 202–203, 204, 205
eating habits
  night work 319
  SAD 394
efferent projections, SCN 41, *43*
the elderly, circadian rhythms 311–312
  abnormalities
    behavioural therapy 332–333
    melatonin replacement therapy 383
    melatonin synthesis and secretion **374–380**
embolism, arterial 194
Emla cream, pharmacokinetics 220
enalapril 198, 200, *201*, 202
enalaprilat *197*
endocrine system
  circadian variations 99
endogenous circadian oscillator 27
endogenous and exogenous components, rhythm 3–7
  field studies **16–27**
endogenous opioid peptide levels **212–213**
endogenous period, generation of 345–346

endorphin concentrations, CSF, chronic pain syndrome 212–213
endothelium-derived nitric oxide 118
enkephalins, circadian variation 212
enteric coatings, time-delay systems 302
entrainment, circadian rhythms
  feeding schedules 67–68
  light 32–41, **43**, 61–62, 371, 400
  GABA, role in 357
  melatonin, effects of 358–359
  see also phase-shifting
enzyme systems, rhythms in **117–121**
EORTC Chronotherapy Group chronotherapy centres **271–273**
eosinophils, nocturnal asthma 160, 161
epithelial cell proliferation 243
ergotropic actions, body clock effects 11
erodible barrier layers, oral time-delay drug delivery 289–290
erodible plugs, time-delay capsules **297–300**
erythropoietin, serum, diurnal rhythmicity, myeloma 244
esomeprazole, GORD, gastric pH *145*
etoposide, chronotoxicity 262
evening allergen exposure, nocturnal asthma 166–167
evening dosing, ACE inhibitors 200
excitatory amino acids (EAA), effects on rhythms **350–352**
exercise, shift-workers' malaise 329–330
exogenous component, circadian rhythm 27
  differences between variables 6–7
extended-release formulations, β_2-adrenoreceptor agonists 171–172
extrinsic factors, nocturnal asthma 167
eye, photoreceptors 63

familial advanced sleep-phase syndrome 68–69
famotidine, gastric acid secretion 139, *140*, *143*

fatigue 15
  night work 319
female aircrew, hormonal and melatonin changes 317
fetal period, melatonin synthesis and secretion 374
FEV_1, asthma 153
  and alveolar eosinophils, correlation 160
  circadian variations 154, *155*
  and glucocorticoid receptor binding affinity 162
  hydrocortisone infusion 164–165
  nocturnal fall
    LTB_4 increase 163
    salmeterol 174
    zileuton 163, 182
field studies, endogenous and exogenous factors, rhythm **16–27**
  field data, purifying **19**, **20–27**
  marker for, choice of **16–19**
floxuridine 245
  hepatic artery infusion 257, 258, 260
fluoropyrimidine derivatives 245, **251–261**
fluorouracil (5-FU) 245
  chronoefficacy 256, 265, 266
  chronotoxicity and chronopharmacology **251–261**
  as combination therapy 250–251, 255, 255–256, 257, *259*, 264–265, 269, 270, 272
flurbiprofen, rheumatoid arthritis 222
fluticasone, inhaled 183
folinic acid with 5-FU (leucovorin; LV) 252, 254, 255, 256, 257, *259*, *261*
food ingestion
  and cholesterol synthesis 121, 319–320
  drug absorption 131
  drug metabolism 99–100
  gastric acid secretion 136–138, 139, *140*, 141, *142*, *143*, *144*
  and H_2RAs, efficacy 143, *144*
  jet lag 324–325
  SAD 394, 395
  and urinary pH, rats 101

food-induced re-synchronisation 67–68
forced desynchronisation protocols 16
formulations
  and physicochemical properties 85–86
  time-delay *see* time-delay formulations
Fos 36
fractional desynchronisation 5–6, 19, 20
  mental and physical performance studies 12–15
FRC, nocturnal asthma 157
free radicals 384–385
free-running period 53
free-running rhythms 8–9, 310, **314**, 336, 371
functional residual capacity (FRC), asthma 157

GABA 32, 357–358
gamma-D-glutamylglycine, rhythm studies 352
gastric acid secretion **134–144**
  pharmacological implications **138–143**
  therapeutic implications **143, 144**
gastric emptying **127–130**, 303
  pharmacological and therapeutic implications 90, **130–131**
gastric mucosal defence **145–147**
gastrin 134
gastrin-releasing peptide (GRP) 32
gastro-oesophageal reflux disease (GORD)
  esomeprazole, omeprazole, pH profile *145*
  nocturnal asthma 167
gastrointestinal cancer
  TNF-α, TNF-α P55 and P57, circadian fluctuations 244
gastrointestinal (GI) tract **127–147**
  blood flow and cardiac output to 90, *92*
  drug release sites 284–285
  gastric acid secretion rhythms **134–144**
  gastric mucosal defence **145–147**
  motility **127–134**

migrating myoelectric complexes, propagation 90, *91*, 129–130, 131
motility rhythms **127–34**
peptic ulcer disease **144, 145**
pH triggered drug delivery 302
pH variability 302
gemcitabine and carboplatin, chronotolerance, advanced NSCLC *261*, 270
geniculohypothalamic tract (GHT) 33, **36**, **38**
glucocorticoids, *per1* expression 67
glucuronidation, temporal variations 97–98
glutamate 35
  5-HT, modulation of 40–41, *42*
  phase-shift studies 37, 351
  RHT transmitter 34, 35
  and substance-P interaction, SCN neurons 35
glutamatergic light-entrainment pathway 35–36
glutathione, reduced 98–99
glyceryl trinitrate, coronary heart disease 202, 203
GRβ expression, nocturnal asthmatics 162
guanylyl cyclase, soluble 119
gut temperature 17–18

H^+K^+-ATPase inhibitors, gastric acid secretion 141, 142, 143, 144, *145*
hamster SCN 32
head and neck cancers chronotherapy 270
heart, β-adrenoreceptors in **111–116**
heart rate
  β-adrenoreceptor antagonists 196, 203, 204
  catecholamines 113
  congestive cardiac failure 115, 116
  day-night variations 193
  as field study marker 17
  related ischaemic episodes 204
Helicobacter pylori 146
hepatic artery infusion, chemotherapy, neoplastic cells in liver 257, *258*, 260

hepatic blood flow 131, 133–134
  circadian variations 90, 92, 99
hepatic cytochrome P450, mono-
    oxygenase system **94–96**
hepatic drug metabolism, rhythm in
    94–100
hepatic first pass effect
  and gastric emptying time 130
hepatic microsomal enzyme activity
    131, 133–134
hepatic transferases 97
hexobarbital oxidase activity and
    sleeping time 94
histamine
  asthma, circadian plasma variation
    184
    gastric acid secretion 134
    rhythms, effects on 360–361
histamine H_2-receptor antagonists
    (H_2RAs) 138, 143, 144
homeostasis 1
hormonal circadian rhythms, altered
    and breast cancer risk 239
    female flight crew 317
5-HT *see* serotonin (5-HT)
hydrochloric acid secretion *see*
    gastric acid secretion
hydrochlorothiazide, BP profile 200
hydrophilic polymers, gellable barrier
    layers, oral drug delivery
    287–289
hydrophilic sandwich capsule,
    time-delay drug delivery
    300–301, *302*
hydroxy-methylglutaryl (HMG)-CoA
    reductase 120–121
hyperphagia, SAD 394, 395
hypersomnia 395
hypertension
  chronopharmacology **195–202**
    ACE inhibitors 118, **198**, 200,
      201
    β-adrenoreceptor antagonists
      195–196
    calcium channel blockers **196**,
      198, *199*
    other antihypertensives **200, 202**
  circadian patterns 193
  nitric oxide synthesis 118
  NOx and cGMP excretion 119

vessel wall stress, NO-cGMP
    pathway 119
hypnotics, jet lag treatment 328
hypothalamic serotonin transporter
    availability *398*

ibuprofen, biphasic release
    288–289
I_{LOT} cationic channel 121
immediate release drug delivery *284*
  and sustained release formulations
    and pharmacokinetics 85, 86,
    87
impermeable capsules, erodible plugs
    297–300
impermeable capsules with separable
    components **295–297**
indocyanine green clearance, hepatic
    blood flow 90, *92*, **133**
indometacin, formulations and
    pharmacokinetics 85, 86
  and side effects 222, 223
infants, circadian rhythms 310, 311
  melatonin synthesis and secretion
    374
inflammatory cells and mediators,
    asthma 153, 159, *160–163*,
    178, *179*
insomnias 312–313
insulin, Chronset formulation 296
insulin sensitivity 12
intercepts, purification by, field
    studies 22, 22–23, *24*, 26
interferon-β
  chronoefficacy and toxicity
    267–268
  circadian rhythmicity, disruption
    265
interferon-β receptor (IFNAR)
  expression, cell cycling 241
interferon-β, chronoefficacy and
    toxicity 267–268
intergeniculate leaflet (IGL) 36, 38
  in nocturnal animals 38
  optic tract projections to 344
intractable pain 215, *216*, 217
ion channel activity, rhythms
    121–122
ipratropium bromide, nocturnal
    asthma 185

irinotecan (CPT-11) 249–250, 261, 264–265
  with 5-FU and folinic acid 257
  with 5-FU, leucovorin and oxaliplatin 255
  chronoefficacy and toxicity 267
  and oxaliplatin 251
isosorbide mononitrate 86, 131, *197*, 203, 289
  formulation and chronokinetics 86
isosorbide dinitrate, coronary heart disease 203
isradipine, 24 hour BP pattern 198, *199*

jet lag 3, **314–317**, 336, 343, 358, 383, 385
  treatment **323**, **324–328**

ketoprofen 77, 223–224

L-NAME, rat blood pressure 119–120
lacidipine, 24 hour BR pattern 198, *199*
lansoprazole, gastric acid secretion 141, 142
'larks', circadian rhythms 309, 310, 336
  night work 320
lesion experiments, biological clocks, identification 53–54
leucovorin *see* folinic acid
leukocytes, bronchoalveolar lavage, asthmatics 160
leukotriene modifiers **182–184**
leukotrienes, nocturnal asthmatics 163
levodopa, time-delay drug delivery 290
lidocaine, erythrocyte to total plasma concentration 94
light
  circadian disorders, treatment 334
  constant, and carcinogenesis 238–239
  entrainment of circadian rhythms **32–41**, 43, 357–358, 371, 400
    per1 and *per2* in 61–62

jet-lag, treatment **325–327**
  melatonin synthesis and secretion 371
  night-time exposure, cell response 347
  phase-shifting *10*, 11, 33, *37*, **325–327**, **346–347**, **400–401**, 402
  SAD, treatment 395, *396*, 397, 401
  shift-workers' malaise, treatment **330–331**
  as zeitgeber **9–11**
light–dark cycle 33
  combined with melatonin secretion 9
  hormone control synchronization 371
lignocaine (lidocaine), pharmacokinetics, time-dependent changes 219, 220
lipid-lowering drugs 121
lipid-soluble drugs 85, 88
5-lipoxygenase enzyme inhibitors 182
local anaesthetics, chronopharmacology **219–221**
long-haul flights *see* jet lag
lovastatin, chronoefficacy 121
low density lipoprotein levels, night work 319–320
lung function
  circadian variation **154–159**, 168, 169
  salmeterol reducing 173

macrophages, nocturnal asthma 160, 162
meals
  jet lag treatment 324–325
  shift-workers' malaise, treatment 329
mechanical coupling, airways and lung parenchyma 157
medial raphe nuclei (MRN) 39, 344
melanopsin 64, 65
melatonin **358–359**
  administration **334**, **335–336**, 372, 374, 384
  jet lag **327–328**

phase-shifting **11**
SAD 400
shift-workers' malaise 329
and ageing **374–380**, *382*
as free radical scavenger **384–385**
and light–dark cycle 9
metabolism 372
nocturnal asthma 165
plasma concentration *373*
SAD **399–400**
seasonal variation *375*, *376*
secretion and control 44, 121–122, **369–372**
retina 55
suppression 64, 238–239
sleep, effects on 15, **381–383**
Melodie brand infusion pumps 272
membrane permeability, chronobiology of 90
membranes, passage of drugs through 94
mental performance
body clock effects 12–15
exogenous factors 15
meperidine *see* pethidine
mepivacaine, chronopharmacology 219
mercaptopurine, chronefficacy 268
met-encephalin neurons, VL-SCN 38
metabolism, body clock, effects of 12
methacholine, inhaled, bronchial reactivity 158–159
methotrexate
chronoefficacy 268
chronokinetics 78
methylprednisolone, asthma 178
methylxanthines, rhythm, effects on 359–360
see also caffeine
metoprolol 203, 204
time-delay formulation 294
migraine *215*, 217
migrating myoelectric complex (MMC) 90, *91*, 129–130, *131*
mitotic activity in presence of tumour 242, 263
mitoxantrone *see* mitozantrone
mitozantrone 245, 246
MK-801 *see* dizocilpine

molsidomine *197*
mometasone furoate dry powder inhaler, dose timing 180–182
mononuclear cells, circadian rhythm in circulating 243–244
monooxygenase activity and composition, rat liver 95–96
montelukast 182, 183, 183–184
morphine, postoperative pain 224–228
mortality, nocturnal asthma 154
motion sensors, wrist, field studies 18, *19*, *20*
mouse genes, circadian timing 59
movement, measurement of, field studies 18, *19*, *20*
multiple infarct dementia
altered sleep/activity phases 313
multiunit recording, SCN neurons 348
muscarinic receptors, phase-shifting studies 359
muscimol, phase-shifting studies 357–358
myocardial infarction
calcium channel blockers 205
early morning rise in 120, 194
and β-adrenoreceptor antagonists 203, 205
late afterdepolarisations 122
pain, biological rhthms of *215*
recurrent, reducing risk of 203

nasopharyngeal human cell line, DNA synthesis 243
natural killer (NK) cells, cytotoxicity 242
nedocromil **184**
neuroanatomy and neurochemistry, biological clock **31–45**
neurotransmitters
and drug effects on rhythms **350–357**
nocturnal asthma **165**
SAD 396
neutrophils, asthmatics 160
N^G-nitro-L-arginine methyl ester, blood pressure effects, rat 119–120

nicotine
  effects on rhythm, *in vitro* 359
  phase-shifting studies 359
nifedipine *197*, 198, *199*, 205
  formulation and chronokinetics 86, **88**
night work 3, **317–323**, 336, 385
  accidents and short-term effects **318–319**
  interindividual differences **320–323**
  long-term effects **319–320**
  melatonin and carcinogenesis 239
  see also shift-workers' malaise
night-shift paralysis 318
nitrates, coronary heart disease 120, **202–203**
nitrendipine, 24 hour BP pattern 198, *199*
nitric oxide donors *see* nitrates, coronary heart disease
nitric oxide (NO)
  in circadian rhythmicity 34–35
  exhaled, nocturnal asthmatics 163
  photic entrainment 36
nitric oxide oxidative products (NOx) 118–119
nitric oxide synthase type III (eNOS) 118
nitric-oxide-cyclic GMP system **118–120**
NMDA receptor stimulation
  NO release 34–35
NMDA, rhythm studies 351–352
nocturnal animals, IGL in 38
nocturnal asthma 153–154
  chronotherapeutic approach *168*
  extrinsic factors 167
  genetic predisposition 166
  lower airway resistance 157, 158
  mechanisms of worsening **159–166**
  sleep disruption 154
  symptoms independent of medication *155*
nocturnal awakenings
  leukotriene modifiers compared to inhaled corticosteroids 183
  mometasone furoate dry powder inhaler 181
nocturnal enuresis 312, 331

non-24-hour sleep syndrome 343
non-Hodgkins lymphoma, circadian rhythms and outcomes 240
non-Q-wave infarction 203
non-selective β-adrenoreceptor antagonists 195
  see also oxprenolol; propranolol
non-small-cell lung cancer
  chronobiological treatment 269–270
  5-FU combined infusional chronotherapy 255–257, *259*, *260*
non-steroidal anti-inflammatory drugs (NSAIDS), chronopharmacology **221, 222–224**
nonphotic phase shifts 38
nonphotic stimulus to CTS, GHT relaying 38
noradrenaline (norepinephrine)
  animal studies
    in heart tissue 113
    turnover 113–114
  SAD 396
noradrenergic control, melatonin synthesis **369, 371**
nurses, night shift 331

obstructive sleep apnoea, nocturnal asthma 167
occupational health
  long-haul travellers **314–317**
  night work *see* night work
omeprazole, gastric acid secretion 141, 142
  GORD *145*
opioid analgesics, temporal variation **224–228**
oral drug administration 86, **88**, *89*, 283
  gastric emptying 130–131
  and gastrointestinal motility 127
  time-dependent differences, pharmacokinetics 75, 78, *79–82*, 134
  see also time-delay formulations, drugs
oral mucosa and skin, cell cycle phases and clock gene expression 236, *237*, 243
oral nitrates **202–203**

oral temperature, mental task performance and subjective alertness *13*, 14
orally administered drugs 134
Oros formulations, drug delivery 285, *286*
osmotic control, time-delay tablet formulation **285–287**
osteoarthritis, pain, rhythms of *215*, 218
'owls', circadian rhythms 309, 310, 336
   night work 320
oxaliplatin 267
oxidase activity, liver microsomal concentration study 95
oxidation, drug metabolism 94
oxitropium bromide, nocturnal asthma 184
oxprenolol *197*

P53, circadian expression 236
P57 soluble receptors, circadian fluctuations 244
P450 isoenzyme, hepatic concentrations 95, 96
pain
   chronobiology **212–219**
   clinical situations, rhythms in 213, 215–219
   evaluation 211
   intensity, circadian variations *216*, 228–229
   rhythms in experimental 213, *214*
   see also analgesics
pancreatic cancer, chronobiological considerations 270–271
pantoprazole, gastric acid secretion 141, 142
paracetamol (acetaminophen) 97, 98
   time-delay formulation 294, 296, 300, *302*
parenteral administration, chronokinetics 78, *83–84*
patient-controlled analgesia (PCA) 224, *225*, 226–227
PEFR, asthmatics 153, 154, *155*, 160
   overnight decline 164, 165, 174
pellets, time-delay formulations **291–293**

pemoline, jet lag treatment 328
peptic ulcer disease 146
   gastric acid secretion *135*
   seasonal rhythms **144, 145**
peptide histidine isoleucine (PHI) 32
PER/CRY heterodimers 60
PER/TIM nuclear translocation 55, 56
performance
   mental and physical *12–15*
   and night work 318
perindopril, 24 hour BP pattern *201*
period genes (*per1*, *per2* and *per3* 55, 58, 60, *61*, 67, 345
   expression, *per1* 67
   induction by light 347, 351–352
   light entrainment pathway 61–62
   mutations affecting 68
peripheral airway resistance, asthma 157–158
peripheral blood mononuclear cells, glucocorticoid receptor binding affinity, asthmatics 162
peripheral resistance reduction, β-adrenoreceptor antagonists 200
peripheral tissues, role of clock genes **65–68**
peristalsis 127–128
peropsin 64–65
pethidine (meperidine), intramuscular, chronokinetics 78
pH triggered drug delivery 302, 303
pharmacodynamics, rhythms and **111–122**
pharmacokinetics
   rhythms in **76–88**
phase response, SCN function 346–347
phase-delay hypothesis, SAD 401
   studies not agreeing with 402
phase-shifting 346–347
   5-HT 40–41
   in vitro 347–348
   in vivo 11, 38, 348–350, 358
   light *10*, 11, 33, *37*, 325–327, 346–347, 400–401, 402
      phase-response curve 10
   melatonin ingestion 11

phase-shifting (*continued*)
  neurotransmitter and drug effects 350–361
  nonphotic 38
  SAD 400–402
phenytoin, plasma levels, free 93
phosphodiesterase activity 114
photic entrainment 32–41, *43*, 61–62, 357–358, 371, 400
photic and nonphotic signals, converging 41
photoperiod and melatonin secretion, SAD 399–400
photopigments
  identification of novel 64
  mediating photoreception **63–65**
physical activity, jet lag treatment 325
physical performance
  and self-chosen work rate 13, *14*
physicochemical properties, drugs 85
pindolol, heart rate 196
pineal gland
  and blood-brain barrier 369
  calcification 380
  melatonin production 349–350, *370*
pinopsin 64
pituitary adenylate cyclase-activating polypeptide (PACAP)
  CTS response to light 35–36
  light and phase-shift, subjective night 37
placebo, analgesic potency, circadian variations 228
plasma 11-hydroxycorticosteroids 2
plasma concentrations and acute side effects 130–131
plasma proteins, drug binding 90, 92–94, 99, 134
platinum derivatives 261–262
  chronotoxicity **246–250**
  see also carboplatin; cisplatin; oxaliplatin
plug separating from permeable capsules, time-delay drug delivery 293–294
PORT system, time-delay capsule formulation 294–295
postoperative patients, analgesia 224, 225, 226

potassium excretion, circadian rhythms 2
prednisone, asthma 178, *179*
pregnancy, β-endorphin levels, time dependent variation 212
press coating technology, continuous sustained-release drug formulation 289
pressure-controlled systems, time-delay drug delivery 300, *301*
primary hypertension 193
  calcium channel blockers 196
Prinzmetal angina *see* variant angina
programmable pump delivery, anti-cancer drugs 254, **271–273**
prolactin response to tryptophan administration 397, *398*
propranolol 114, 196, *197*, 202–203, 204
  absorption, evening administration 130
  fast or slow eroding tablets *299*
  melatonin secretion reduction 371
  pharmacokinetics, time-dependent variations 77
  physicochemical and pharmacokinetic variations 85
  time-delay capsule formulation 297, *299*
proton pump 134
proton pump inhibitors
  food ingestion and 144
  gastric acid secretion 141, 142, 143
  see also omeprazole
pulsatile release drug formulation
  β$_2$-adrenoreceptor agonists 171, *172*
  pellet 292, *293*, *294*
  or sustained release *284*
Pulsincap formulation, time-delay capsule 295–296, *297*
pumps, programmable portable *254*, **271–273**
purification of data, field studies, rhythms 19, 20
  accuracy 23, 25
  early models **20–21**
  recent purification methods **21–27**
purinergic, effects on rhythms 359–360

quinapril 198, 200, *201*
quipazine, phase shift studies 353, 355–356

ramipril 198, 200, *201*
ranitidine, gastric acid secretion 139, *141*, *142*, *144*
raphe nuclei 33, 39, 344
raphe-hypothalamic tract (RaHT) 33, **39–41**
receptors and signalling pathways, rhythms in **111–117**
rectal and oral drug administration, chronokinetics **86**, **88**, *89*
rectal temperature 17, *18*
  circadian rhythm 309, 310
  long haul travel *316*
recurrent brief SAD 393
reduced glutathione 98–99
renal drug excretion **100–101**
renal reabsorption 100
renin activity 117
renin-angiotensin system **117–118**
rest–activity cycles 44
  alterations and tumour growth 263
  SAD, phase delay 401–402
resynchronisation, biological clock and zeitgebers 384, 385–386
retina
  afferents to SCN 32, 33
    release of glutamate, substance-P and PACAP 36, *37*
  photoreceptors 63
retinal-binding G protein-coupled receptor 64
retinal-raphe-SCN pathway 41, 344
retinohypothalamic tract (RHT) 32, **33–36**, 63, 344
rheumatoid arthritis
  NSAIDs 221, 222
  pain, circadian rhythm *215*, 218
rhythm amplitude **346**
rhythm, endogenous and exogenous components **6–7**
rifampicin, chronokinetics study 76, 78
rostral suprachiasmatic nuclei projections 44
routes of administration and rhythms **86**, 88

salbutamol sulfate (albuterol sulfate), extended release 171, 175–176
salbutamol, time-delay formulations 285, *286*, 290, 295
salmeterol **172–176**
  comparison with
    albuterol sulphate 175–176
    bambuterol 176
    theophylline 174–175
seasonal affective disorder (SAD) **393–403**
  biological rhythms **396–402**
  demographic and clinical characteristics *394*
  diagnosis **393–396**
  photoperiod, association with 399
seasonal pattern 393–394
seasonal rhythms
  melatonin, plasma concentrations *375*, *376*
  peptic ulcer disease **144**, **145**
  reproduction, photic response 346–347
seasonality 393
  brain 5-HT function **396–398**
secondary hypertension, calcium channel blockers 198
senile dementia
  light therapy, circadian rhythm disorder 334
  multiple infarct dementia 313
  see also Alzheimer's disease
serotonergic system
  ascending pathway 39
  destruction of 39–40
serotonin (5-HT) **352–357**
  glutamate, regulating release during light phase *42*
  nonphotic information, relay of 39
  phase shifts 40–41
  photic responses, modulation of 40, 41
  SAD 395, **396–398**
  seasonality 103
serotonin receptor subtypes
  in CNS **116–117**
  potential drug targets
    5-HT_{1A} receptors 353–354
    and light 354

serotonin receptor subtypes
    (*continued*)
    5-HT$_{1B}$ receptors 354–355
    5-HT$_2$ receptors 355–356
    5-HT$_{5A}$ receptors 356
    5-HT$_7$ receptors 356–357
Shaggy/GSK-3, TIM phosphorylation 55–56
shift-work
    and breast cancer 239
    see also night work
shift-workers' malaise 3, **317–323**, 336, 343
    treatment **329–331**
    see also night work
side effects
    acute, plasma concentrations and 130–131
    theophylline and salmeterol, comparison 175
    timing to minimize 223, 229
sigmoidal release profile, pellet formulation **291–292**
simvastatin, chronoefficacy 121
sinoatrial node, enhanced parasympathetic drive, asthmatics 165
sleep
    body clock effects 15
    and lung function 156
    melatonin, effects of **381–383**
    night work and sleep habits 320
    theophylline administration 171
sleep disorders
    advanced sleep phase syndrome 313, 343
    delayed sleep phase syndrome 313, 334, 358, 385
    melatonin secretion, impaired 381–383
    see also hypersomnia; insomnias
sleep disruption, nocturnal asthma 154
    oral β$_2$-adrenoreceptor agonists 172
    salmeterol 173, 175
sleep-wake cycle 1, 3, 343
    ageing and 381, 382–383
sleeping at work, night shift 318
sodium channel, drugs acting on 122

sodium cromoglicate, asthma **184**
steroid resistance, nocturnal asthma 165
stroke 194
sub-syndromal SAD 393
subparaventricular zone (SPV) 44
substance-P
    light entrainment pathway 35
    light and phase-shift, subjective night 37
sudden cardiac death 194, 203
sulfamethoxazole, time-dependent pharmacokinetic study 76, *78*
sulfation, temporal variations 97–98
6-sulfatoxymelatonin 372
suprachiasmatic nuclei (SCN) 31, **32**, 45, 53
    afferent pathways **32–41**, 41, *43*
    body clock, effects of 11
    degeneration, old age 312
    efferent projections 41, *43*
    function **343–362**
        methodology **347–350**
        neurotransmitter and drug effects **350–361**
    innervation 344–345
    intracellular Ca^{2+} concentration 122
    light as zeitgeber 10, 11
    melatonin receptors 11
    murine model, TGF-α rhythmic expression 241
    neurochemistry 32
    neurons, communication between 44
    and peripheral clocks, hierarchy between 236
    photic input response, 5-HT modulation 117
    rodent neurochemistry 33
    serotonergic imput 352–353
sustained release drug delivery *284*
    food ingestion and gastric emptying 131, *132*
    and immediate-release formulations and pharmcokinetics 85, 86, *87*
sympathetic tone, early morning increase in 205
sympathomimetic/agonist activity, compounds with intrinsic 195

tablets
  time-delay formulations **285–291**
  fast or slow eroding tablets *299*
tau mutant hamsters 31–32, 56
  retinal circadian melatonin release 66
temazepam, jet lag treatment 328
temperature *see* core temperature
tenoxicam, spondylarthritis 222
terbutaline, nocturnal asthma 157–158
testosterone hydroxylation, circadian variation 95, 96
TFN-α P55, circadian fluctuations 244
theophylline **169–171**
  absorption, evening administration 130
  comparison with salmeterol 174–175
  formulation and chronokinetics 86, *87*
  once-daily or twice-daily 169–171, 175
thermoregulatory reflexes 1
three-dimensional printing techniques, time-delay drug delivery 291
thymidylate synthase activity 236, 243
Time-Clock formulation 289
Time-Controlled Explosion System 292
time-delay formulations, drugs 284
  capsules 293–*301*
  pellets **291–293**
  tablets **285–291**
  see also extended-release formulations, β$_2$-adrenoreceptor agonists
time-dependent changes, pharmacokinetics
  oral administration 78, *79–82*
  parenteral administration **83–84**
time-dependent hydrophilic barriers, tablet formulation 288
time-free environments **8–9**
timeless (*tim*)/TIMELESS, gene and protein 55, 56, 59
timeout, *Drosophila melanogaster* 59
tiotropium 185

TNF-α *see* tumour necrosis factor α (TNF-α)
toothache, biological rhythms, pain 213, 215, *216*
transcutaneous passage, local anaesthetics 220, 221
transforming growth factor α (TGF-α) rhythmic expression in SCN, murine 241
transplantation of candidate tissues, identification of biological clocks 53, 54
travel fatigue 315
triamcinolone acetonide, asthma 180
triazolam, phase-shifting studies 38, 358
trophotropic actions, body clock effects 11–12
tryptophan 369
tuberous sclerosis 383
tumour, biological rhythms in host or **240–244**
tumour development
  influence on host circadian rhythmicity 238, **262–264**
tumour markers 263
tumour necrosis factor α (TNF-α) 250
  circadian fluctuations 244
  circadian rhythmicity, disruption 265
tumour outcome, circadian coordination and carcinogenesis 238–240

ulcer healing 143
upper airway inflammation
  nocturnal asthma 167
urinary drug excretion *see* renal drug excretion
urinary pH
  and drug excretion 101
  and renal toxicity, aminoglycoside 101, *102*
urine production at night 312, 331

vagal mechanism, gastric acid secretion 136, 141
valproic acid
  binding in plasma samples 92–93
  plasma levels, free 93

valproic acid (*continued*)
  rectal administration and chronopharmacokinetic variation 86, *89*
variant angina 120, 194, 202
vasodilation, NO-cGMP pathway 119
vasointestinal polypeptide (VIP) 32
vasomotor tone, temporal variations 202
ventral-lateral suprachiasmatic nuclei (VL-SCN) 32
  efferents 41, *43*
  light-induced phase shifts 36
  PACAP 35
  rodent, neurochemistry *33*
  serotonergic innervation 39
ventricular arrhythmias, following myocardial infarction 194, 204
verapamil 196, *197*
  sustained-release formulation 287
vinorelbine 248
  body temperature and activity rhythms, mouse models 264
P-388 leukaemia, mice, chrono-efficacy and toxicity 266
vrille (*vri*) 56

waking, spontaneous 15
water-soluble drugs
  temporal variations, pharmacokinetics 85
wax compression coating, time-delay drug delivery 291
wheel-running activity, rodents 346, 348–349
Wnt signal transduction pathway 55
wrist activity 18, *19*
  fractional desynchronisation *20*

zafirlukast 182, 183
zeitgebers 9, 27, 309, 371
  and body clock, abnormal phase relationship 313
  in humans **9–11**
  and night work 321, 322, 323
zileuton 182, 183
  FEV_1, nocturnal decline 163
zolpidem, jet lag treatment 328